T0319243

The Safety Relief Valve Handbook

The Safety Relief Valve Handbook

Design and Use of Process Safety Valves to ASME and International Codes and Standards

Marc Hellemans

www.icheme.org

IChemE

heart of the process

ELSEVIER

AMSTERDAM • BOSTON • HEIDELBERG • LONDON • NEW YORK • OXFORD
PARIS • SAN DIEGO • SAN FRANCISCO • SINGAPORE • SYDNEY • TOKYO

Butterworth-Heinemann is an imprint of Elsevier

Butterworth-Heinemann is an imprint of Elsevier
Linacre House, Jordan Hill, Oxford OX2 8DP, UK
30 Corporate Drive, Suite 400, Burlington, MA 01803, USA

First edition 2009

British Library Cataloguing-in-Publication Data
A catalogue record for this book is available from the British Library

Library of Congress Cataloging-in-Publication Data
A catalog record for this book is available from the Library of Congress

ISBN: 978-1-85617-712-2

Typeset by Charon Tec Ltd (A Macmillan Company), Chennai, India
www.charontec.com

Printed and bound in Great Britain

09 10 11 12 10 9 8 7 6 5 4 3 2 1

Contents

Preface

> The expert is not the genius who knows everything, but the one who knows where to find expertise.

The process industry represents a huge part of many gross national products in the world. It includes the oil and gas, petrochemical, power, pulp and paper, pharmaceuticals, food and many other industries that produce goods which enhance the quality of life.

The primary purpose of a safety relief valve (SRV) is the last protection of life, environment and property in this process industry by safely venting process fluid from an overpressurized vessel. It is not the purpose of a safety valve to control or regulate the pressure in the vessel or system that the valve protects, and it does not take the place of a control or regulating valve.

SRVs are used in a wide variety of process conditions, ranging from clean service to heavily corrosive and toxic process fluids, from very cold to very hot temperatures. This results in a correspondingly large number of damage mechanisms that can prevent them from working if they are not selected, inspected, maintained and tested correctly. This book will guide one to correctly select the equipment for the correct application by understanding the fundamentals of the SRV taking into account the corresponding codes that apply worldwide. Risk-based inspection procedures are introduced in this book as a method of minimizing the chances of failure, and therefore maintaining the highest levels of safety.

SRVs should be taken very seriously. Manufactured from castings, they may not look very sophisticated, but in their design, accuracy and function, they resemble a delicate instrument whilst performing an essential role. Self-contained and self-operating devices, they must always accurately respond to system conditions and prevent catastrophic failure when other instruments and control systems fail to adequately control process limits.

Although it would be ideal that SRVs would never have to work, practice proves, however, that operational or other process-related accidents do occur in our current competitive process industries and that SRVs work more than we think and actually act as the silent sentinels of our processes and avoid many major accidents.

Unfortunately, because of the lack of knowledge, industry's competitiveness and disappearing expertise worldwide, the quality of selection and utilization of SRVs is often insufficient. This possibly jeopardizes the safety of each of us living or working in the neighbourhood of a process plant (Figure P.1).

FIGURE P.1
Accident caused by a failing SRV

More than 25+ years of experience in advising designers, end-users and maintenance people in the selection, handling and maintenance of safe safety relief systems, together with independent studies, described in detail in Chapter 12, have shown that more than half of the pressure-containing equipment installed in the process industry has a small to serious pressure relief system deficiency as compared to widely accepted engineering practices and even legal codes. The types of deficiencies are roughly split between absent and/or undersized pressure relief devices, wrongly selected valves and improperly installed SRVs.

Correct SRV sizing, selection, manufacture, assembly, testing, installation and maintenance as described in this book are all critical for the optimal protection of the pressure vessel, system, property and life. This book explains the fundamental terminology, design and codes to allow most engineers to make the correct decisions in applying SRVs in the process industry and to improve the safety to higher levels.

The challenge for today's process, design or maintenance engineer is to gain knowledge and expertise on a wide array of components in the process industry of which the SRV is the last component that is not always considered as one contributing much to the process in itself and therefore many times is forgotten in literature about instrumentation, process control or valves and actuators.

In addition, the range of information covered makes the book especially well suited for use by technical high schools, colleges, universities and post-scholar education. The book is also intended to be a valuable reference tool for the professional working with SRVs, be it as process designer, end-user, inspector or for all those in a consultancy or sales role.

The information contained in this book is offered as a guide and not as ultimate solution to any problem. The actual selection of valves and valve products is dependent on numerous factors and should be made only after consultation with qualified personnel of the user who is the only one knowing his detailed process conditions and also with the manufacturer who is the holder of the design of the device that is going to protect the system.

This SRV handbook contains vital technical information relating to specifically spring- and pilot-operated SRVs used in the process industry for positive pressure applications above 0.5 Barg and as such for devices subject to both ASME VIII and PED, hence which are subject to law. Most publications currently available are exclusively based on US norms and codes. This book also addresses in detail the European requirements.

I hope that the knowledge gained from this book will enable readers to perform their jobs more effectively, to understand the manufacturers' literature better and understand their jargon, to become more aware of the potential benefits and pitfalls of the currently available technology and to make more informed and creative decisions in the selection and use of SRVs and their applications.

Note: The information in this book is not to be used for ASME Section III nuclear applications, as some of the requirements are somewhat unique and complex. Nor will this book handle the applications for tank venting under 0.5 Barg, otherwise also known as breather valves or conservation vents as they are not subject to any law worldwide and follow mainly the API 2000 guidelines.

Marc Hellemans

Acknowledgements

I wish to acknowledge my previous colleagues from whom I have learned so much throughout my career. They are all passionate about the world of Safety Relief Valves and passed to me their enthusiasm for our field of expertise: Gary Emerson, Paul Marinshaw, Don Papa, Jean Paul Boyer and Randall Miller.

- Anderson Greenwood Crosby (Tyco Flow Control)
- Fike Corporation
- Consolidated (Dresser)
- Emerson process management
- Bopp & Reuther GmbH
- Weir group
- Farris (Curtis Wright)
- Shand & Jurs
- Spirax Sarco
- National board
- Chyioda Engineering Co.

CHAPTER 1

History

It is believed that the French scientist Denis Papin was the inventor of the Safety Valve, which he first applied to his newly developed steam digester at the end of the seventeenth century. Safety Valves were indeed designed and used for many years mainly for steam applications or distillation installations throughout Europe (Figure 1.1).

FIGURE 1.1
Denis Papin

The Safety Valve was kept closed by means of a lever and a movable weight; sliding the weight along the lever enabled Papin to keep the valve in place and regulate the steam pressure.

The device worked satisfactorily for many years and was even commercialised until the beginning of the twentieth century (Figure 1.2).

Some believe, however, that Papin was only the inventor of some improvements and that Safety Valves were already being used some 50 years earlier on a steam digester designed by Rudolf Glauber, a German-Dutch alchemist. In his *Practice on Philosophical Furnaces,* translated into English in 1651, Glauber describes the modes by which he prevents retorts and stills from bursting from an excessive pressure. A sort of conical valve was fitted, being ground airtight to its seat, and loaded with a 'cap of lead', so that when the vapour became too 'high', it slightly raised the valve and a portion escaped; the valve then closed again on itself, 'being pressed down by the loaded cap'.

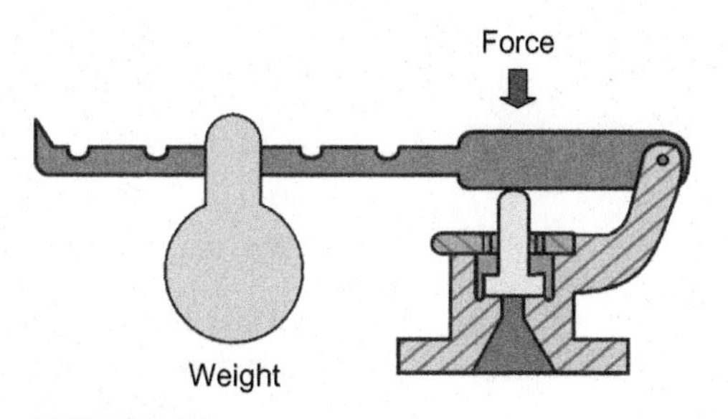

FIGURE 1.2
Early 20th century-type weight-loaded Safety Relief Valve

The idea was followed by others, and we find in *The Art Of Distillation,* by John French, published in London; the following concerning the action of Safety Valves:

> Upon the top of a stubble (valve) there may be fastened some lead, that if the spirit pressure be too strong, it will only heave up the stubble and let it fall down.

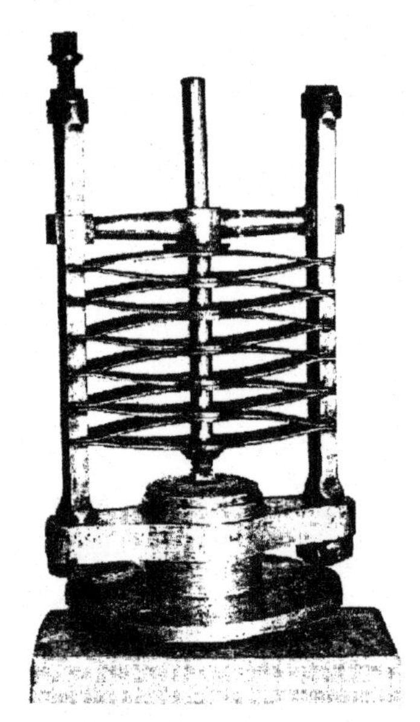

FIGURE 1.3
First open-ended spring-operated Safety Relief Valve

It should be realized that the word *steam*, for which application safety valves were later further developed, was still unknown at the time, being of later coinage.

Around 1830 Timothy Hackworth developed an open-ended Safety Valve for the steam trains and boilers that were first being built around that time, which was the start of the Safety Valve design as we know it today. However, the steam installations didn't really become much safer with the safety devices then in use (Figure 1.3).

Because of the number of boiler explosions and related fatalities in Europe, a select committee of the British House of Commons, looking into the explosions on steam ships reported in June 1817:

> Boilers – should have two safety valves, they shall be inspected and penalties be inflicted on unauthorised persons interfering with the Safety Valves.

Many explosions were caused by inadequate boiler design or by people rendering the Safety Valves inoperative in order to increase the boiler pressure. Due to further explosions, 1882 saw the passing of the Boiler Explosion Act, in which a boiler was defined as

> Any closed vessel used for generating steam or for heating water, or for heating other liquids or into which steam is admitted for heating, steaming, boiling or other similar purposes.

In Great Britain, voluntary bodies such as the Steam Users Association supplied reports to the government beginning in 1854. In the period from 1881 to 1907, there were still a total of 1871 boiler explosions investigated by the Board of Trade. These explosions accounted for 732 fatalities and 1563 non-fatal injuries.

In the United States, the safety records were just as bad. In the period from 1906 to 1911, there were 1700 boiler explosions in the New England area alone, accounting for 1300 fatalities.

In 1901 Parliament passed the Factories and Workshop Act further regulating steam boilers. Among the improvements were

> A steam gauge and water gauge are to be fitted to the boiler and the boiler and associate safety devices are to be inspected every 14 months.

The American Society of Mechanical Engineers (ASME), was asked by the government to formulate a design code, and developed the famous Boiler and Pressure Vessel Code between 1911 and 1914 as a set of safety rules to address the serious problem of boiler explosions in the United States. Average steam pressure in those days had reached only about 300 PSI (20 bar). Europe and other parts of the world used the code as a basis for their own safety rules.

The ASME Boiler and Pressure Vessel Code, Section I, became a mandatory requirement in all states that 'recognized the need for legislation'.

This code included rules for the overpressure protection of boilers, based on the best industry practice of the time. The principles of today's code rules for overpressure protection is little changed from the first code.

With the expansion of the process industries, the need for a code that would be applicable to 'unfired' vessels (roughly, every pressure-containing vessel that is not a boiler) was identified, which gave rise to the Section VIII of the ASME code. Today, the ASME Boiler and Pressure Vessel Code is composed of 12 sections:

Section I = Power Boilers
Section II = Materials
Section III = Rules for the Construction of Nuclear Power Plant Components
Section IV = Heating Boilers
Section V = Nondestructive Examination
Section VI = Recommended Rules for the Care and Operations of Heating Boilers
Section VII = Recommended Guidelines for the Care of Power Boilers
Section VIII = Pressure Vessels – Division I
Section IX = Welding and Brazing Qualifications
Section X = Fiber-Reinforced Plastic Pressure Vessels
Section XI = Rules for Inservice Inspection of Nuclear Power Plant Components
Section XII = Rules for the Construction & Continued Service of Transport Tanks

With the growth of the petroleum and petrochemical industries, the American Petroleum Institute (API) sought uniformity of the dimensional and physical characteristics of pressure-relieving devices. To date, the API has published the following internationally acknowledged documents:

RP* 520 = Sizing, Selection and Installation of Pressure-Relieving Devices in Refineries (Part 1: Sizing and Selection; Part 2: Installation)
RP* 521 = Guide for Pressure-Relieving and Depressurising Systems
Std 526 = Flanged Steel Pressure Relief Valves
Std 527 = Seat Tightness of Pressure-Relief Valves
RP* 576 = Inspection of Pressure-Relieving Devices
Std 200 = Venting Low-Pressure Storage Tanks
*RP means recommended practices

While the ASME code was the law, the API practices became the internationally recognized recommendations still used today.

Today the ASME codes are still mandatory in the United States and Canada. Both ASME and API are applied worldwide. Many European countries also developed their own national rules for the protection against overpressure of process equipment and these remained in force well into the twentieth century. Most were based on the ASME code, but they were sometimes also developed to protect national trade (see also Appendix M).

To allow free circulation of goods in the European Community, EU member states were prohibited from making new technical rules and from updating the existing ones. Instead, they agreed to a new overall directive, the Pressure Equipment Directive (PED), which was published in 1997. The PED has become compulsory for equipment 'put in the market' after 29 May 2002 (refer to Article 20, paragraph 3 of the PED).

Today in Europe, the term *safety valve* is used to describe Safety Valves, Safety/Relief-Valves and Relief Valves. This term is now used in European Norms (EN) and ISO 4126 descriptions. Safety Valves are included as 'Safety Accessories' in the PED (Article 1, paragraph 2.1.3) and are classified in risk category IV (the highest). As with ASME, the legislation texts are complex and possibly open to interpretation. In this book, we have distilled all parts directly related to Safety (Relief) Valves and tried to make them comprehensive and practically usable.

In order for a manufacturer to mark their product Conformité Européene (CE), each product group and type must undergo a conformity assessment comprising the EC type or design examination and the assurance of the production quality system.

FIGURE 1.4
Official CE marking

The procedures to certify conformity to the PED are audited by a notified body of the Member States of the European Community. With the completion of the above, the manufacturer may stamp the CE mark on their product. The users, by law, must install devices that carry the CE mark for all equipment put into the market since 29 May 2002 (Figure 1.4).

However, because these directives have a very large scope, they cannot be very specific as to the details of the goods they address. To set guidelines on how to address the requirements of the directives, the European Committee for Standardisation (CEN) was empowered by the European Council to draw European Standards. After almost 20 years of efforts on the part of end users and manufacturers who were part of the committee, CEN released in 2004 a set of standards particularly for Safety Valves, EN 4126 Parts 1 through 7, 'Safety Devices for Protection Against Excessive Pressure' (similar to API):

EN 4126 Part 1 = Safety Valves (spring loaded)
EN 4126 Part 2 = Bursting Discs Safety Devices
EN 4126 Part 3 = Safety Valves and Bursting Discs in Combination

EN 4126 Part 4 = Pilot-Operated Safety Valves
EN 4126 Part 5 = Controlled Safety Pressure Relief Systems (CSPRS)
EN 4126 Part 6 = Application, Installation of Bursting Discs
EN 4126 Part 7 = Common Data (steam tables, etc.)

This standard now replaces any equivalent standard that existed before in each country of the European Union. In the EU, as in the United States, the CE mark is law and the EN 4126 is a standard (recommended) practice.

Since the economy has become global, most major manufacturers must comply with both the U.S. and European codes, but it is important to know that ASME and CE are both mandated by law, whereas API and EN(ISO) are recommendations. We will expand on both in Chapter 5. Some nations outside the United States and Europe still have their own national codes (GOST – Russia, JIS – Japan, SQL – China, UDT – Poland, etc.), but most basic practices are rather similar.

1.1 ACRONYMS, ABBREVIATIONS

API = American Petroleum Institute
API RP = API Recommended Practice
API Std = API Standard
ASME = American Society of Mechanical Engineers
BS = British Standards
CEN = Commité Européen de Normalisation (European Committee for Standardisation)
EN = European Normalisation (European Standard)
ISO = International Standard Organisation
NB = National Board of Pressure Vessel Inspectors

CHAPTER 2

Overpressure Protection

In this chapter, we will define what is considered a potential overpressure scenario in process systems and where the safety relief valves (SRVs) are needed.

2.1 GENERAL DEFINITION OF AN SRV

First, we need to define generally what we are talking about: A pressure relief device is any device that can purge a system from an overpressure condition. More particularly, an SRV is a pressure relief device that is self-actuated, and whose primary purpose is the protection of life and equipment. Through a controlled discharge of a required (rated) amount of fluid at a predetermined pressure, an SRV must prevent overpressure in pressurized vessels and systems, and it operates within limits which are determined by international codes. An SRV is often the final control device in the prevention of accidents or explosions caused by overpressure.

The SRV must close at a predetermined pressure when the system pressure has returned to a safe level at values determined by the codes.

SRVs must be designed with materials compatible with many process fluids, from simple air and water to the most corrosive and toxic media. They must also be designed to operate in a consistently smooth manner on a variety of fluids and fluid phases. These design parameters lead to a wide array of SRV products available in the market today, with the one constant being that they all must comply with the internationally recognized codes.

2.2 WHERE DO SRVs FIT IN THE PROCESS?

Every industrial process system is designed to work against a certain maximum pressure and temperature called its rating or design pressure. It is in the economic interest of the users to work as close as possible towards the maximum limits of this design pressure in order to optimize the process output, hence increase the profitability of the system.

Nowadays, pressures and flow in the process industry are controlled by electronic process systems and highly sophisticated instrumentation devices. Almost all control systems are powered by an outside power source (electric, pneumatic, hydraulic). The law requires that when everything fails regardless of the built-in redundancies, there is still an independent working device powered only by the medium it protects. This is the function of the SRV, which, when everything else works correctly in the system, should never have to work. However, practice proves the contrary, and there are a variety of incidents which will allow the system pressure to exceed the design pressure.

Although many pressure relief devices are called SRVs, not every SRV has the same characteristics or operational precision. Only the choice of the correct pressure safety device for the right application will assure the safety of the system and allow the user to maximize process output and minimize downtime for maintenance purposes. Making the correct choice also means avoiding interference between the process instrumentation set points in the control loop and the pressure relief device limits selected. These SRV operational limits can vary greatly even when all are complying with the codes.

2.3 WHERE DO SRVs ACT WITHIN THE PROCESS?

Let's consider a typical basic process control loop (Figure 2.1). A pressure-indicating transmitter (PIT) sends the pressure signal to a proportional integral derivative controller (PID), which sends a signal to the control valve to

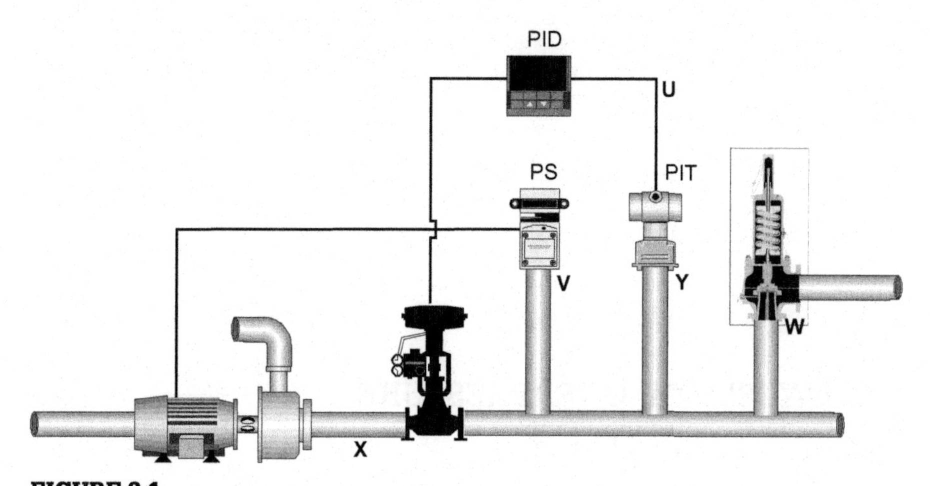

FIGURE 2.1
Traditional control loop

regulate the flow or pressure in the system. This PID can trip the system on at least Hi-Lo alarms. A pressure sensor/switch is usually built in as a redundancy to switch off the pump in case of excessive pressure, for instance. The SRV is installed in case all these powered devices fail or, for instance, when a downstream valve is shut off in an uncontrolled manner.

Each instrument in this control loop has its tolerances, for instance +/ − 5%. To ensure smooth operation, tolerances should never interfere with each other. Also, the SRV should be selected so that it does not start to open under the highest pressure switch setting plus its tolerance. Therefore, it is important to know the tolerance of the pressure relief device, or in this case the SRV, and the same applies for the SRV closure. In short, tolerances should never interfere.

■ Example

Pressure switch is set at 10 barg with an accuracy of 5%. Therefore, the switch could trip at 10.5 barg. In this case, an SRV should be selected that does not start leaking under this pressure at the minimum.

Let's assume that a user has made an investment of an installation with a design pressure of 110 barg; a typical traditional ASME VIII–API 520-type spring-loaded SRV is used and the instruments in the control loop have 5% accuracy.

SRV set pressure needs to be 100 barg (as it will need 10% overpressure to flow its nominal flow – see details later). It is allowed by law to leak at 90% of set pressure = 90 barg.

PS is set at minimum 5% before SRV first leak point = 85.5 barg

PS low point is 5% under its set point = 81.22 barg

PIT control point and hence the process pressure is minimum 5% under the PS low point. Therefore, maximum recommended process pressure and hence maximum output is at 77.16 barg. However, the user is paying for a design pressure investment of 110 barg. Therefore, the user is paying 30% for safety settings and tolerances which will not bring any output and is giving no economic advantage.

A choice of higher accuracy instrumentation and higher accuracy SRVs will allow the user to obtain higher output levels, as will be demonstrated later. ■

2.4 CAUSES OF OVERPRESSURE

2.4.1 Blocked discharge

The full system input flow continues to feed the vessel while the outlet is partially or totally blocked due to personnel error, valve failure, actuator failure, lack of power to operate the valve or an operational upset in the control loops (Figure 2.2).

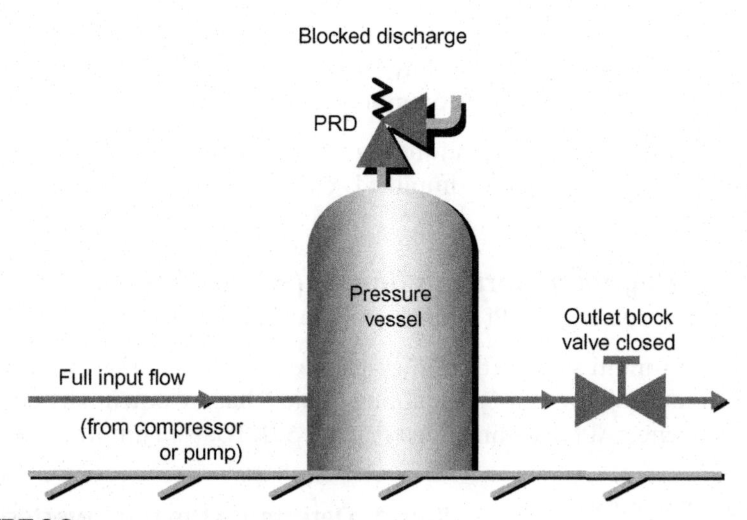

FIGURE 2.2
Blocked discharge

The worst case is obviously for the outlet valve to be fully closed. A less severe case could be for the outlet valve not to be fully open, with the system input being greater than the flow through the outlet valve, ultimately causing a system overpressure.

In any case, if we can see in a P+ID (piping and instrumentation diagram) the possibility of a blocked discharge, the rated capacity of the SRV protecting the system should be based on fully closed valve(s) and take into consideration the maximum flow capacity of the device(s) (pumps, compressors, ventilators, etc.) feeding the system.

2.4.2 Fire case

This case covers the event that the pressure vessel is exposed to external fire, which would cause the system to heat up quickly. Vapours would expand; hence there would be a faster increase in pressure above the system design pressure (Figure 2.3).

In case of liquid storage, the liquid would flash into vapour, causing a significant pressure increase.

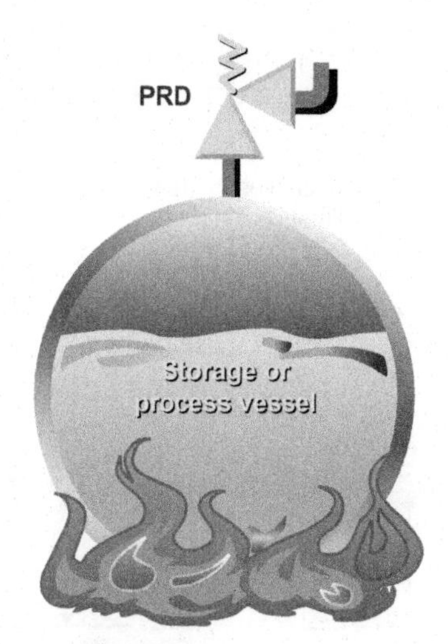

FIGURE 2.3
Fire case

Together with the blocked outlet case, external fire cases are probably one of the most common cases where SRVs are required within the modern process industry.

The procedure used for fire sizing depends sometimes on the codes and engineering practices applied in each installation and determined by the end users. The following sizing procedure, according to API RP 520 Part 1 (see codes in Chapter 4), is the most commonly used.

We will here detail the calculations according to the code and will later also give the (conservative) simplified empiric method used by some professionals.

2.4.2.1 Sizing for vaporizing liquids (wetted vessels)

Formulas are according to API and therefore in English units.

The following multi-step method may be used for calculating the required orifice area for SRVs on vessels containing liquids that are potentially exposed to fire. (Reference: API Recommended Practice 521, Fourth Edition.)

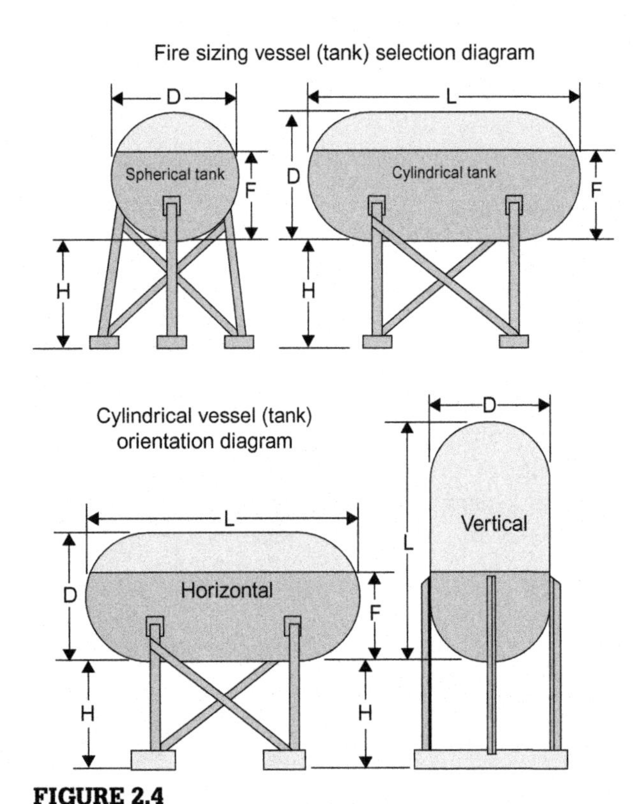

FIGURE 2.4
Tank selection diagram

Step 1: Determine the total wetted surface area.
The type of tanks taken into account in the formulas is represented in these schematics (Figure 2.4).

The following formulas are used to determine the wetted surface area of a vessel (potential surface in contact with the stored liquid). They use the logic as stated in API RP 521, Fourth Edition, table 4: 'Effects of Fire on the Wetted Surfaces of a Vessel'.

Wetted surface area A_{wet} in ft^2.

Sphere:

$$A_{wet} = \pi E_s D$$

Horizontal cylinder with flat ends:

$$A_{wet} = \left(\frac{\pi DB}{180}\right)\left(L + \frac{D}{2}\right) - D\left(\left(\frac{D}{2} - E\right)\sin B\right)$$

Horizontal cylinder with spherical ends:

$$A_{wet} = \pi D\left(E + \frac{B(L-D)}{180}\right)$$

Vertical cylinder with flat ends:

$$\text{If } E < L, \text{ then: } A_{wet} = \pi D\left(\frac{D}{4} + E\right)$$

$$\text{If } E = L, \text{ then: } A_{wet} = \pi D\left(\frac{D}{2} + E\right)$$

Vertical cylinder with spherical ends:

$$A_{wet} = \pi ED$$

where

A_{wet} = Wetted area (ft^2)

E= Effective liquid level in feet, up to 25 ft from the flame source (usually ground level); reference logic diagram effective liquid level (Figure 2.5)

E_s = Effective spherical liquid level in feet, up to a maximum horizontal diameter or up to a height of 25 ft, whichever is greater; reference logic diagram effective liquid level (Figure 2.5)

D = Vessel diameter in feet (see Figure 2.4)

B = Effective liquid level, angle degrees = $\cos^{-1}\left(1 - \frac{2E}{D}\right)$

L = Vessel end-to-end length in feet (see Figure 2.4)

where in Figure 2.5

K = Effective total height of liquid surface (ft)
K_1 = Total height of liquid surface (ft)
H = Vessel elevation (ft)

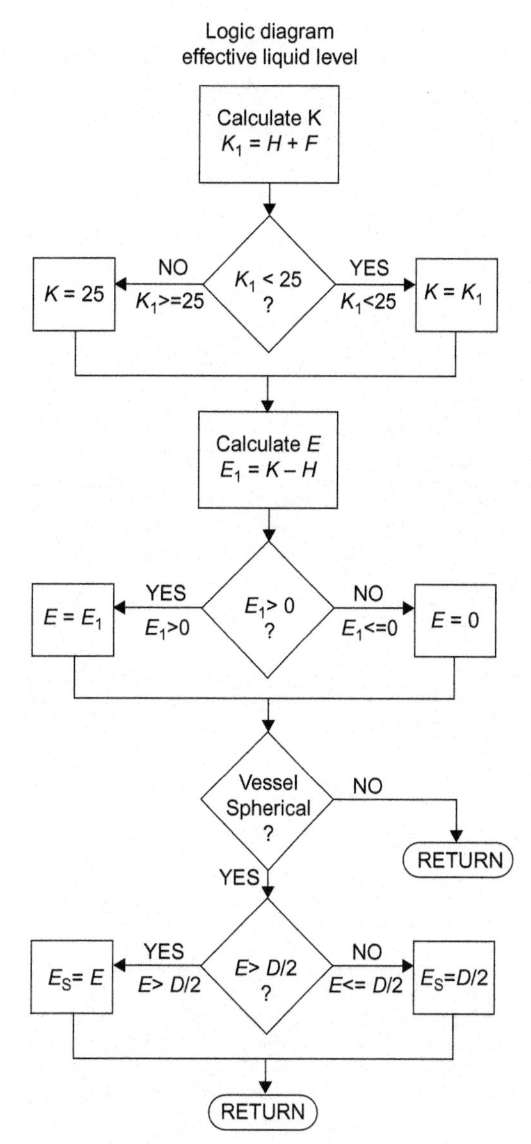

FIGURE 2.5
Logic diagram: Effective liquid level

F = Liquid depth in vessel (ft)
E = Effective liquid level (ft)
E_1 = Initial liquid level (ft)
E_s = Effective spherical liquid level (ft)

Step 2: Determine the total heat absorption.
A determination is made depending on the firefighting accommodations on the site.

When prompt firefighting efforts and adequate drainage exist,

$$Q = 21,000 \times F \times (A_{wet})^{0.82}$$

When prompt firefighting efforts and adequate drainage do not exist,

$$Q = 34,500 \times F \times (A_{wet})^{0.82}$$

where

Q = Total heat absorption to the wetted surface in BTU/h

F = Environmental factor (see Figure 2.6)

A_{wet} = Total wetted surface area in square feet (see calculations in Step 1)

Equipment Type	Factor F[1]
Bare Vessel	1.0
Insulated Vessel[2] (These arbitrary insulation conductance values are shown as examples and are in BTU's per hour square foot per degree Fahrenheit):	
4	0.3
2	0.15
1	0.075
0.67	0.05
0.50	0.0375
0.40	0.03
0.33	0.026
Water application facilites, on bare vessel[3]	1.0
Depressurizing and emptying facilities[4]	1.0

Notes

(1) These are suggested values assumed for the conditions in API Recommended Practice 521, Section 3.15.2. When these conditions do not exist, engineering judgement should be exercised either in selecting a higher factor or in means of protecting vessels from fire exposure in API Recommended Practice 521, Section 3.15.4 - 3.15.5.

(2) Insulation shall resist dislodgement by fire hose streams. Reference API Recommended Practice 521, Table D-5 for further explanation.

(3) Reference API Recommended Practice 521, Section 3.15.4.2.

(4) Reference API Recommended Practice 521, Section 3.15.4.3.

FIGURE 2.6
Table with environmental factors

Step 3: Determine the rate of vapour or gas vaporized from the liquid.

$$W = \frac{Q}{H_{vap}}$$

where

W = Mass flow in lbs/h
Q = Total heat absorption of the wetted surface in BTU/h
H_{vap} = Latent heat of vaporization in BTU/lb

Step 4: Calculate the minimum required relieving area.
If the valve is used as a supplemental device for vessels which may be exposed to fire, an overpressure factor of 21% may be used. However, allowable overpressure may vary according to local regulations. Specific application requirements should be referenced for the allowable overpressure.

The minimum required relieving area (in^2) can now be calculated using the following equation for gas and vapour relief valve sizing (lbs/h):

$$A = \frac{W\sqrt{TZ}}{CKP_1K_b\sqrt{M}}$$

where

A = Minimum required effective discharge area (in^2).

C = Coefficient determined from an expression of the ratio of specific heats of the gas or vapour at standard conditions (see Appendix D). Use $C = 315$ if value is unknown.

K = Effective coefficient of discharge of the used SRV, typical
$K = 0.975$ for gas (depending on the manufacturer's approval – check with manufacturer).

K_b = Capacity correction factor due to backpressure. For standard valves with superimposed (constant) backpressure exceeding critical see Appendix B. For bellows valves with superimposed or variable backpressure see Appendix B. For pilot-operated valves see note below.

M = Molecular weight of the gas or vapour obtained from standard tables or tables in Appendix N.

P_1 = Relieving pressure in psia. This is the set pressure (psig) + overpressure (psi) + atmospheric pressure (psia).

T = Absolute temperature of the fluid at the valve inlet, degrees Rankine (°F + 460).

W = Required relieving capacity (lbs/h).

Z = Compressibility factor (see Appendix C). Use $Z = 1$ if value is unknown.

Note:

Pilot-operated valves: Snap-acting

Backpressure has no effect on the set pressure or flow capacity of pilot-operated pressure relief valves except when the flow is subcritical (ratio of absolute backpressure to absolute relieving pressure exceeds 55%). In this case, the flow correction factor *Kb* (see Appendix B) must be applied. If the ratio of absolute backpressure to absolute relieving pressure is less than 55%, no correction factor is required, $Kb = 1$.

Pilot-operated valves: Modulating

The pilot exhaust is normally vented to the main valve outlet. Set pressure and operability are unaffected by backpressure up to 70% of set pressure, provided that a backflow preventer is used whenever backpressure is expected to exceed inlet pressure during operation (consult the manufacturer for backpressures greater than 70% of set pressure). The capacity is affected, however, when flow is subcritical (ratio of absolute backpressure to absolute relieving pressure exceeds 55%). In this case, the flow correction factor *Kb* (see Appendix B) must be applied. If the ratio of absolute backpressure to absolute relieving pressure is less than 55%, no correction factor is required, $Kb = 1$.

While the above is the current official method, various manufacturers and end users use simplified methods which are considered rather conservative and safe. They reference the old version API RP520, Part 1, D.5.

$$W = \frac{Q}{V}$$

where

W = Required valve capacity in lbs/h
V = Latent heat of evaporization in BTU/lb

Some examples	
Ammonia	589
Benzene	169
Butane	166
CO_2	150
Ethane	210
Ethylene	208
Methane	219
Propane	183
Water	970

Q = Total heat input to wetted surface of vessel.

$$Q = 21000FA^{0.82}$$

where

A = Total vessel wetted surface in ft^2 (up to 25 ft maximum above ground level or in the case of a sphere to the elevation of the largest diameter, whichever is greatest) (Figure 2.7).

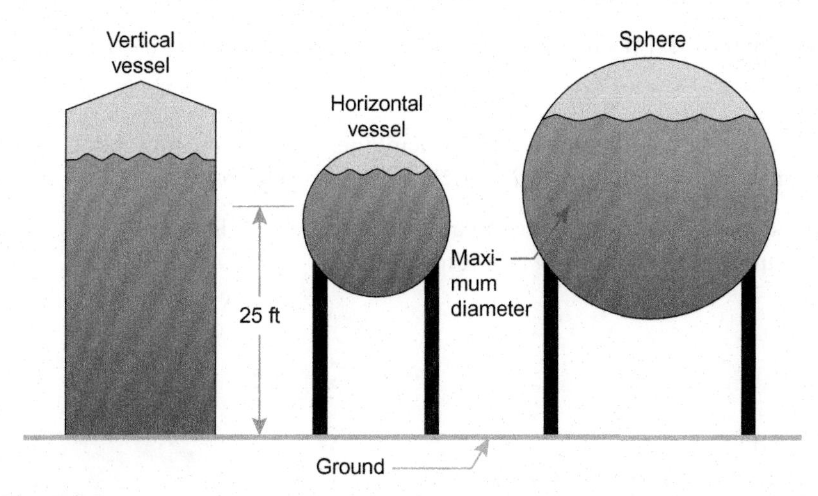

FIGURE 2.7
Heights of wetted surfaces

$F = 1$ if assumed that there is no vessel insulation.

2.4.2.2 Sizing for vessels containing gases and vapours only (unwetted vessels)

Calculations are in English units and based on API recommendations.

The following method may be used for calculating the required orifice area for SRVs on vessels containing gases that are exposed to fire. Reference API Recommended Practice 521, Fourth Edition.

$$A = \frac{F'A'}{\sqrt{P_1}}$$

where

A = Minimum required effective discharge area (in^2).

A' = Exposed surface area of the vessel (ft^2).

P_1 = Relieving pressure, psi, absolute (set pressure [psig] + overpressure [psi] + atmospheric pressure [psia]).

$$F' = \frac{0.1406(T_w - T_1)^{1.25}}{CKT_1^{0.6506}}$$

The recommended minimum value of F' is 0.01.

When the minimum value is unknown, $F' = 0.045$ should be used.

where

T_w = Vessel wall temperature, degrees Rankine. The API recommended maximum wall temperature of 1100°F for carbon steel vessels.

T_1 = Gas temperature at the upstream pressure, degrees Rankine, as determined by the following relationship:

$$T_1 = \frac{P_1 T_n}{P_n}$$

T_n = Normal operating gas temperature, degrees Rankine.

P_n = Normal operating gas pressure, pounds per square inch , absolute (normal operating gas pressure [psig] + atmospheric pressure [psia]).

C = Coefficient from Appendix D.

K = Effective coefficient of discharge.

While the above is the current official method, various manufacturers and end users use simplified methods, which are considered rather conservative and safe and reference the old API RP520, Part 1, D.5 (Figure 2.8).

$$A = \frac{F' A_s}{\sqrt{P}}$$

where

A = Calculated safety relief valve (SRV) orifice area (in^2)

A_s = Total exposed surface area of vessel(ft^2)

P_1 = Set pressure – inlet pressure loss + allowable overpressure (21%) + 14.7 psia

F'= 0.042 is a conservative number which relates to bare vessel metal temperature at relief

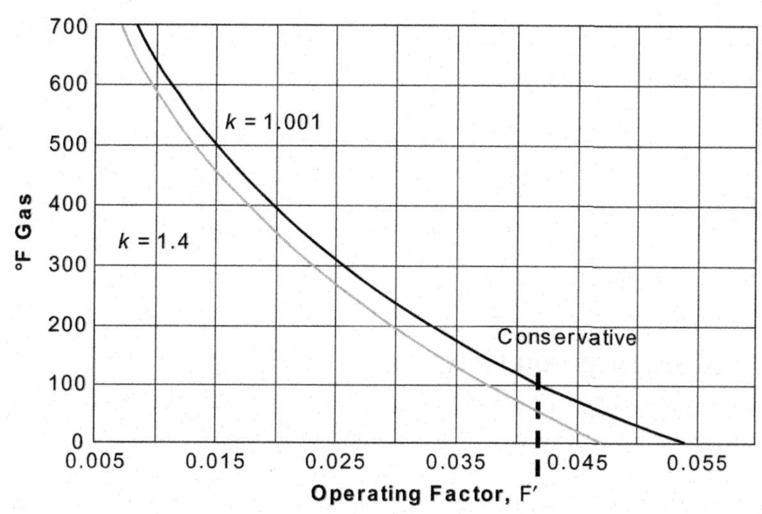

FIGURE 2.8
Operating factor F′ in function of gas temperature

2.4.3 Thermal expansion

When a pipe or vessel is totally filled with a liquid which can be blocked in, for instance, by closing two isolation valves, the liquid in the pipe or pressure vessel can expand very slowly due to heat gain by the sun or an uncontrolled heating system. This will result in tremendous internal hydraulic forces inside the pipe or pressure vessel, as the liquid is non-compressible and needs to be evacuated. This section of pipe then needs thermal relief (Figure 2.9).

The flows required for thermal relief are very small, and there are special thermal relief valves on the market that accommodate this specific application. Oversizing a thermal relief valve is never a good idea, and orifice sizes preferably below API orifice D are recommended.

Per API 521, Section 3.14 the following formula can be used:

$$Q = \frac{BH}{500GC}$$

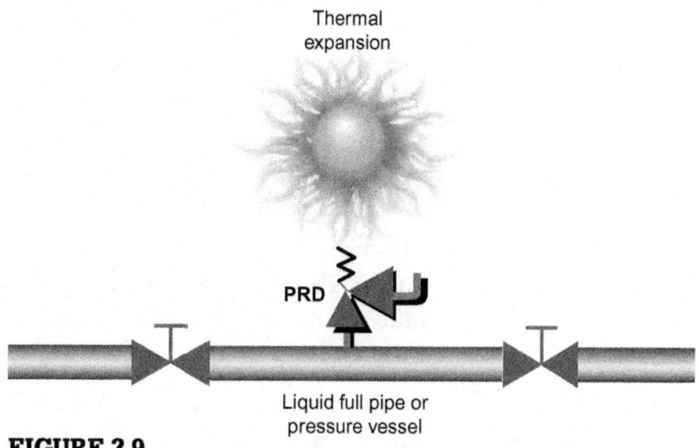

FIGURE 2.9
Thermal expansion

where

Q = Flow rate (gpm – gallons per minute)
B = Cubicle expansion coefficient per °F
H = Total heat transfer rate (BTU/h)
Heat transfer rate (BTU flow formula can be used and can be found in standard tables)
G = Specific gravity
C = Specific heat (BTU/lbs/°F)

2.4.4 Runaway reaction

If the possibility exists of runaway chemical reactions, SRVs should be provided and it is up to the chemical engineers to determine the different scenarios and provide models that take into account the amount of vapour that can be produced by the chemical runaway reaction.

2.4.5 Tube rupture in heat exchangers

If a tube ruptures in a heat exchanger, it creates a massive overpressure because, under pressure, the fluids will evaporate rapidly when exposed to atmosphere. Conservatively, the capacity of the heat exchanger can be taken as the required flow for the SRV. In that respect, ASME and API are not very specific in their recommendations and state the following:

ASME VIII Division 13, Paragraph UG-133 (d):

> *Heat exchangers and similar vessels shall be protected with a relieving device of sufficient capacity to avoid overpressure in the case of an internal failure.*

API RP 521 Section 3.18 states:

> *Complete tube rupture, in which a large quantity of high pressure fluid will flow to the lower pressure exchanger side, is a remote but possible contingency.*

API has, however, long used the 'two-thirds rule' to identify tube rupture scenarios. This rule states that tube rupture protection is not required when the ratio of the low pressure to high pressure side design pressure is greater than two-thirds.

Basically, it remains up to the design engineer and/or end user to determine whether and which type of SRV is to be used when doing a hazard and operability analysis (HAZOP).

2.5 DETERMINE OVERPRESSURE PROTECTION REQUIREMENTS

Despite all safety precautions, equipment failure, human error and other external events can sometimes lead to increased pressures beyond the safe levels, resulting in a relief event. These possible events are described above, but what are the potential lines of defence and why use relief systems which go beyond the simple use of an SRV? The SRV is in fact only a part of the relief system and definitely the most important one.

The different lines of defence against overpressure are as follows:

- *Use an inherently safe design*: This basically means to use a low-pressure design where overpressure accidents would be minimal, but this still cannot eliminate every possible pressure hazard. Even low pressure accidents can cause serious damage, as can be seen here following a pneumatic pressure test on a low-pressure storage tank (Figure 2.10).
- *Passive control*: One can overdesign the process equipment, but this can become exceedingly expensive. In this case, one would select a piping or pressure vessel rating which is far beyond the normal process pressures. Cases are known where this initially happened, but later, for economical reasons during revamps process, pressures were increased to reach the rating limits, which eliminates the initial intention.
- *Active control*: This involves selecting correct relief systems for the process by analysing different possible overpressure scenarios.

FIGURE 2.10
Accident during pneumatic test of storage tanks

> Ultimately, relief systems are the safest and most economic method, and by law they are designed to work correctly if the proper devices are selected and codes respected.

A relief system is a combination of a pressure relief device and the associated lines and process equipment that are used to safely handle the fluid.

The relief design methodology needs to be considered in the correct order, as follows:

1. Locate potential relief points.
   a. On all pressure containing vessels (liquid or gas)
   b. Blocked in sections of cool liquid lines that could be exposed to heat in any shape or form.
   c. Discharge sides of positive displacement pumps, compressors, turbines, etc.
   d. Vessel steam jackets
   e. Chemical reactors
   f. Heat exchangers

2. Choose the general type of the relief device.
   a. Rupture discs
   b. Spring-operated SRVs
   c. Pilot-operated SRVs
   d. A combination of the above

Depending on the specific application, it will be determined later which specific configuration of the above general types we should use in order to get optimal efficiency.

3. Develop the different possible overpressure scenarios for a specific pressure-containing vessel.
   a. Description of one specific relief event
   b. Usually each possible relief has more than one relief event and multiple scenarios. For example:
      i. Overfilling
      ii. Fire
      iii. Runaway reaction
      iv. Blocked lines with subsequent expansion
   c. Developed through a PHA (process hazard analysis)

It is our experience that this should be done in concert with both the design engineers and the end users, and eventually the process engineers. In many instances still today, each works separately, resulting in oversights of some possible relief events.

Overpressure Scenario

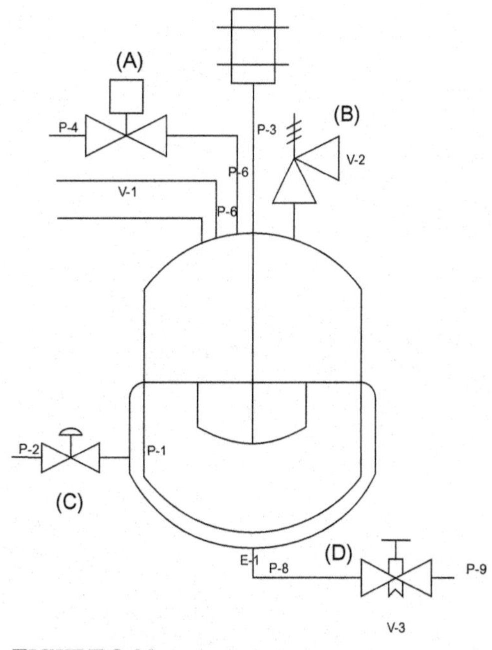

FIGURE 2.11
A reactor with organic substrate catalyst nitric acid

1. Control valve (A) on the nitric acid feed line can be stuck open while the manual outlet valve (D) at the bottom is closed, and the vessel can overfill (Figure 2.11).
2. The steam regulator (C) to the steam jacket can fail and cause overpressure in the vessel.
3. Coolant system could fail, which can cause a runaway reaction as a result.

Then we can start sizing the system for the necessary relief (B)

d. Determining the necessary relief rates (see Chapter 2)
e. Determining the relief vent area (see Chapter 7)

In any case, always use the worst case as the necessary relief flow scenario.

4. Design the complete relief system. A relief system entails more than just installing an SRV or a rupture disc on a pressure vessel; it also includes the following:
   a. Look for the necessity of a back-up relief device(s) – evaluate the necessity for eventual redundancy (possibly for maintenance reasons). This can be two SRVs, two rupture discs or a rupture disc in parallel with an SRV. It is recommended to stagger the settings slightly, having the SRV open first.
   b. Design the correct piping leading to the relief device(s) – avoid excessive inlet pressure drops (see Chapter 6).
   c. The environmental conditioning of the relief devices – can they discharge to atmosphere or not?
   d. Design the discharge piping/headers – avoid the unnecessary creation of backpressure on the safety valve, or determine the correct backpressure so it can be taken into account when sizing and selecting the relief device (see Chapter 6).
   e. Design a blowdown drum.
   f. Design the condensers, flare stacks or scrubbers (if any).

2.6 OVERPRESSURE RELIEF DEVICES

In our current process industry, we use different types of relief devices as primary protection, which can be divided as per attached schematic (Figure 2.12).

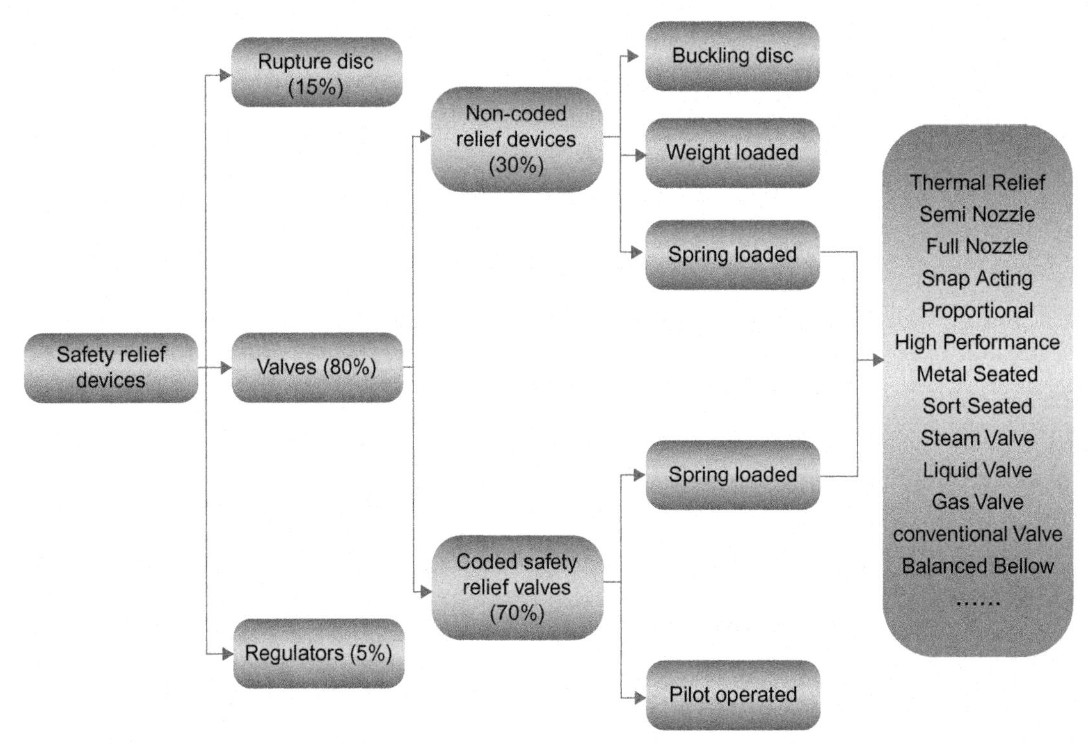

FIGURE 2.12
The market of pressure relief devices

The industry basically uses the following main groups of pressure relief devices to ensure overpressure protection:

- Reclosing devices
- Non-reclosing devices
- Combinations of reclosing and non-reclosing devices

The choice for the best solution is driven by a number of individual parameters, both technical and economical.

Generally spoken, the use of non-reclosing pressure relief devices will offer, in most cases, the lower cost solution but requires that the process is shut down or redirected through alternative safety systems to allow for replacement of the burst device. Subsequently non-reclosing pressure relief devices will only be selected as primary relief solutions in cases where loss of process media or shutdown for repair is tolerated or possible (see Section 5.4).

The selection of non-reclosing devices as a secondary or backup system is, however, a more widely accepted solution.

Reclosing devices allow for continued operation of the process, even when spurious overpressures occur. Consequently, reclosing devices will be preferred for primary relief applications where long-term opening of the process equipment can not be tolerated. The potential for leakage, fouling, plugging or icing can, however, render these critical devices sometimes inefficient, and great care is needed in making the correct selection of the device, taking into account both the application and the local codes. This also accounts for the wide variety of valves available on the market; they are all there for a reason.

Reclosing relief devices are mainly safety valves, relief valves or SRVs (either weight, spring or pilot operated), whereas the non-reclosing relief devices are bursting disc or buckling pin devices.

Combinations of SRVs and bursting discs are also relatively popular as they offer the best of both individual solutions (Figure 2.13).

The most commonly used combination will be a design where the bursting disc device is installed upstream of the safety valve. In such a configuration, the bursting disc device will provide a pressure and chemical seal between the process and the downstream valve, resulting in a better safety factor and reduced operational and maintenance cost (leakage, repair, corrosion, etc.) on the SRV in case of dirty, corrosive or polymerizing fluids.

The use of bursting disc devices on the downstream side of safety valves may be considered in cases where corrosion or fouling of the valve trim may be a concern (a common problem in systems using common headers to evacuate process media).

In all cases where combinations of bursting disc devices with SRVs are used, measures must be taken to ensure that the space between the valve seat and the bursting disc is kept at atmospheric pressure. Any increase of pressure in this cavity due to, for example, temperature changes, minute pressure leaks, and so forth, will result in a dramatic and uncontrolled change in opening pressure of the safety system. Also, with this combination the SRV must be 10% oversized to accommodate the eventual pressure drop over the bursting disc.

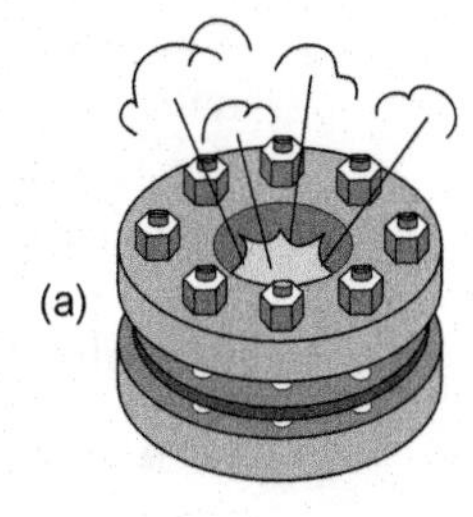

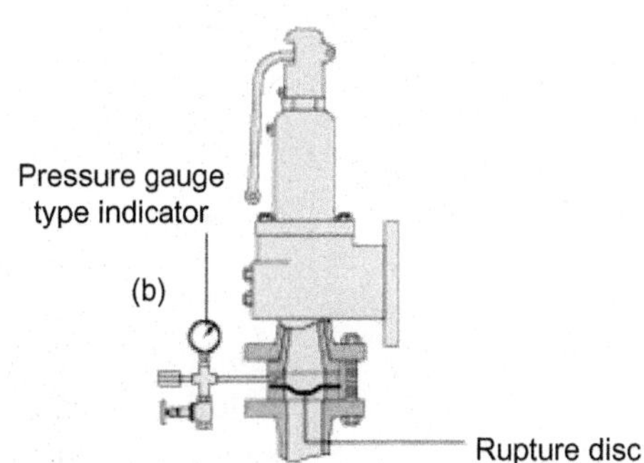

FIGURE 2.13
Rupture disc and SRV combination

Alternatively, in cases where pressure relief cannot be applied due to environmental or process issues, the use of controlled safety pressure relief systems (CSPRS) or safety-related measurement, control and regulating (SRMCR) devices may be evaluated. They are used primarily on steam systems. Such systems will generally be developed to interact with the process to avoid the occurrence of situations possibly leading to unsafe conditions. Such systems have to be carefully selected, taking into account guidance regarding safety redundancy specified in design documents such as IEC 61508 'Functional safety of electrical/electronic/ programmable electronic safety-related systems', IEC 61511 and ANSI/ISA S84.01.

2.7 RISK ASSESSMENT AND REDUCTION

One of the most critical steps in establishing the appropriate role and settings of the individual safety systems will be the risk assessment analysis, the process in which engineers consider and analyse all possible conditions in order to select the most appropriate safety concept, which ensures safe operation under all possible circumstances and scenarios (see Section 13.4).

Identifying the potential hazards (PHA, process hazard analysis, or HAZOP, hazard and operability analysis) during operation must be done from a wide-angle approach; dangerous situations can occur due to many root-cause situations other than those specified by, for instance, ASME or PED. Based on the results of the risk assessment, the pressure equipment can be correctly designed and the most effective safety system selected.

Basically, the process equipment shall be designed to:

- Eliminate or reduce the potential hazards as defined.
- Provide adequate protection measures if the hazards can not be eliminated.
- Inform the system user of the existence of eventual residual hazards.
- Indicate the appropriate protection measures used and prevent misuse of safety systems as applied.

Under all circumstances, preference will be given to inherently safe design solutions. Safety systems should be designed to operate independent of any other functions and should operate reliably under all conditions determined by the risk analysis (including start-up, shutdown and maintenance and repair situations).

The most commonly used pressure relief devices in the industry are:

1. Rupture discs (non-reclosing devices)

2. SRVs
   a. For low pressure, the conservation vent, or breather valve (not a subject for this book as other codes and recommendations apply for pressures under 0.5 barg)

   b. For pressures above 0.5 barg – spring-operated SRVs

c. Pilot-operated SRVs

In this book, we will only elaborate on pressure relief devices used above 0.5 barg and therefore subject to local legislations.

CHAPTER 3

Terminology

The terminology in pressure relief devices is very specific and therefore the terms used in this field are explained in order for the reader to better understand the literature on the subject. Some definitions are as they are given in the API.

Pressure relief device is the general term for a device designed to prevent pressure or vacuum from exceeding a predetermined value in a pressure vessel by the transfer of a fluid during emergency or abnormal pressure conditions. There are, however, different definitions for specific devices, their testing and their operating characteristics.

3.1 TESTING

Bench or test stand testing: Testing of a pressure relief device on a test stand using an external pressure source with or without an auxiliary lift device, to determine some or all of its operating characteristics, without necessarily flowing the rated capacity. This is required on a regular basis when the valve is taken into the maintenance cycle (see Chapter 10) at least to see that there is no shift on the set pressure and that the valve would open correctly during a pressure upset.

Flow capacity testing: The usually special testing of a pressure relief device to determine its operating characteristics, including measured relieving capacity. This tests whether the valve flows the capacity as stated in the literature or as per given flow coefficients, or to simply determine the flow coefficient of the valve as such. This is done on a spot-check basis by independent notified bodies in limited locations worldwide especially approved for that purpose.

Hydrostatic testing: Before assembly, each valve body is hydrostatically tested at the manufacturer at standard 1.5 times its maximum rating, typically during a period of 1–3 minutes. This is also called the shell test and reveals eventual deficiencies in the castings. Several manufacturers have different procedures, which can usually be obtained for review.

In-place testing: Testing of a pressure relief device installed on but not protecting a system, using an external pressure source, with or without an auxiliary lift device to determine or check some or all of its operating characteristics; mainly opening. Also, set pressure can sometimes be obtained by calculation.

In-service testing: Testing of a pressure relief device installed on and protecting a system using system pressure or an external pressure source, with or without an auxiliary lift device to determine or check some or all of its operating characteristics; mainly opening and set pressure. Usually this requires about 75% of set pressure present under the valve while testing (also known under commercial denominations such as Trevitest, Sesitest, etc.).

Leak test pressure: The specified inlet static pressure at which a quantitative seat leakage test is performed in accordance with a standard procedure (e.g. API 527, see Section 4.2).

Pre-start up testing: It is highly recommended that all valves be visually inspected before installation for dirt and particles, and the same goes for the system the valve will be installed upon. Especially new installations are prone to contain welding beads, pipe scale and other foreign objects, which are inadvertently trapped during construction. These foreign materials are devastating for the valve, and it is recommended that the system be purged carefully before installing the safety relief valves (SRVs) as these are very destructive when the valve opens. Also, caution should be taken that all protective materials, such as flange protectors, are removed before installing the valve.

It is also recommended that the valve be isolated or gagged during pressure testing of the system, but make sure the gag is removed after testing.

Some companies or local customs require the valves to be tested just before start-up, but normally the valves have already been set and sealed correctly at the manufacturer. This is not a recommended practice, but if needed on spring-operated valves, crack pressure can be checked by applying a suitable pressure source at the inlet of the valve. However, in the usual case on site, the volumetric capacity upstream is insufficient, and therefore a false reseat pressure (usually lower than actual) will be obtained.

On non-flowing-type pilot valves with a field test connection, the set pressure can be easily checked. It is recommended that the manufacturer's instructions, which should accompany the valve, be carefully followed.

Shell test: See hydrostatic testing.

3.2 TYPE OF DEVICES

3.2.1 Reclosing pressure-relieving devices

Reclosing pressure relief devices have a variety of names, although there used to be a clear definition based on the US market and API. However, when the Pressure Equipment Directive (PED) came into effect in 1997, this somewhat added to the confusion, as PED uses the overall term *safety valve* for every pressure-relieving device subject to the PED code. Originally the following were the definitions for the different terms per API and are still in use today.

Relief valve (RV): Spring- or weight-loaded pressure relief valve (PRV) actuated by the static pressure of the fluid. An RV opens normally in proportion to the pressure increase and is used primarily on incompressible fluids (liquids) (per PED it is called a *safety valve*).

Safety valve (SV): Spring-loaded PRV actuated by the static pressure of the fluid and characterized by rapid opening or pop action. Normally used with compressible fluids (gas, vapours and steam), the SV is also the general denomination in PED.

Safety relief valve (SRV): Spring-loaded PRV that may be used as either a safety or a relief valve, depending on the application. The SRV works within well-determined operational limits (per PED it is called *safety valve*).

Pressure relief valve (PRV): A more general term for a device that is designed to be actuated by the medium it protects, based on the inlet static pressure, and to reclose after normal and safe conditions have been restored within certain predetermined limits. It may be one of the following types and have one or more of the following design features. Since PRV is the general term, we have a large number of valve-type denominations that can be called a PRV.

- *Low-lift PRV*: A PRV in which the actual discharge area is the curtain area. It will not necessarily lift open fully to its capacity and acts more or less proportionally to the pressure increase but is usually relatively unstable during its relief cycle.
- *Full-lift PRV*: A PRV in which the actual discharge area is the bore area.
- *Reduced bore PRV*: A PRV in which the flow path area below the seat is smaller than the flow area at the inlet to the valve, creating a venturi effect.
- *Full bore PRV*: A PRV in which the bore area is equal to the flow area at the inlet to the valve, and there are no protrusions in the bore area between the inlet and flow area.
- *Pilot-operated PRV*: A PRV in which the disc or piston is held closed by system pressure, and the holding pressure of that piston is controlled by a separate pilot valve actuated by system pressure (see Section 5.3).

- *Conventional direct spring-loaded PRV*: A direct spring-loaded PRV which is held closed by a spring force which can be adjusted within a certain range and whose operational characteristics are directly affected by changes in the backpressure which is exercised at the outlet of the valve (see Section 5.2).
- *Balanced direct spring-loaded PRV*: The same as a conventional direct spring-loaded PRV, but which incorporates the means (typically a bellow) of minimizing the effect of backpressure on the operational characteristics (opening pressure, closing pressure and relieving capacity) (see Section 5.2).
- *Power-actuated PRV*: A PRV which can act independently using the force exercised by a spring, and which is additionally actuated by an externally powered control device (see Section 5.2.6.8).

3.2.2 Non-reclosing pressure relief device

This pressure relief device is designed to actuate by means of the process fluid and remains open after operation. A manual resetting or replacement will need to be provided after a pressure upset. The different design types include:

- *Rupture disc device*: A device that contains a disc which ruptures when the static differential pressure between the upstream and downstream side of the disc reaches a predetermined value. A rupture disc device includes a rupture disc and may include a rupture disc holder with eventual accessories (e.g. pressure gauge) (Figure 3.1).

FIGURE 3.1
Bursting disc or rupture disc in disc holder

- *Buckling or rupture pin device*: A device actuated by static differential or static inlet pressure and designed to function by the breakage of a load-carrying section of a pin or to function by the buckling of an axially loaded compressive pin that supports a pressure-containing member (Figure 3.2).

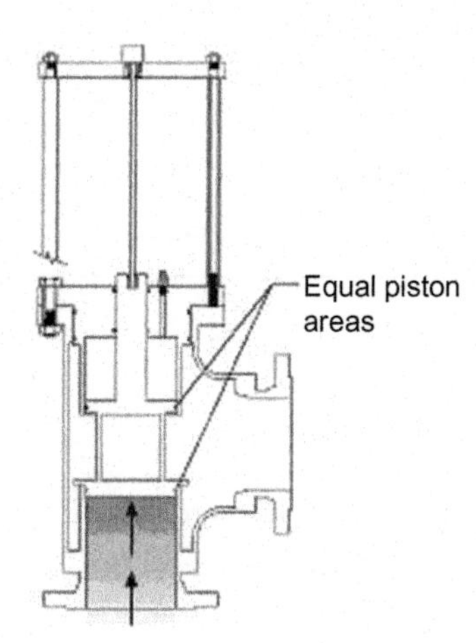

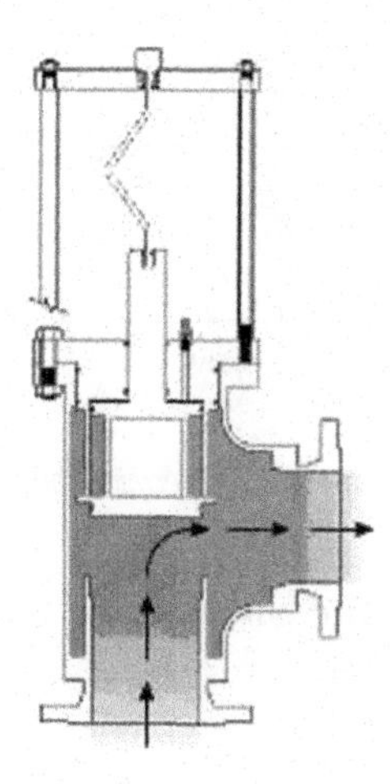

FIGURE 3.2
Buckling pin valve

- *Fusible plug device*: A device designed to function by the yielding or melting of a plug, at a predetermined temperature, which supports a pressure-containing member or contains pressure by itself. This device does not act on pressure but on temperature and can operate to both open or shut a valve. Usually used on lower pressure systems only. Fusible plugs are used regularly to protect internally fired steam boilers. If overheating occurs due to low water conditions, the plugs are designed to allow pressure to reduce, thereby preventing collapse of the boiler. They are also used to protect compressed air systems from the risk of an explosion occurring due to ignition of oil vapour, or they are used to protect air receivers from the risk of an explosion occurring due to external fire. They also usually provide an audible signal when opening (Figure 3.3).

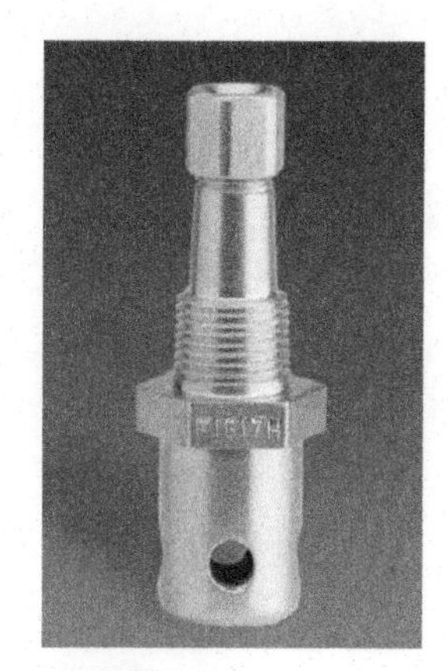

FIGURE 3.3
Fusible plug device

3.3 DIMENSIONAL TERMS

Actual discharge area: The lesser of the curtain and effective discharge or flow areas. The measured minimum net discharge area determines the flow through a valve (Figure 3.4).

Curtain area: Effectively the area of the cylindrical or conical discharge opening between the seating surfaces created by the lift of the disc above

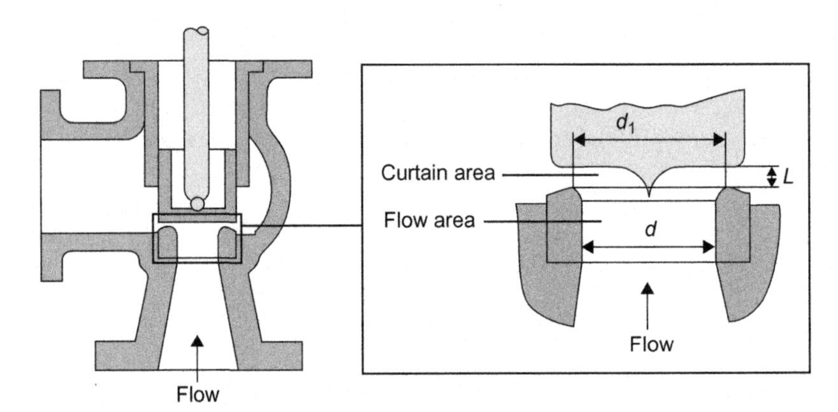

FIGURE 3.4

Different sections which determine the flow of the valve

the seat. The diameter of the curtain area is represented by dimension d_1 in Figure 3.4.

$$\text{Curtain area} = \pi d_1 L$$

Effective discharge or flow area: The minimum cross-sectional area between the inlet and the seat, at its narrowest point. The diameter of the flow area is represented by dimension d in Figure 3.4. It is a nominal or computed area of flow through an SRV, differing from the actual discharge area, and is used in recognized flow formulas to determine the capacity of an SRV. So, in short, it is the computed area based on flow formulas.

> *API 520 Definition: A nominal or computed area used with an effective discharge coefficient to calculate the minimum required relieving capacity for a pressure relief valve per the preliminary sizing equations contained in API Standard 526. API Standard 526 provides effective discharge area for a range of sizes in terms of letter designations D through T.*

$$\text{Flow area} = \frac{\pi d^2}{4}$$

Inlet size: The nominal pipe size of the inlet of an SRV, unless otherwise designated.

Outlet size: The nominal pipe size of the outlet of an SRV, unless otherwise designated.

Lift: The actual travel of the disc of the SRV from its nozzle away from the closed position when a valve is relieving.

Nozzle or bore diameter: The minimum diameter of a nozzle.

Orifice area: See effective discharge area (Figure 3.5). API has determined standard orifice areas from *D* to *T*, each with a corresponding orifice size (in^2 or cm^2) (See Section 4.2.3).

Seat diameter: The smallest diameter of contact between the fixed (nozzle) and moving portions of the pressure-containing elements of an SRV.

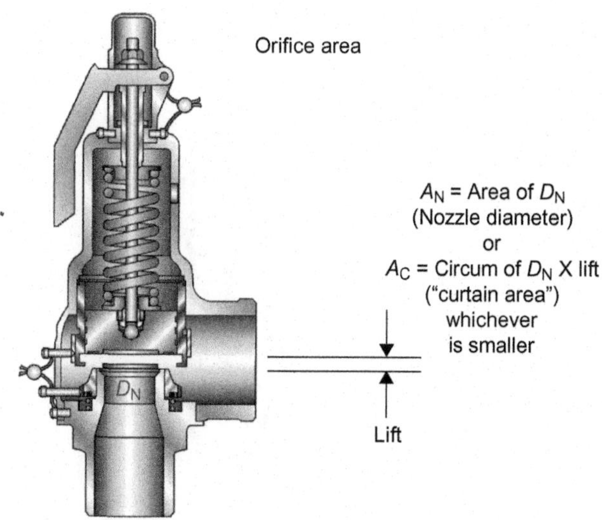

FIGURE 3.5
Determination of the orifice area

3.4 OPERATIONAL TERMS

Accumulation: The pressure increase over and above the MAWP (maximum allowable working pressure) during the discharge of the pressure relief device. Expressed in pressure units or as a percentage of set pressure. Maximum allowable accumulation is established by applicable codes for operating and fire contingencies (see Section 3.6).

Backpressure: The static pressure which exists at the outlet of an SRV due to existing pressure in the discharge system. It is the sum of superimposed and built-up backpressure, and potentially influences the set pressure and certainly the operation of the valve (Figure 3.6).

Based on backpressure existing at the outlet of a spring-operated SRV, we use the *conventional* SRV for backpressures typically under 10% of set pressure. We must use a *balanced bellows SRV* or a *pilot-operated SRV* for instable backpressures or backpressures over 10% (Figure 3.7).

We must differentiate two types of pressures existing at the SRV outlet prior to opening:

- Built-up backpressure (variable)
- Superimposed back pressure
  a. Constant
  b. Variable

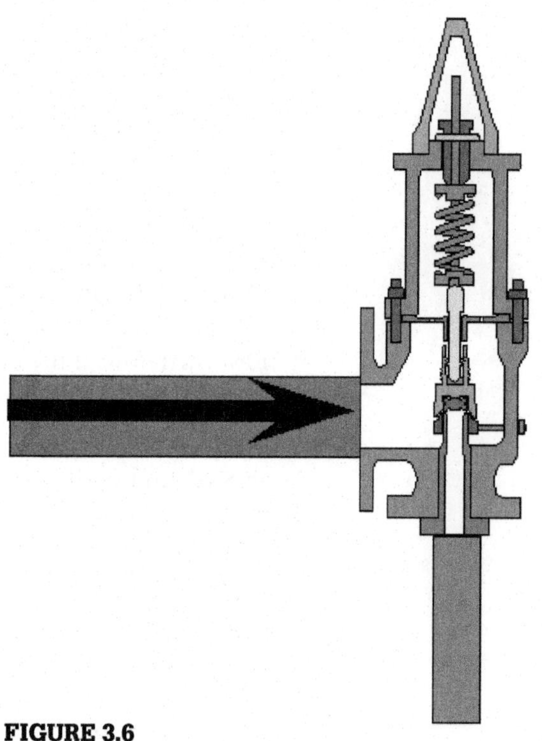

FIGURE 3.6
Backpressure

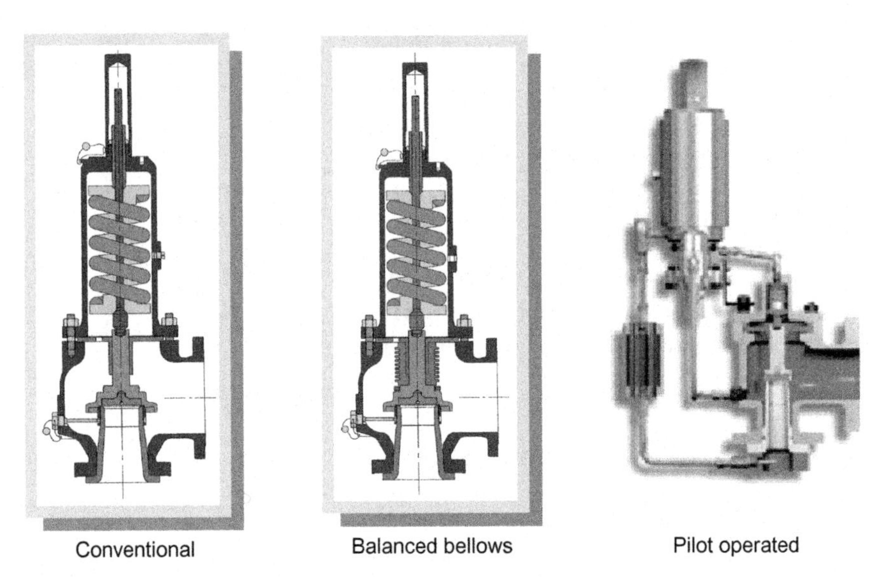

FIGURE 3.7
Conventional balanced bellows and pilot operated safety relief valves

Built-up backpressure: Occurs when the safety valve is open and flowing due to:

- The rate of flow through the PRV
- The size and/or configuration of the discharge piping
- Other sources of pressure into the discharge header

Built-up backpressure is always variable and usually occurs due to friction (vortices/turbulences) and pressure drops through the discharge piping. It is allowed to be up to 10% (of set pressure) on conventional design SRVs but will cause reduced capacity and unstable operation if the pressure gets over 10%. Therefore, if the built-up backpressure is greater than 10%, we will need to use a balanced bellows – or pilot-operated design SRV.

Superimposed backpressure: Superimposed backpressures acting on the outlet of an SRV can be either constant or variable. Superimposed backpressure occurs when the valve is closed and pressure already exists at the outlet of the valve. This is due to existing constant and/or variable pressures which exist in the discharge system.

Constant superimposed backpressure: Usually backpressures that occur when a safety valve outlet is connected to a static pressure source and doesn't change appreciably under any conditions of operation. In this case, conventional valves may be used if the valve spring setting is reduced by the amount of the constant backpressure (Figure 3.8).

In case of constant backpressure: Actual set pressure = bench set + backpressure

Example:

Required set pressure: 10 bars
constant backpressure: 2 bars
Bench set pressure: 10 – 2 = 8 bars

Variable superimposed backpressure: Usually the result of one or more SRVs discharging into a common header. The backpressures may be different at each moment and at each relief cycle. Bellows or pilot design is always required since no predetermined set pressure is possible when the outlet pressure is acting on the trim of the valve, therefore directly influencing the set pressure, and the set point will vary with backpressure (Figure 3.9).

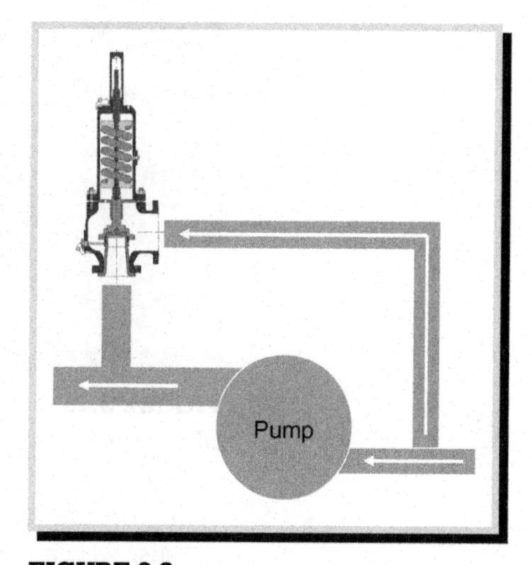

FIGURE 3.8
Constant superimposed backpressure

The typical effects of backpressure are:

- Increase of the set pressure in unbalanced valves
- Reduced flow capacity
- Instability of the valve resulting in chatter
- Corrosive attack to the spring chamber components

As we indicated, we can compensate for the effects of backpressure by selecting the correct valves such as balanced bellows – or pilot-operated valves but the backpressure also has an effect on the rated capacity of the valve and therefore larger valves could be required if backpressure exists (Figure 3.10).

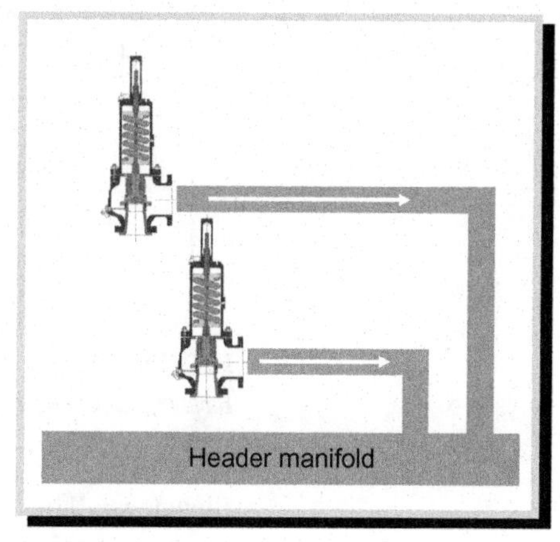

FIGURE 3.9
Variable superimposed backpressure

This illustrates why a conventional SRV is best suited for simple discharge via a tail pipe into atmosphere. Its ability to tolerate built-up backpressure is very limited.

The loss of lift and resultant capacity at higher levels of backpressure is caused by the backpressure acting on the external surfaces of the bellows, attempting to lengthen it, which produces an increased spring rate of the bellows. To maintain the bellows' structural integrity and resist instability, they are normally limited to 50% of backpressure (as a percentage of set pressure) or less. Above that value of backpressure, pilot valves should be considered.

Balanced bellows design: Design used for SRVs subject to backpressures over and above 10%. The effective area of the bellows must be the same as the nozzle seat area ($A_B = A_N$), this way the bellow prevents backpressure from

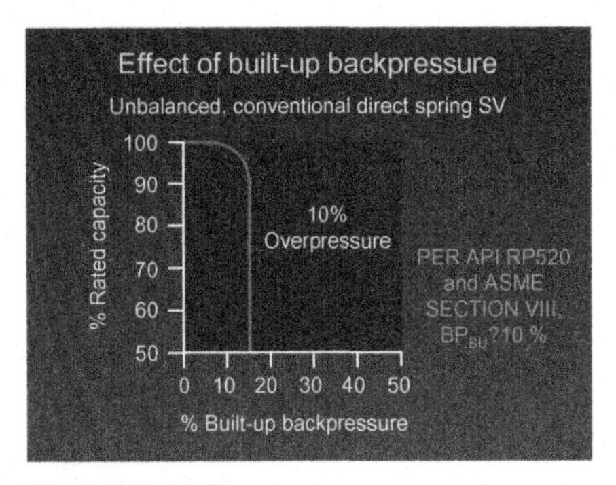

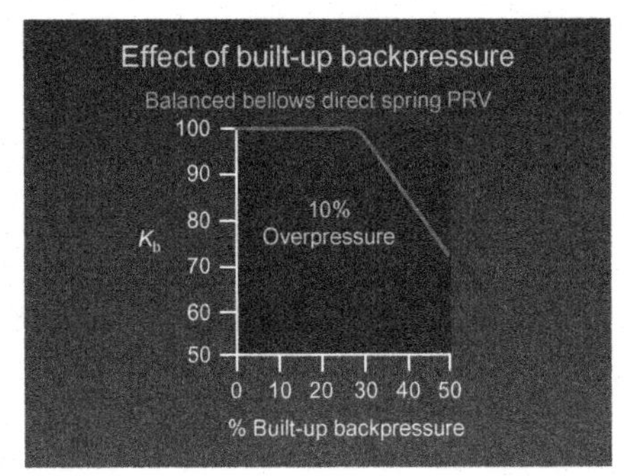

FIGURE 3.10
Effects of backpressure on flows

acting on the top area of the disc and cancels the effects of backpressure on the disc which results in a stable set pressure (Figure 3.11).

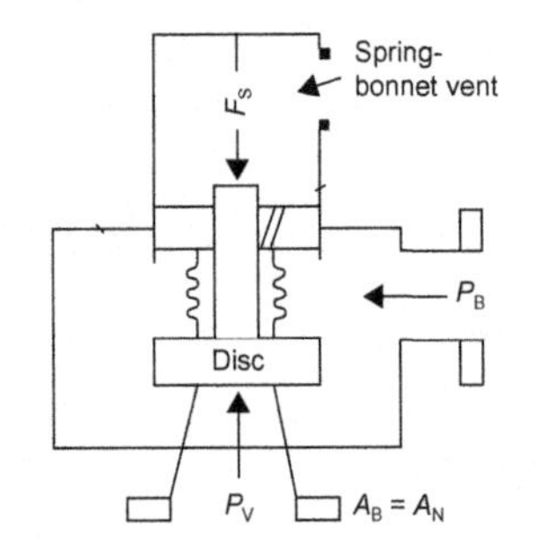

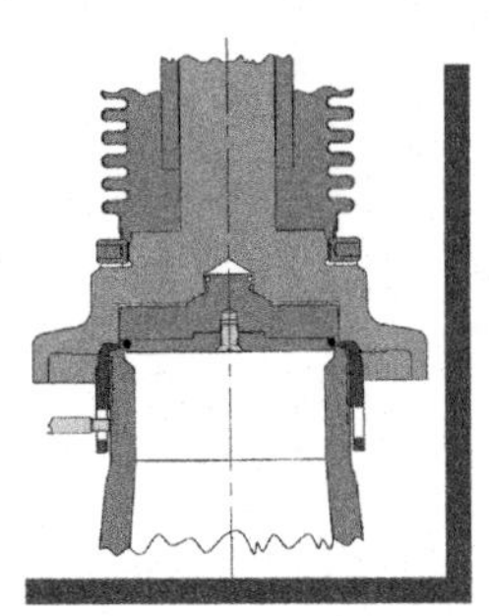

FIGURE 3.11
Balanced bellow designs

Blowdown: The difference between the actual set pressure of an SRV and the actual reseating pressure, expressed as a percentage of set pressure or in pressure units.

Chatter, simmer or flutter: Abnormal, rapid reciprocating motion of the movable parts of a PRV in which the disc makes rapid contacts with the seat. This results in audible and/or visible escape of compressible fluid between the seat and the disc at an inlet static pressure around the set pressure and at no measurable capacity, damaging the valve rapidly.

Closing pressure: The value of decreasing inlet static pressure at which the valve disc re-establishes contact with the seat or at which the disc lift becomes zero.

Coefficient of discharge: Also called the 'K' factor, the ratio of the measured relieving capacity to the theoretical relieving capacity. It determines the flow capacity of the SRV and can be slightly different from type of valve or from manufacturer to manufacturer. It is, however, not at all a rating for the valve quality. Not necessarily all valves with a high 'K' factor are high-quality valves (see Section 7.1).

Cold differential set pressure: The inlet static pressure at which an SRV is adjusted to open on the test stand. This test pressure includes corrections for service conditions of superimposed backpressure and/or low or high temperature.

Conventional SRVs: Spring-operated SRVs which can be used up to backpressures of 10%.

Crack(ing) pressure: See opening pressure.

Design (gauge) pressure, rating: Most severe condition of a temperature and pressure combination expected during operation. May also be used instead of the MAWP (in all cases where the MAWP has not been established). Design pressure is equal to or less than the MAWP.

Dome pressure: The pressure at the dome connection of a pilot-operated valve, which is usually the same as the inlet pressure. It is the pressure that is exercised on top of the unbalanced piston in the main valve and which, in normal operating conditions, is the force that keeps the valve closed.

Huddling chamber: An annular pressure/boosting chamber in an SRV located above the seat area for the purpose of generating a rapid opening (Figure 3.12).

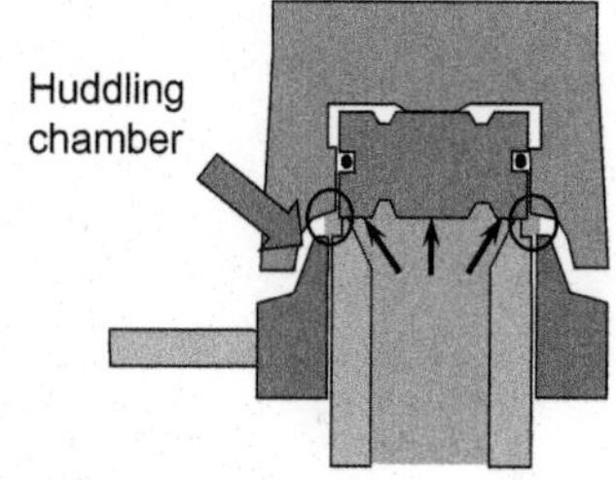

FIGURE 3.12
Huddling chamber

Leak pressure: The value of increasing inlet static pressure at which the first bubble occurs when an SRV is tested by means of air under the valve disc and a specified water seal on the outlet.

Maximum allowable working pressure (MAWP): The maximum permissible gauge pressure of a vessel in its operating position at a designated temperature. The pressure is based on calculations for each element in a vessel, using nominal thickness exclusive of additional metal thickness allowed for corrosion. The MAWP is the basis for the upper limit of pressure setting of the safety relief devices that protect the vessel (see Section 3.6).

Maximum operating pressure (MOP): Maximum pressure expected during normal system operation.

Modulating action: A gradual opening and closing characteristic of some SRVs, particularly some pilot-operated types, in which the main valve opens in proportion to the relief demand at that time. This proportionality is not necessarily linear.

Opening pressure: The value of increasing static pressure of an SRV at which there is a measurable lift and the disc is fully loose from the nozzle, or at which the discharge becomes continuous as determined by seeing, feeling or hearing.

Overpressure: The pressure increase over the set pressure of an SRV at which the valve will flow nominal or rated flow, usually expressed as a percentage of the set pressure. (= Accumulation when the relieving device is set at the MAWP and no inlet pipe losses exist to the relieving device.)

Pop action: An opening and closing characteristic of an SRV in which the valve immediately snaps open into high lift and closes with equal abruptness.

Popping pressure: The inlet static pressure at which the disc starts moving in the opening direction.

Primary pressure: The pressure at the inlet of an SRV.

Rated relieving capacity: The measured relieving capacity approved by the applicable code or regulation, to be used as a basis for the application of an SRV on a system which requires compliance with the code.

Rated coefficient of discharge (API): The coefficient of discharge determined in accordance with the applicable code or regulation which is used together with the actual discharge area to calculate the rated flow capacity of an SRV (see Section 7.1).

Relieving conditions: The inlet pressure and temperature on an SRV that occur during an overpressure condition. The relieving pressure is equal to the valve set pressure or the rupture disc burst pressure plus the allowed overpressure for that application. The temperature of the flowing fluid at relieving conditions may be and usually is higher or lower than the operating temperature, depending on the fluid.

Relieving pressure: Set pressure plus overpressure.

Resealing pressure: The value of decreasing inlet static pressure at which no further leakage is detected after closing. The method of detection may be a specified water seal on the outlet (API 527) or other means appropriate for this application (see Section 4.2.3).

Reseating pressure: See 'resealing pressure'.

Set pressure: The value of increasing inlet static pressure at which an SRV displays one of the operational characteristics as defined under opening pressure, popping pressure, start-to-leak pressure. (The applicable operating characteristic for a specific design is specified by the manufacturer.)

Superimposed backpressure: The static pressure existing at the outlet of a pressure relief device at the time the device is required to operate. It is the result of pressure in the discharge system from other sources and may be constant or variable. Also see backpressure.

Test pressure: See relieving pressure.

Theoretical relieving capacity: The computed capacity expressed in gravimetric or volumetric units of a theoretically perfect nozzle having a minimum cross-sectional flow area equal to the actual discharge area of an SRV or the net flow area of a non-reclosing pressure relief device.

Variable backpressure: A superimposed backpressure that will vary with time. Also see Backpressure.

3.5 COMPONENT TERMS

Adjusting ring/blowdown ring/control ring/nozzle ring: A ring assembled to the nozzle or guide (or a ring for each) of a direct spring-operated SRV, used to control the opening characteristics and/or the reseat pressure (blowdown).

The adjusting/blowdown or control ring or rings (in case of ASME I valves) are the nozzle and guide rings through which controllable spring valve actions (accurate opening, full-lift and proper blowdown) are obtained in meeting varying conditions. Correct valve operation will depend upon correct control ring settings.

Most manufacturers have recommendations on how to set the nozzle ring(s) in order to obtain a correct opening and closing characteristic of their SRV design. Incorrect ring adjustment may cause valves to:

- *Have too long or too short a blowdown:* If the nozzle ring is set higher than the recommended setting, the valve may fail to reseat correctly while the system is operating. The blowdown can be too long and the valve might have problems closing correctly causing excessive damage to the seat and nozzle area.
- *Have too long a simmer:* Dual ring control, valves (ASME I) may fail to achieve full-lift at allowable overpressure limit if the nozzle ring is set lower than that recommended, a situation which could lead to a catastrophic event as it could take too long for the valve to react on the overpressure condition.

To ensure correct control ring settings, each reputable manufacturer issues nozzle ring and guide ring settings for its specific range of SRVs. These settings are taken from datum points. Generally, this datum is the face which the nozzle ring touches when it is in its highest position.

Body: A pressure-retaining or -containing member of an SRV that supports the parts of the valve assembly and has provision(s) for connecting to the primary and/or secondary pressure source.

Bonnet: A component of a direct spring-operated SRV or of a pilot on a pilot-operated SRV that supports and houses the spring. It may or may not be pressure containing, depending on the design and/or application.

Cap: A component on top of the bonnet used to restrict access and/or protect the set pressure adjustment screw. It may or may not be a pressure-containing part. In operation it should always be sealed with a leaded wire.

Disc/seat: The pressure-containing, movable element of an SRV which effects closure.

Flowing pilot: A pilot design that continuously discharges the fluid through the pilot throughout the relieving cycle of the pilot-operated safety relief valve (POSRV).

Lifting device/lift lever: A device for manually opening an SRV by the application of an external force to lessen the spring loading which holds the SRV closed. This allows the SRV to open manually below set pressure.

Non-flowing pilot: A pilot in which the fluid flows only just enough to enable a change in the position of the piston in the main valve it controls. During the relief cycle, there is no fluid going through the pilot.

Nozzle: A pressure-containing inlet flow passage that includes the fixed portion of the disc/seat closure combination responsible for the tightness of the valve. The capacity of a full-lift SRV is determined by the precision diameter of the nozzle bore. Generally there are two types of nozzle design. In most ANSI designs, the full-nozzle design is used, while most DIN designs are semi-nozzle (Figure 3.13).

Pilot: An auxiliary valve assembly utilized on pilot-operated SRVs to determine the opening pressure, the closing pressure and the opening and closing characteristics of the main valve.

Pilot-operated safety relief valve (POSRV): A self-actuated SRV comprising a main valve and a pilot. The operations of the main valve are controlled by the pilot, which responds to the pressure of the fluid. The main valve assures the quantity (capacity) of the valve, while the pilot assures the quality (accuracy) of the operation(Figure 3.14).

Piston: Utilized in the main valve assembly of most POSRVs, its movement and position control the main valve opening, flow and valve closure.

Power actuated/assisted safety valves (CSPRS – controlled safety pressure relief system): A spring-operated safety valve actuated or assisted by an externally powered control device which can be hydraulic, pneumatic or electric (Figure 3.15).

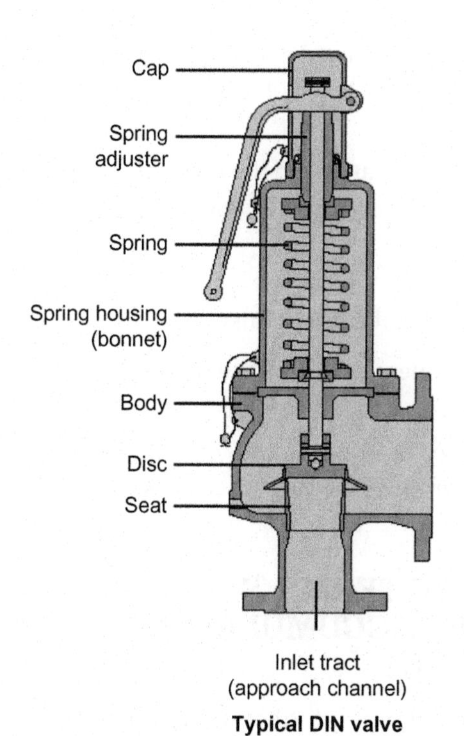

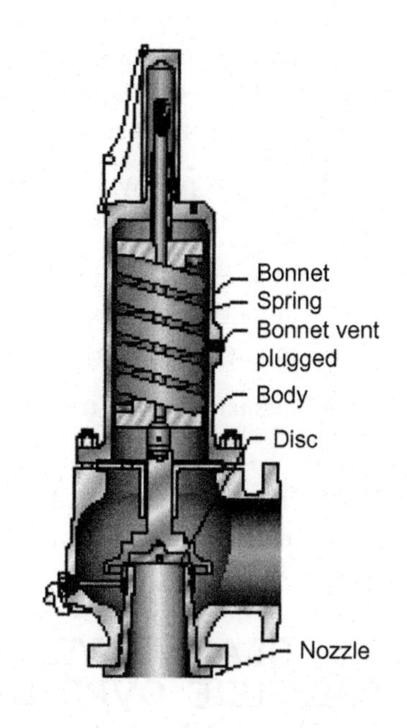

FIGURE 3.13
Semi- and full-nozzle designs

FIGURE 3.14
Pilot-operated SRV

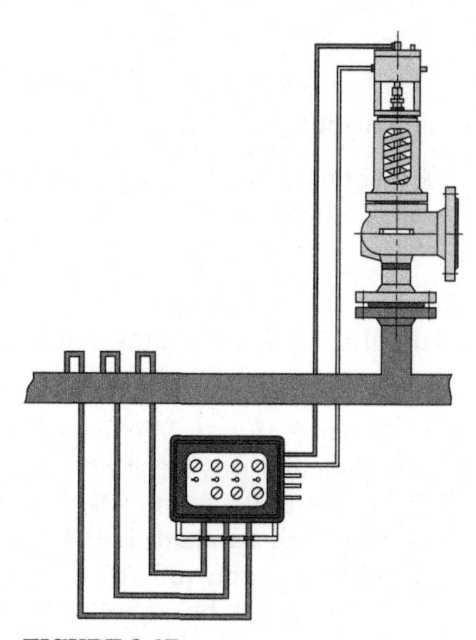

FIGURE 3.15
Possible CSPRS system set-up

Seat: The pressure-containing contact between the fixed and moving portions of the pressure-containing elements of a valve. It may be made of metallic, plastic or elastomere material.

Spindle/stem: A part whose axial orientation is parallel to the travel of the disc/seat. It may be used in one or more of the following functions:

- Assist in alignment
- Guide disc/seat travel
- Transfer of internal or external forces to the discs/seats

Spring washer: A load-transferring component in an SRV that supports the spring.

Spring: The element in a direct spring SRV or pilot that provides the force to keep the disc/seat on the nozzle and determines the valve's set pressure.

3.6 CLARIFICATION OF THE TERMS: SET PRESSURE, OVERPRESSURE, ACCUMULATION, MAWP AND DESIGN PRESSURE

While we will cover some of the subjects above in further detail, it has to be noted that in some instances many users and even manufacturers get confused by the terminology of pressure relief devices, in particular, the important terms *overpressure, accumulation, MAWP* and *design pressure* (see also Appendix H for a visualization).

Even today, an important point of discussion among end users, inspectors and manufacturers is the exact definition of *set pressure.*

A lot of users state that if you want to reduce the size of a relief device for cost savings, then you have to design it at a higher set pressure, while not ignoring the weakest link – the MAWP.

Others, amongst whom are many manufacturers, state that this is not true and that for a certain MAWP, the capacity of the relief device is not a function of its set point, but of MAWP alone.

For example, for an MAWP of 100 barg, the relief valve capacity will be the same whether it is set at 80 barg or 100 barg. In both cases, the maximum relieving pressure for the ASME non-fire case (or, for instance, the BS 5500 fire case – British standard) is 124.7 barg, and the discharge capacity will remain identical. The only difference is that if the set point is 80 barg, the allowable overpressure will be 37.5%, while, at the same for a set point of 100 barg, it will be only 10%. For the ASME fire case, the values will be 51.25% and 21% respectively. These values are defined very clearly in API 520, tables 2–6.

So among specialists there are apparently different interpretations or understandings on what API 520 and the ASME Boiler and Pressure Vessel Code, Section VIII, Division 1, are really saying, while it is these organizations which are setting the codes that determine how pressure vessels are to be designed and protected. Remember that these codes are law and must be followed.

Most SRVs especially installed in the oil, gas and petrochemical industry both in the United States and in Europe have been designed around this ASME code and the API recommendations. Of course most also comply with the European PED, which is also not specific on this particular subject.

Although the code was established in the early 1900s, today the question of what exactly is set pressure remains a point of discussion among different parties.

I'm sure many people dealing with SRVs have had this discussion at least once in their lives.

It starts when it comes to setting the SRV, where ASME simply refers to set pressure as the 'first audible leak'. Others refer to the set pressure as being when there is the first lift of the disc from the nozzle without any touching or simmer. This is what most manufacturers always applied as the rule, but it is actually difficult to physically measure.

The second big confusion exists around MAWP and design pressure. In paragraph 1.2.3.2 (b), API 520 defines MAWP as

> *... the maximum gauge pressure permissible at the top of a completed vessel in its normal operating position at the designated coincident temperature specified for that pressure.*

Probably the operative word here is 'completed'. The vessel is completed when a fabricator, according to the code laid down by ASME, has designed it.

It is the vessel fabricator and not the process engineer who determines MAWP. Some may try to stretch the definition of 'completed' to mean that the vessel is also erected in place. Not quite, because the certified vessel drawings, which are delivered long before the vessel is erected, already must contain this information.

In the same paragraph, API 520 says that the MAWP is normally higher than design pressure. The process engineer usually sets the design pressure as the value obtained after adding a margin to the most severe pressure expected during normal operation at a coincident temperature.

Depending upon the company, this margin is typically a maximum of 25 psig or 10%, whichever is greatest. The vessel specification sheet contains the design pressure, along with the design temperature, size, normal operating conditions and material of construction among others. It is this document that will eventually end up in a fabricator's lap and from which the mechanical design is made.

Unfortunately, project schedules may require that SRV sizing be carried out long before the fabricator has finished the mechanical design and certified the MAWP of his vessel. The process engineer must then use some pressure on which to base the relieving rate calculations. Paragraph 1.2.3.2 (c), API 520 states that the design pressure may be used in place of the MAWP in all cases where the MAWP has not been established.

Guess what pressure the process engineer usually would set the SRVs at? He will, of course, use his familiar design pressure. There are even times when the relief valve is set lower than the design pressure. For example, a high design pressure may be desirable for mechanical integrity, but an SRV set pressure at the design pressure could end up with a coincidental temperature that would require the use of exotic construction materials or one that promotes decomposition and/or runaway reactions.

So, why the confusion?

The confusion is due to a number of reasons. First is the way ASME does not relate the maximum allowable pressure (MAP) limits to SRV capacity. Throughout the entire document, ASME, Section VIII, Division 1, refers to MAWP when talking about SRV set pressure and allowable overpressure.

We could interpret what is stated (in part) in paragraph UG-125 of ASME Section VIII, Division 1:

> *....All pressure vessels other than unfired steam boilers shall be protected by a pressure relief device that shall prevent the pressure from rising more than 10% or 3 psi, whichever is greater, above the maximum allowable working pressure except as permitted in (1) and (2) below...*

Sub-paragraphs (1) and (2) mention cases where the pressure rise may be higher.

However, when ASME talks about certifying the capacity of a relief device, MAWP is never mentioned. ASME Section VIII, Division 1, clearly states in paragraph UG-131 (c)(1) that:

> *.....Capacity certification tests shall be conducted at a pressure which does not exceed the pressure for which the pressure relief valve is set to operate by more than 10% or 3 psi, whichever is greater, except as provided in (c)(2)...*

Sub-paragraph (c) (2) covers a fire case. Again, capacity certification is based only on the set pressure of the SRV and is unrelated to MAWP, unless of course the set pressure is MAWP.

Another area of confusion might involve the definition of *capacity* and how the term is used in ASME and API. Relieving rates are determined from 'what can go wrong' scenarios and, if allowed to go unchecked, would overpressure the vessel.

Once the process engineer determines the controlling relieving rate from all the scenarios, the required SRV orifice size is determined usually by the manufacturer using the appropriate equation given in API or one of our manufacturer's sizing programs, which usually use API or EN formulas. Once the required SRV orifice size is calculated, an actual orifice size equal to or greater than the calculated orifice size is chosen from a particular manufacturer's available selection. The maximum flow through this actual valve will be the valve's capacity.

The problem and solution can be summarized as follows:

Misinterpretation of code:

Capacity based on MAWP + Allowable Overpressure

Code as written:

Capacity based on Set Pressure + Allowable Overpressure

So, basically it is important to note that design pressure is not the same as MAWP and that the timing on which each is determined is important in determining the set pressure of the SRV.

In any case, the code clearly requires that the SRV's capacity be based solely on set pressure and not on the vessel's MAWP.

Indeed, as shown above, if the SRV's capacity were based on MAWP, then code might even force the process engineer into an unsafe design.

A good analogy is highway speed limits. In some European countries, many highway speed limits are set for 120 km/h. This does not mean a driver cannot travel more slowly and, under certain conditions for safety, it is almost a necessity that one does.

If it is safe to do so and the protected vessel can be allowed to pressurize to a greater extent, the SRV set pressure can be increased, thereby reducing the SRV's size and cost. Remember also that there is piping and possibly downstream equipment to 'catch' and process the relieving fluid associated with the SRV, which may also benefit by this reduction.

One way of accomplishing a reduction in the SRV size is by increasing the vessel's design pressure.

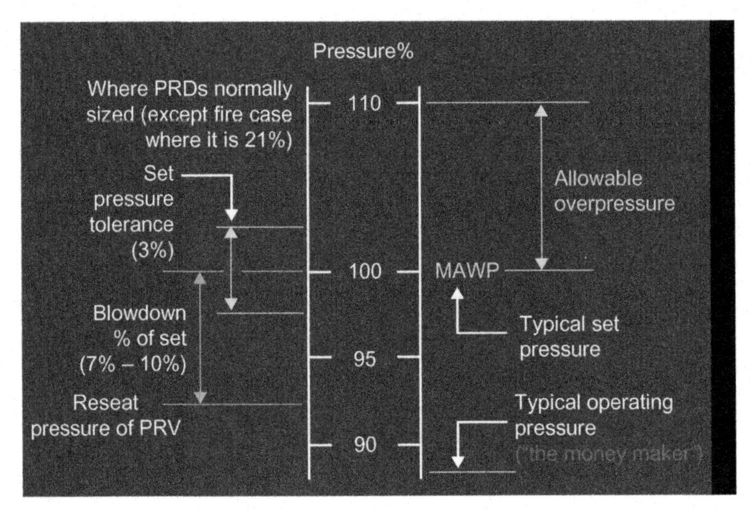

FIGURE 3.16
Pressure table

There is an economic trade-off here as the vessel's cost can increase above what you may save by reducing the size of the valve.

Another frequently used approach is to consider increasing the SRV's set pressure right up to MAWP after receiving the certified vessel drawings. However, depending on project schedule, the cost savings may be offset by the high costs associated with late design changes.

Presented here is a schematics (Figure 3.16) determining the code requirements for an SRV where:

- MAWP is typically set pressure and is determined by the pressure vessel supplier, approved by a notified body.
- The tolerance on set pressure per ASME is +/−3%.
- Design pressure can be under or (usually) above MAWP.
- SRV must flow nominal flow (maximum capacity) at 10% overpressure above its set pressure.
- SRV must be bubble tight up to 90% of set pressure.
- The valve must be adjustable to reclose between 7% and 10% under set pressure (blowdown).

3.6.1 PED versus ASME

Here we will explain the terms *accumulation* and *overpressure* relative to the European Pressure Equipment Directive PED 97/23/EC versus as used by ASME.

3.6.1.1 *Overpressure versus accumulation*

Every pressure vessel installed in the European Union, and the SRV(s) protecting it, must comply with the PED. In fact, in Europe they must comply with any and all European Directives that may apply, for example, the ATEX Directive, which we will discuss in more detail in Chapter 4 on the 'codes'.

Like other pressure vessel codes, the PED recognizes that most SRVs need some increase in pressure above their set pressure to be fully opened and relieve their full rated capacity. This is what we call *overpressure.*

The PED, as does the ASME VIII, accepts that the SRV is set at the MAP or PS that is the exact equivalent to MAWP in the ASME code. Also in PED, MAP or PS are the design pressure of the weakest component of the equipment used in a pressure vessel, which needs protecting from potential overpressures. This pressure vessel may be an assembly of different components – pipes, flanges, nozzles, shells, and so on – and each may have a different design pressure. It establishes the limit of the pressure vessel for very short and exceptional increases of pressure above MAP, specifically to enable the SRV to operate properly and reach its rated capacity. This exceptional increase is called *accumulation.* So accumulation is specific to the individual pressure vessel and does not relate to the SRV. The pressure increase of the SRV to nominal flow is called overpressure.

3.6.1.2 *ASME VIII on accumulation*

Some codes have established different levels of accumulations, depending on the situation of the equipment. For example, the ASME Section VIII imposes the following accumulations on pressure vessels:

ASME Section VIII Case Accumulation

ASME Section VIII	Case	Accumulation
Protection with ONE safety relief valve	Fire case	21% of MAWP
Other cases		10% of MAWP
Protection with MULTIPLE safety relief valves	Fire case	21% of MAWP
Other cases		16% of MAWP

Note: To emphasize the difference between overpressure (safety relief valve characteristic) and accumulation (code limitation on the pressure vessel), safety relief valves installed for fire cases will have an overpressure of 10% like most safety relief valves, even if the allowed accumulation on the pressure vessel is 21% in the case of ASME VIII.

3.6.1.3 *Multiple valves*

In the case of multiple valves protecting the equipment, at least one valve must always be set at no more than the MAWP. Of course, it is always allowed to set the safety valve at a set pressure lower than MAWP.

Many codes allow that the other valves be set higher than the MAWP, typically up to 5% above the MAWP, this is to allow the 'staggering' of the set pressures to avoid interaction between the valves.

Staggering the valves means that the settings of the valves are not all set at the same set point so that not all valves open at the same moment but in a sequence.

Practical example following the directions of ASME VIII – UG125 for multiple SRVs installation:

Multiple valves are used for a wide variety of reasons but most common are:

- The total required capacity is too high for even the largest API lettered valve.
- Multiple smaller valves are preferred because they are easier to handle for later maintenance or the weight of the valves needs to be divided over a larger surface because of support reasons.
- The reaction force of one valve would become too high and would require too much support structure.
- And others such as isometry, place, insulation, etc.

One SRV set at or below MAWP. Balance may be staggered, set with the highest being no more than 105% of MAWP.

Vessel MAWP 200 barg
Normal operating pressure 175 barg
Quantity of valves 4
1 set at 195 barg (i.e. 10% operating/set)
1 set at 200 barg (MAWP)
1 set at 205 barg (2.5% above MAWP)
1 set at 210 barg (maximum 5% above MAWP)

3.6.1.4 PED on accumulation

According to the PED, the allowed accumulation for pressure equipment is 10% in all cases but fire.

PED 97/23/EC	Case	Setting of SRV	Accumulation
Protection with ONE SRV	Fire case	Below or at MAP	10% above MAP or higher if proved safe by the vessel designer
	All other cases		10% above MAP
Protection with MULTIPLE SRVs	Fire case	One valve below or at MAP and other valves up to MAP + 5%	10% above MAP or higher if proved safe by the vessel designer
	All other cases		10% above MAP

The PED annexes below give the references to support this table.

3.6.1.5 *All cases except fire*

The maximum accumulation on the equipment must be no higher than 10% above MAP, even in the case of multiple safety valves.

See PED Annex I, clause 7.3.

> *Pressure limiting devices, particularly for pressure vessels: The momentary pressure surge referred to in 2.11.2 must be kept to 10% of the maximum allowable pressure.*

In the case of multiple valves, only one valve needs to be set at no more than MAP. The others can be set up to MAP +5% (inclusive). In any case, the required capacity must always be relieved at no more than MAP +10%. As taken from the harmonized standard EN 764-7 paragraphs:

> *To support the PED, many European (EN) standards are now 'harmonised' with the PED. The PED has been introduced as the pressure vessel code in each and every country of the European Union. Likewise, the new EN standards have been introduced and they replace all the local standards. For example, the standard for safety valves, EN 4126, is now the German standard DIN EN 4126, or the French standard NF EN 4126, etc... This harmonised standard has an annex (called 'Annex ZA') that lists the paragraphs which address the requirements of the PED. By following these harmonised standards, one is sure therefore to comply with the PED clauses supported by the paragraphs listed in the Annex ZA. (However, we may need to follow several harmonised standards to cover all the possibilities!)*
>
> *This Annex ZA is reviewed and approved by the PED experts of the European Commission, and so can be considered as an official interpretation about how to comply with some of the PED requirements.*
>
> *It is always important to remember that these harmonised standards remain 'standards': they are not compulsory. Only because they support the PED, they give useful guidelines to comply with the PED.*
>
> *Other of such standards which are useful are the EN 764-7 and EN 12952-10: Water-tube boilers and auxiliary installations – Requirements for safeguards against excessive pressure. Although its Annex ZA shows that it supports the PED clauses 2.11 and 7.3, this standard does not give any indication on the set pressure and the overpressure of the Safety Relief Valves, in which case one can refer back to the above.*

According to PED, Pressure Equipment, Part 7 – Safety systems for unfired pressure equipment, its Annex ZA lists the paragraphs which address some of the PED Annex I clauses.

PED Annex I clauses 2.11.2 & 7.3:

Supporting EN 764-7, paragraphs:

6.1.4 Pressure limit: Pressure limiting devices shall be effective at a pressure such that the pressure in the equipment is prevented from exceeding 1.1 times the maximum allowable pressure PS with the exception of external fire (see 7.2).

6.2.2.1 Safety valves shall have a set pressure not exceeding the maximum allowable pressure PS of the equipment, except as permitted in 6.2.2.2 or 6.2.2.3.

6.2.2.2 If the required discharge capacity is provided by more than one safety valve, only one of the valves needs to be set as specified in 6.2.2.1. The additional valve or valves may be set at a pressure not more than 5% in excess of the maximum allowable pressure PS providing the requirements of 6.1.4 are met.

6.2.2.3 Alternatively the safety valve set pressure may be above the maximum allowable pressure PS providing that:

- *the valve(s) can attain the certified capacity at 5% overpressure or less; and*
- *the requirements of 6.1.4 are met; and*
- *an additional pressure limiter is fitted to ensure that the permitted maximum allowable pressure PS is not exceeded (including peak values) during continuous operation.*

PED Annex I clause 2.12:

Supporting EN 764-7 paragraph:

7.2 External fire: Where there is a potential risk for external hazards, such as fire or impact, the pressure equipment shall be protected against them in order to keep the equipment within safe limits.

Note: Protection against over-pressurization during external fire should be based on a detailed thermal response evaluation similar to the risk evaluation. Pressures higher than 1.1 PS can be permitted depending on the damage limitation requirement. Following fire attack the equipment should not be returned to service without a thorough review of its fitness for service.

Exception to the above: It is permitted to have a single valve or all safety valves set higher than MAP (but still not more than MAP +5%) if all valves are certified with a maximum overpressure of 5% or less and that the sizing is done at MAP +10% and there is an additional pressure limiter (can be a control valve, etc., but only installed for this unique purpose) that ensure MAP is never exceeded (from standard EN 764-7 harmonized paragraphs).

3.6.1.6 Fire case only

The unique safety valve or at least one of multiple safety valves protecting the equipment must be set at a pressure not higher than MAP of the equipment (PED Annex I, clause 2.11.2).

The 10% maximum accumulation can be exceeded, as long as this is safe. This means that the equipment designer must be able to prove to the authorities that an accumulation above 10% will not create additional risks or hazards (PED Annex I clause 2.12 and official guideline 5/2 June 2000).

3.6.1.7 Fired vessels, boilers

While the ASME has two different sections, Section I for fired vessels and Section VIII for unfired vessels, the PED encompass all these conventional vessels (excluding nuclear applications). Therefore, according to PED, unlike with ASME, these above considerations on accumulation and multiple valves installation also apply to boilers and other fired vessels as well as for unfired vessels as long as the pressures are higher than 0.5 barg.

3.6.1.8 Practical applications

Note that in the multiple valves case, where one valve is set at PS and the others set staggered above but not more than 5% above PS, it is not said that the valves must be certified at 5% or less overpressure. It is only said that one needs to be sure that the increase pressure will not exceed 10%.

But how can we be sure if the valves are certified for such low overpressure? So far, most of the pressure vessel codes, particularly the ASME Section VIII, certified Safety Valves with 10% overpressure, no less.

However, some select manufacturers can supply safety valves with capacities certified for overpressures lower than 10%.

With these valves, the set pressures of multiple valves can be staggered up to 5% accumulation and still can respect the 10% maximum accumulation for full capacity.

This is a rather new issue, and it is always wise to contact the manufacturer to ask for their certification, but in any case each application will need to be reviewed first. Furthermore, the flow coefficient to use for the sizing of the valve may be different than the usual ones (established at 10% overpressure).

3.6.1.9 Conclusion

For single safety valve applications on cases other than fire: no change.

For fire case, basically the PED does not dictate any particular accumulation, and this must be reviewed by the designer of the pressure equipment and this engages his/her responsibility. This should not exclusively be decided by the safety valve characteristics, manufacturer or the user.

For multiple valves cases, in principle, each application would need to be reviewed prior to any commitment. It is important to note here that the flow coefficient to be used may be different than the usual ones published in manufacturers' catalogues.

This is a very recent issue since 2006, and many users are still not aware of this and also very few manufacturers promote this issue.

CHAPTER 4

Codes and Standards

Codes and standards relevant to safety relief valves (SRVs) can vary quite considerably in format around the world, and many are sections within codes relevant to boilers or pressure-containing vessels. Some will only outline performance requirements, tolerances and essential constructional detail, but give no guidance on dimensions, orifice sizes and so forth. Others will be related to installation and application. It is quite common within many markets to use several codes in conjunction with one another and it is not uncommon that specifications call for sections taken from several codes, which makes compliance by manufacturers complex and uneconomical. An overview of most common worldwide codes and standards is given in Appendix M.

As already mentioned in the previous chapter, SRVs are completely governed by local codes and regulations. However, since 2002 the two major worldwide codes are ASME and PED. Both are laws and are, in any case, the basis of most international codes. There might be detailed but usually irrelevant differences, but if it complies with either or both ASME and PED, it is my opinion that your system is safe. The main problem is that a lot of installations do not comply with ASME, PED or local codes because of misinterpretations of these codes, which we will try to address and clarify further in this handbook. The worldwide governing standards and recommended practices are API 520 and EN4126, and here the reasoning is the same as with the codes.

ASME	American Society of Mechanical Engineers
API	American Petroleum Institute
PED	Pressure Equipment Directive
EN	European Normalization

While the US ASME and European PED codes are very similar, they also unfortunately have distinct differences which make things sometimes difficult in a global economic environment. However, the codes are the law and must be adhered to when equipment is installed in that specific region. Therefore, most reputable SRV manufacturers have both organizations' approvals. However, compliance with the codes is not limited to selecting an approved supplier; that is only a part of compliance.

The PED has definitely the merit that it supersedes, from a legal perspective, the very many old local codes in all European member states, codes such as BS (UK), ISPESL (Italy), TUV (Germany), Stoomwezen (the Netherlands), and UDT (Poland). Compliance with PED allows the manufacturer to CE mark their product as required by the European Union (EU) and is an assurance for the end-users that the selected material to protect their systems is in accordance to the law. Manufacturers' approvals, however, are limited in time and need to be renewed regularly, which is an additional guarantee. Therefore, even with the most reputable manufacturers, it is always wise to check the status of the approvals to make sure they have not expired.

A lot of European users still also require their local code in addition to the PED. This is their prerogative but it is not legally required and adds to the costs. It has to be noted that the maintenance of all the certification on top of PED and ASME is a huge cost to the manufacturers and is significantly influencing the prices of the SRVs on the market.

The world of codes and standards is an absolute labyrinth, so the emphasis here will be only on the main issues from ASME and PED, which cover about 80% of all worldwide requirements. The exceptions are China and India, who follow their own guidelines, although they are mostly focused on boiler applications. Also, we will concentrate on the industrial process applications and only refer to these applicable codes.

While local codes have different ways of presenting things (even sizing formulas), it can be mathematically proven that their results are practically the same, which is normal given the fact they are all based on hydrodynamic and thermodynamic fundamentals and that the differences are mainly due to the use of different units. (See Appendix A, 'Relevant Tables And References,' where an example shows that the ASME calculations are virtually the same as those required by GOST, the Russian standard.)

However, it is the API 520 which incorporates more detail on correction factors (more conservative) for backpressure and viscosity, for example, than any other code known today. Therefore, most people worldwide use API for their sizing.

European PED does not differentiate between directly and indirectly fired pressure vessels (steam boilers), nor nuclear nor any other, while ASME differentiates among the following:

ASME Section I	Direct fired pressure vessels
ASME Section III	Nuclear power plants
ASME Section IV	Heating boilers
ASME Section VIII	Unfired pressure vessels

API recommended practices and standards have been an important guidance for users and engineering companies for a long time, while in many European countries, national rules for protection against overpressure in process equipment were developed and remained in force until the beginning of the twenty-first century. We will review the API recommendations, judged to be conservative, in a little more detail.

In order to allow free circulation of goods in the European Community, member states were prohibited from making new technical rules and from updating existing ones after PED became law in 2002. Instead, they have to conform to the PED, published in 1997. PED has become compulsory for equipment 'put in the market' after 29 May 2002 (refer to Article 20. paragraph 3 of the PED).

To be able to EC mark a product, the manufacturer must undergo, for each product and type of valve, a conformity assessment comprising the EC type or design examination and the assurance of the production quality system. The manufacturer must also demonstrate the quality compliances of all sub-suppliers and ensure that all critical parts (or at least pressure-retaining parts) are fully traceable and accompanied by a material certificate. Procedures to certify conformity to PED are carried out by a notified body approved by the member states of the European Community. With completion of the assessment, the manufacturer may stamp the EC mark on the product.

The main codes and standards and their regulatory organizations relative to SRVs are summarized in the table in Appendix J.

These codes provide all the rules for the design, fabrication, testing, materials and certification of boilers and unfired pressure vessels. The rules include requirements for the pressure-relieving devices to be installed on every boiler or other pressure vessels.

All pressure relief valves (PRVs) set at 15 psi (1.03 barg) or greater will be manufactured and certified according to ASME for the United States and Canada or set at or above 0.5 barg for European member states.

4.1 OVERVIEW OPERATIONAL REQUIREMENTS

The following table offers a condensed overview of the main operational code requirements according to the current worldwide codes applicable today:

Function	ASME I	ASME VIII	PED
Set pressure tolerance	2 psi (0.14 bar) ≤ 70 psi (4.76 bar)	2 psi (0.14 bar) ≤ 70 psi (4.76 bar)	
	3% above 70 psi (4.76 bar)	3% above 70 psi (4.76 bar)	
	10 psi (0.68 bar) between 300 psi (20.4 bar) and 1000 psi (68.03 bar)		
	1% above 1000 psi (68.03 bar)		
Blowdown	< 67 psi (4.62 bar) = 4 psi	Gas/vapour: 7%–10%	No code requirement but 7%–10% is the industry standard
		Liquid: No requirement	
	≥ 67 psi (4.62 bar) and ≤ 250 psi (17.24 bar) = 6%	Neither are code requirements, but recommendations	
	>250 psi (17.24 bar) and < 375 psi (25.86 bar) = 15 psi		
	4% above 375 psi (25.86 bar)		
Overpressure	2 psi (0.14 bar) or 3%, whichever is greater	3 psi (0.2 bar) or 10% whichever is greater	10% above accumulation
		Multiple valves: 16%	
		Fire case: 21%	

4.2 ASME AND API CODES AND STANDARDS – CLARIFICATIONS

ASME/ANSI B16.34, valves – Flanged, threaded and welding ends: This standard covers pressure/temperature ratings, dimensions, tolerances, materials, non-destructive examination requirements, testing and marking for cast, forged and manufactured flanged, threaded and welding end valves. This standard is not specifically applicable to PRVs but is often used by manufacturers as 'good engineering practice'.

ASME/ANSI B16.5, pipe flanges and flanged fittings: This standard provides allowable materials, pressure/temperature limits and flange dimensions for standard ANSI flanges. Most ANSI flanged-ended PRVs will conform to these requirements.

ASME stamps: When approved by ASME, the valve manufacturer can ASME stamp or mark the valve. It then also needs to comply with specific requirements.

ASME Section I approved valves carry the V stamp.

ASME Section III valves (for nuclear applications) must carry the N, NV and NPT stamp.

ASME Section IV (for heating boilers) valves carry the H stamp.

ASME Section VIII (unfired pressure vessels in the process industry) valves carry the UV stamp.

4.2.1 National Board approval

ASME in itself does not approve nor certify the safety devices; this is done by the National Board (NB). The NB certifies the valve's capacity and verifies the valve's compliance with the ASME code. The NB maintains and publishes the 'red book' – NB-18, which contains all manufacturers and products approved according to ASME. It also publishes the true flow coefficients as measured and approved by them. So, when in doubt, one can always consult the NB-18 document at the website: http://www.nationalboard.org/SiteDocuments/NB18/PDFs/NB18ToC.pdf

NB's method for certifying the capacity (and flow coefficient) is very similar to that of other notified bodies such as PED.

- At an ASME-approved flow facility, a total of 9 valves of a particular valve design or range are flow tested at 10% overpressure above set pressure. They select three valve sizes and three set pressures. This way they establish the K_D factor of each test valve, considering the flow conditions and each measured orifice area.
- K_D is established by dividing the flow of the test valve by the flow of a 'perfect nozzle'.
- Then they calculate the average K_D for the 9 test valves.
- No test valve(s) K_D can deviate more than $+/-$ 5% from that calculated average.

- Then the ASME K flow factor of the valve is established by de-rating the K_D factor with 10%: K = Average $K_D \times 0.9$.
- The actual K_D, K and A (orifice area) values are published for all code stamped relief valves in the NB-18 (red book).

Ever since this 10% de-rating rule was established in 1962, it has been a cause of confusion. Manufacturers' catalogues do not always show the same coefficients as those published in the red book, making it extremely confusing for end users, who do not know which coefficients to use without verifying the NB-18 each time for every supplier and each valve range.

To eliminate the need for new capacity tables, revised catalogues and so on, the ASME/NB allowed manufacturers to use the K_D figures as K values on the condition that the relief valve flow areas would be increased by at least 10%. The manufacturer can show any K and any A (orifice area) as long as their advertised KA is equal or smaller to the certified ones.

Of course, the capacity (W) of the valve is directly proportional to the KA.

$$W = CKAP\sqrt{\frac{M}{TZ}}$$

Since 1962 most SRV manufacturers have overstated their K values and understated their A values.

For example a perfect nozzle has a $K_D = 1$ and a $K = 0.9$. Yet some manufacturers show their $K = 0.975$ or 0.95, which theoretically is impossible. To compensate for this, the manufacturer must furnish actual SRV orifice areas larger than those published in their brochure.

Example: Gas service – J orifice (API A = 8.303 cm^2)

	National Board			Vendor catalogue		
	K	*A* (cm^2)	*KA*	*K*	*A* (cm^2)	*KA*
Consolidated spring valve	0.855	3.774	3.227	0.95	3.269	3.106
AGCO pilot valve	0.830	3.393	2.816	0.830	3.393	2.816

4.2.2 Main paragraph excerpts from ASME VIII

A. *ASME VIII – UG-125:* Operating to set pressure ratio. Set pressure may not exceed MAWP and system pressure must be below the blowdown of the used valve (Figure 4.1).

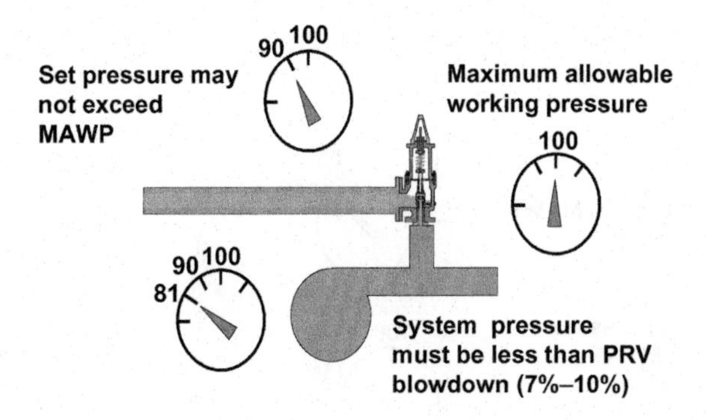

FIGURE 4.1
Operating to set pressure ratios

B. *ASME VIII – UG-125:* Full capacity must be achieved at 10% above set pressure (Figure 4.2).

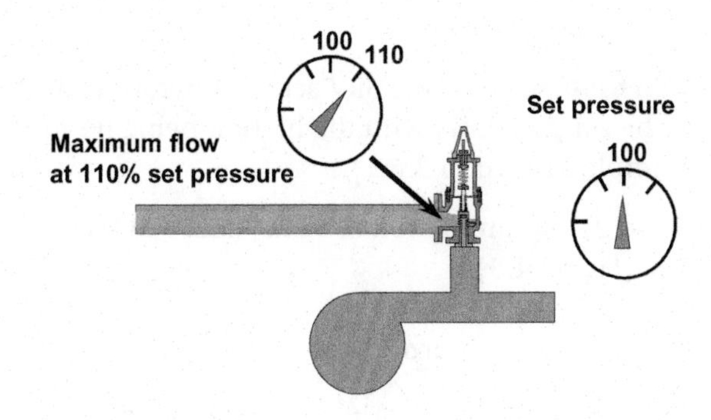

FIGURE 4.2
Full capacity at 10% overpressure over set

C. *ASME VIII – G-125:* Accumulation of the pressure vessel above MAWP for multiple valves is 16% or 0.27 barg (Figure 4.3).

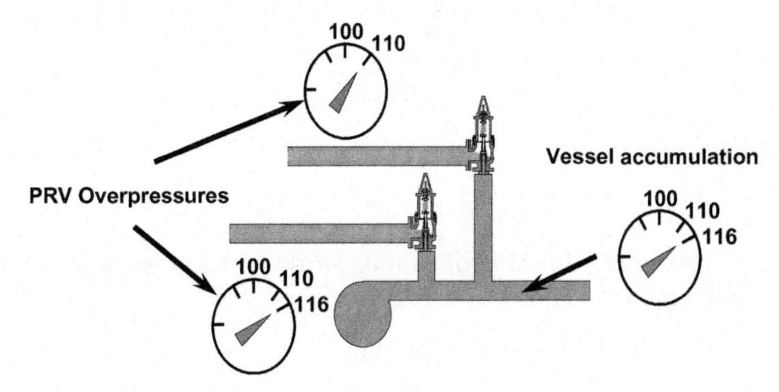

FIGURE 4.3
Accumulation for multiple valves

D. *ASME VIII – UG-125:* Relationship between set pressure, overpressure, accumulation and MAWP on multiple valves (Figure 4.4).

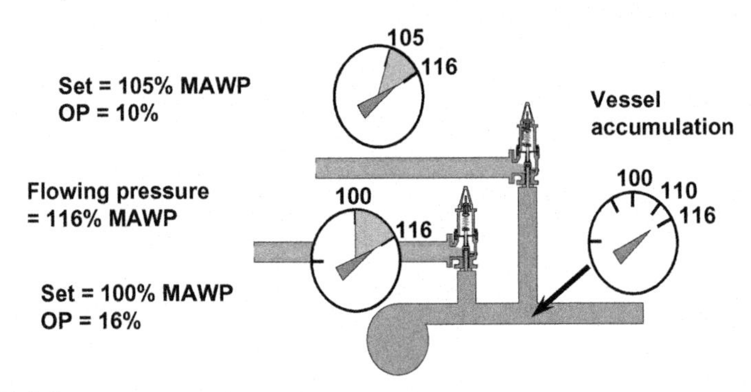

FIGURE 4.4
Multiple-valve scenario

Multiple Safety Valves Installation: One SRV set at or below MAWP. The balance of the valves may be staggered, set with the highest being no more than 105% of MAWP (see example in Section 3.6.1).

E. *ASME VII – UG-125:* Accumulation of vessel above MAWP for process and fire case is 21% (Figure 4.5).

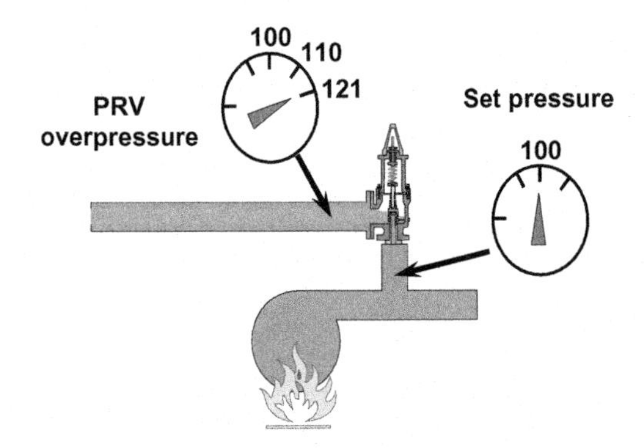

FIGURE 4.5
Fire case accumulation

For a full summary of allowable operating, working, relief, set and blowdown pressures, see Appendix H.

F. *ASME VII – UG-129:* Nameplate marking.

The nameplate is a mandatory feature of every (approved) SRV. It is, as it were, the 'passport' of the valve. Records need to be kept to ensure that all maintenance and/or eventual changes are traceable. The nameplate has to contain the following minimum information:

- Manufacturer
- Model number
- Inlet size
- Set pressure
- CDTP (cold differential set pressure)
- Orifice
- Capacity
- Year built
- Valve serial number
- ASME symbol

NB UV

ANDERSON GREENWOOD AND CO. STAFFORD, TX
PRESSURE RELIEF VALVE NAMEPLATE

MODEL NO.D-30TS124ALS0100
INLET: 250 MM ORIFICE: 167.74 cm2
SET PRESSURE: 6.89 BARG
CAPACITY: 139,961 Nm3/hr AIR
CDTP: 6.20 BARG BACKPRESSURE:. 68 BARG
SERIAL NUMBER: 96 – 10324 YEAR BUILT: 1996

G. *ASME VIII – UG-136*: Miscellaneous recommendations:

- Blowdown is not specifically addressed in the code, and therefore valves are not tested on blowdown at the manufacturers, but 7%–10% is the suggested industry standard. Note that to set blowdown, the valve would have to flow its rated capacity. Worldwide, very few manufacturers are equipped to full flow their valves on all media. If setting blowdown is a requirement, the capacity of the test stands needs to be checked, or in some instances the test can be performed during the process itself.

- Springs must be corrosion resistant or have a corrosion-resistant coating.

- Cast iron seats are not permitted.

- A safety factor of 4:1 needs to be applied for pressure-containing components – body wall thickness, closed bonnet wall thickness, bonnet bolts and bonnet bolt threads in body. These components can be stressed to maximum 25% of their nominal maximum tensile strength.

- A lift lever is required on steam, hot water and air above 60 °C (in Europe only required on steam).

- Hydrotesting is required of primary pressure parts (usually nozzle, body and bonnet) at 1.5 times the design rating if the nozzle is larger than 25 mm or pressures from 20.7 barg and if it is cast or welded.

- The secondary pressure zone (outlet) needs to be tested at 2.07 barg with air.

- Set pressure testing is required.

- Seat leakage testing is required according to API 527 (see p. 65).

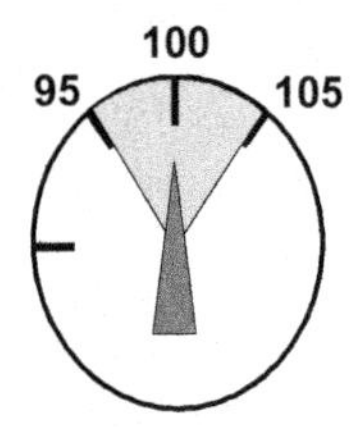

FIGURE 4.6
Spring adjustment tolerances

H. *ASME VIII – UG-126*

The field spring adjustment (Figure 4.6) needs to be within +/− 5% of set pressure. This means that the SRV must be capable of being reset differently from factory nameplate set pressure by +/− 5% with no change of any parts or impaired performance (instability, not full lift at 10% overpressure, blowdown, etc.)

Note that the valve must be set within the spring's set pressure range. However, it is important to note that once the valve is shipped, it becomes a 'used valve' and it may be reset +/− 5% outside the published spring range (Figure 4.7).

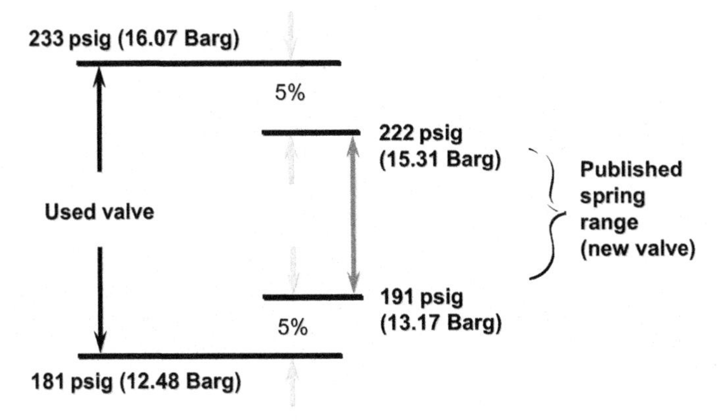

FIGURE 4.7
Typical spring values for a full nozzle 3K4 spring valve

Although theoretically the valve could probably be set outside the 5% range, there are a number of possible consequences of not following this 5% rule:

- In case of a normal spring selection by the manufacturer, the SRV can probably be reset 15% higher than nameplate set, but then the coils of the spring will be compressed more closely together during the resetting so that the spring may stack solid before full lift is achieved.
- The SRV could also probably be reset 15% lower than the nameplate set. But at a lower pressure level, there will probably be insufficient energy to overcome the too stiff spring rate, requiring an overpressure higher than that allowable to achieve full valve lift. As a result the valve would no longer comply with the codes.

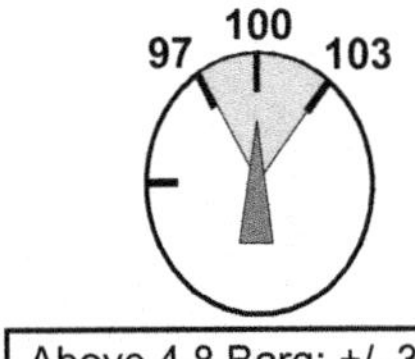

FIGURE 4.8
Set pressure tolerance

The set pressure tolerance is 3% for set pressures above 4.8 barg (Figure 4.8).

I. *ASME VIII:* Appendix M

Appendix M is recommendations only:

M5 and M6 – Permits the use of inlet and outlet block valves provided that:

- The used block valves are full bore.
- They are locked open in normal operation.
- If inlet and outlet valves are used, both valves must be interlocked so that they are both open or closed and that they can both be locked in open position (Figure 4.9).

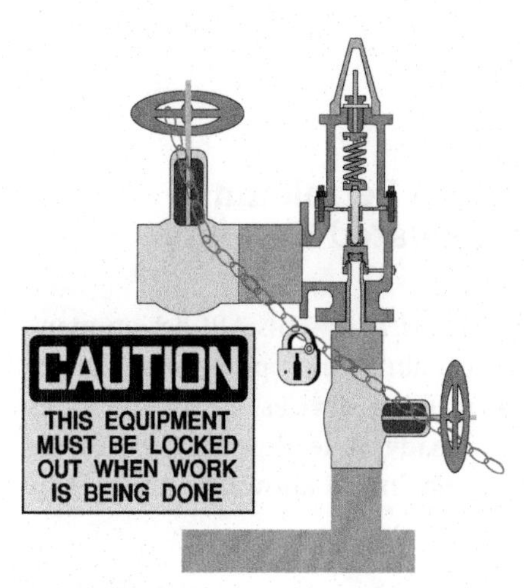

FIGURE 4.9
Use of isolation valves before and after SRVs

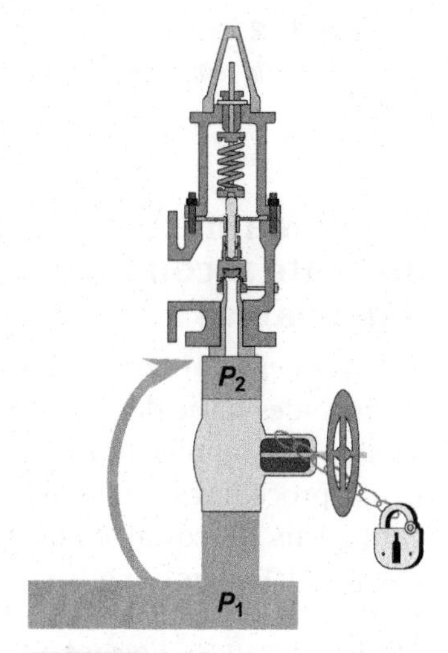

FIGURE 4.10
Maximum pressure inlet drop (3%)

M7 – The inlet pressure drop between the protected pressure point and the relief valve (between P_1 and P_2) must be 3% or less (Figure 4.10).

M8 – Discharge lines (Figure 4.11) may be used provided that:

- They do not reduce in anyway the SRV capacity below the required one.
- It may not adversely affect the SRV operation.
- The vent pipe must be as short as possible.
- The vent pipe must have a vertical riser.
- It must have a long radius elbow (see later in installation in Section 6.1)
- It must have adequate drainage.

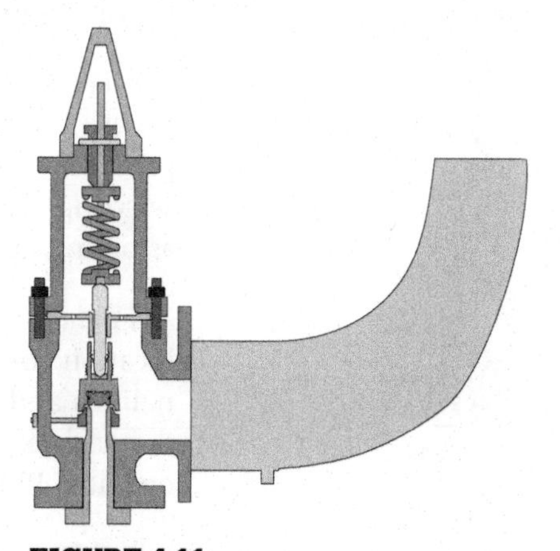

FIGURE 4.11
Discharge piping

M12 – A SRV must be mounted vertically (Figure 4.12).

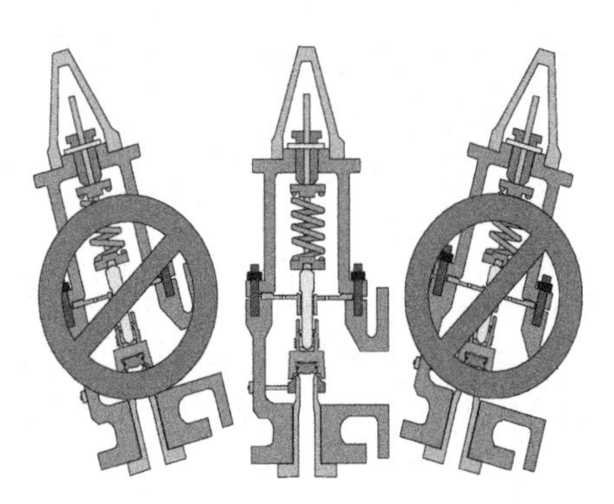

FIGURE 4.12
Mounting SRVs in vertical position only

4.2.3 Main excerpts from American Petroleum Institute recommended practices related to safety relief valves

API Recommended Practice 520 Part I, Sizing and Selection: This API design manual includes basic definitions and information about the operational characteristics and applications of various pressure relief devices. It also includes sizing procedures and methods based on steady state flow of Newtonian fluids. This RP covers equipment that has a maximum allowable pressure of 15 psig (1.03 barg) or greater.

API Recommended Practice 520 Part II, Installation: This part covers methods of installation for pressure relief devices, including recommended piping practices, reaction force calculations and precautions on pre-installation, handling and inspection.

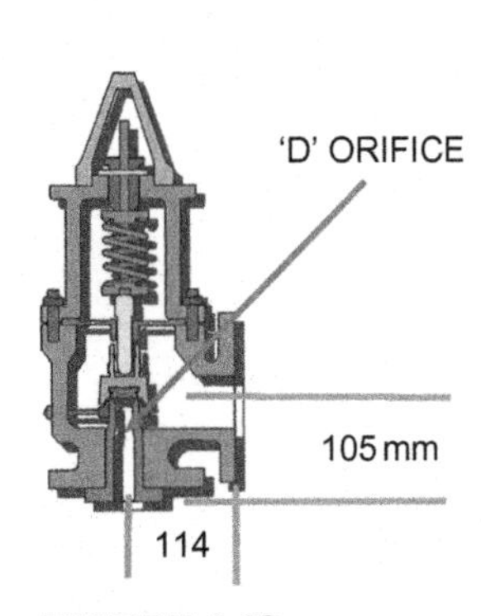

FIGURE 4.13
API 526 determines standard connections

API Recommended Practice 521, Guide for Pressure Relief and Depressurizing Systems: This recommended practice provides guidelines for examining the principal causes of overpressure, determining individual relieving rates and selecting and designing disposal systems, including such component parts as piping, vessels, flares and vent stacks.

API Standard 526, Flanged Steel Pressure Relief Valves: This is a purchase specification for flanged steel PRVs. Basic requirements, such as orifice designations and area, materials, pressure–temperature limits and centre-to-face dimensions, inlet and outlet, are given for both spring-operated and pilot-operated PRVs (Figure 4.13).

API 526 also determines standard orifice areas and their respective letter denomination for sizing purposes.

Orifice letter	in^2	cm^2
D	0.110	0.71
E	0.196	1.26
F	0.307	1.98
G	0.503	3.25
H	0.785	5.06
J	1.287	8.30
K	1.838	11.85
L	2.853	18.40
M	3.60	23.22
N	4.34	28.00
P	6.38	41.15
Q	11.05	71.27
R	16	103.20
T	26	167.74

When an orifice size is calculated per the formulas in Chapter 7, one should use a valve with the next letter size up, as most manufacturers only supply their 'API' valves in the above-mentioned orifices. Some manufacturers are able to adapt their valve to customized orifices (i.e., as close as possible to the calculated value) but this comes at a premium price.

API Standard 527, Seat Tightness of Pressure Relief Valves: This standard describes tests with air, steam and water to determine the seat tightness of metal- and soft-seated PRVs. Valves of conventional, bellows- and pilot-operated designs are covered. Acceptable leakage rates are defined for gas, steam and liquid.

It is interesting to notice here that a larger orifice valve usually has a lower potential leak rate than a smaller orifice valve even though the perimeter of the seal is larger. This is because the unit force per centimetre of circumference is directly proportional to the sealing diameter.

$$\text{Circumference} = \pi D$$

$$\text{Area} = \frac{\pi D^2}{4}$$

$$\text{Force} = PD$$

$$\text{Unit force} = \frac{P\pi D^2}{4} \times \frac{1}{\pi D} = \frac{PD}{4}$$

To improve the sealing characteristics of small orifice valves, the seat sealing area is sometimes made larger than the through bore. For soft-seated valves, a softer elastomer is also used as long as the temperatures are not too high. Here the manufacturer has got to find the correct compromise.

The recommended practices to test for leakage are as follows:

4.2.3.1 Gas

The set-up is shown in Figure 4.14. With the valve always mounted vertically, the leakage rate in bubbles per minute shall be determined with an inlet pressure held at 90% of the set pressure after the valve has been set and popped, except for valves set at or below 3.4 barg (50 psig), in which case the pressure shall be held at 0.34 barg (5 psig) below set pressure immediately after popping. The test pressure with air at atmospheric temperature shall be applied as follows:

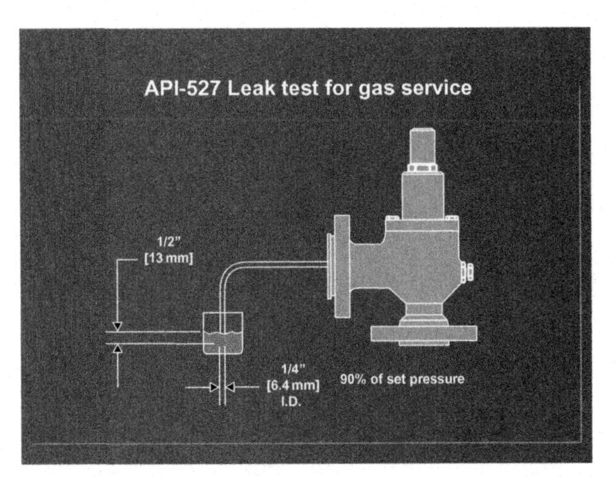

FIGURE 4.14
API 527 test set-up

Inlet size	Duration of test
½–2 in. (up to 50 mm)	1 min
2 ½–4 in. (65–100 mm)	2 min
6 in. (150 mm) and up	5 min

The following table shows the maximum allowable seat leakage rates for new valves, having never undergone the rigours of transit to the jobsite and any mishandling during transportation. A metal-seated valve is clearly always more vulnerable to damage during transportation.

Set pressure		Metal seat leakage at 90% of set pressure in bubbles/minute after 1 pop	
Barg	**psig**	**Orifices D–F**	**Orifices G–T**
1.03–69	15–1000	40	20
103	1500	60	30
138	2000	80	40
172	2500	100	50
207	3000	100	60
276	4000	100	80
414	6000	100	100
		Soft seat leakage at 90% of set pressure in bubbles/minute after 1 pop	
1.03–414	15–6000	0	

Usually this test is also applicable for set pressures up to 700 barg. In that case, the bubble count of 414 barg is used.

4.2.3.2 Steam

Set pressure must be demonstrated first and then pressure is again lowered. Then the inlet pressure is again raised to 90% of set pressure (hold pressure is 0.34 barg below set pressure if the set pressure is lower than 3.44 barg). The pressure needs to be held at 90% as follows:

Inlet size	Duration of test
½–2 in. (up to 50 mm)	1 min
2 ½– 4 in. (65–100 mm)	2 min
6 in. (150 mm) and up	5 min

Neither audible nor visual leak may occur at 90% of set pressure during this period.

4.2.3.3 Water

Set pressure must be demonstrated first and then pressure is again lowered. Then the inlet pressure is raised to 90% of set pressure (hold pressure is 0.34 barg below set pressure if set pressure is lower than 3.44 barg) and needs to be held at 90% for all sizes for 1 minute. Air can be used as test medium (Figure 4.15).

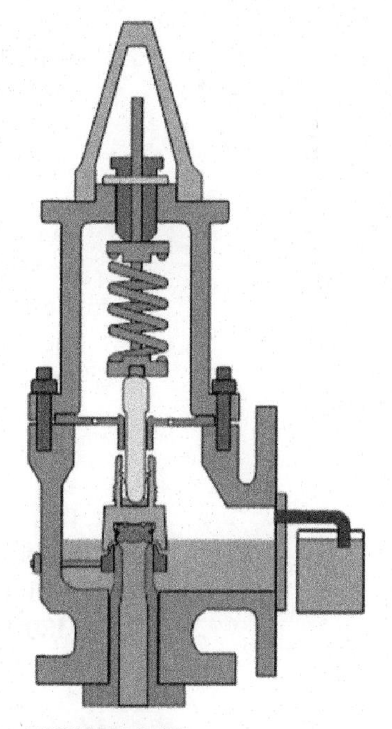

FIGURE 4.15
API 527 test set-up

Set pressure		Seat leakage at 90% of set pressure in bubbles/minute	
Barg	**psig**	**Orifices D–F**	**Orifices G–T**
0.34–69	5–1000	20	10
103	1500	30	15
138	2000	40	20
172	2500	50	25
207	3000	50	30
276	4000	50	40
414	6000	50	50

4.3 NACE

The National Association of Corrosion Engineers (NACE) makes recommendations to the industry on how to protect installations against all sorts of corrosion. It was established to protect people, assets and the environment against the effects of corrosion.

Compliance with NACE is a frequent requirement for valves, especially in the modern oil and gas industry. Because NACE has already changed a couple of times in the early twenty-first century, there tends to be some confusion as to which code exactly to apply where.

To understand the present, we need to know the past. NACE MR0175 (issued in 1975) allowed manufacturers to recommend certain materials based on the code for applications involving sour gas, or hydrogen sulphide (H_2S). It specified material requirements to protect the valves against sulphide stress cracking. It was and is not a performance nor a design specification nor a standard. Until recently it was known as NACE MR0175-2002 – Sulphide Stress Cracking-Resistant Metallic Materials for Oilfield Equipment.

This standard was and still is widely used by operators, either in the oil and gas production fields or refineries. Most SRV manufacturers have established a standard 'Bill of Materials' that complies with the minimum requirements of NACE MR0175, taking into account certain specific valve material selections based on their hardness and under the condition of the following operating conditions:

- Sour gas present in systems operating > 3.44 barg
- Partial pressure of $H_2S > 0.0034$ barg

- Sour oil and multi-phases
  a. Gas to ratio is > 141.6 NM/BBL
  b. Gas phase > 15% H_2S
  c. Partial pressure of H_2S > 0.689 barg
  d. Surface operating pressure > 17.24 barg

A revision of the NACE MR0175 was issued in 2002, which became NACE MR0175-2002. Some important definitions derived from this 2002 issue are as follows:

Partial pressure – Total absolute pressure multiplied by the mole fraction of the component in the mixture. (In an ideal gas, the mole fraction is equal to the volume fraction of the component.)

Pressure-containing parts – Those valve parts whose failure to function as intended would result in a release of retained fluid to the atmosphere.

Sour environment – Environments containing water and H_2S as per the above-mentioned criteria.

Stress corrosion cracking – Cracking of a material produced by the combined action of corrosion and tensile stress (residual or applied).

Sulphide stress cracking – Cracking of a metal under the combined action of tensile stress and corrosion in the presence of water and H_2S.

The 2002 issue was fundamentally not very different from the NACE MR0175 except for some details. The confusion, however, started when, in 2003, NACE issued no less than two new revisions in the same year:

NACE MR0175-2003 – 'Metals for Sulphide Stress Cracking and Stress Corrosion Cracking resistance in Sour Oilfield Environments' (published by NACE), a standard which is now already withdrawn.

and

NACE MR0175/ISO 15156 Parts 1 to 3 – 'Materials for use in H_2S-Containing Environments in Oil and Gas Production'. This is a joint NACE and ISO standard which has become the current standard.

These revised standards differ from the 2002 version as they now consider the effects of chlorides and free sulphur along with H_2S, which was not the case before.

The major impacts of the new 2003 issue of the NACE MR0175/ISO 15156, Parts 1–3, are the following:

- Part 1 – 'General principles for selection of cracking-resistant materials'.
- Part 2 – 'Cracking-resistant carbon and low-alloy steels, and the use of cast irons'.
- Part 3 – 'Cracking-resistant, corrosion-resistant alloys and other alloys'.

The NACE MR0175-2002 was treating only the level of H_2S in fluids and, as mentioned, most SRV manufacturers simply established a 'NACE bill of material' for their products which complied with the requirements of that NACE edition.

The NACE MR0175/ISO 15156 (1–3), however, is more difficult to interpret and does not address SRVs specifically. From this edition, different manufacturers (and users) may and will select different (more standard and/or exotic) materials, while both groups claim to comply with the standard based on their own interpretation.

Some specifics about this new NACE version:

- It only addresses metals.
- It may not apply to refineries, crude oil storage operating below 4.3 barg (65 psig).
- Carbon and low-alloy steels (per Part 2) are not really applicable if the partial pressure of H_2S is smaller than 3 mbara (0.05 psia).
- For other alloys (per Part 3) no general lower limit in H_2S concentration is given.

Generally with the data provided in a typical enquiry for an SRV, it is very unlikely that a responsible manufacturer would or could select 'standard' materials (316 SS, 17-4PH, etc.) and claim to meet this new standard. A safe and conservative attitude would be to offer very exotic materials as listed in the standard with the danger of strongly overengineering the valve and possibly increasing the cost of the valve(s) significantly due to the possible interpretations.

It is therefore of great importance to understand and agree upon a material selection based on additional and very detailed end-user process data. The minimum additional required data to make a reasonable material selection is:

- Maximum service temperature
- Maximum H_2S partial pressure
- If applicable, the partial pressure of CO_2
- Maximum percentage of elemental (free) sulphur
- Maximum chloride concentration

With the above data, the right materials (and tests as required by the standard) can be selected – that is, the most economical grades can be offered to the end user. End users can also provide their own selection of required materials and establish a list themselves. However, due to their lack of experience in valve building, this also can require complete re-engineering of the valve and become not very cost effective.

Manufacturers are mostly waiving the responsibility of compliance with NACE and are putting that responsibility exclusively with end users; or they propose to comply with the old NACE MR0175-2002, for which a material list exists, which can be approved by end users.

The new NACE standard indeed encourages dialogue between user and material suppliers. This is even stressed within the standard per Part 1, paragraph 5 – 'General principles':

> The equipment supplier may need to exchange information with the equipment manufacturer, the materials supplier and/or the materials manufacturer.
>
> The equipment user shall determine whether or not the service conditions are such that this part of NACE MR0175/ISO 15156 applies. If necessary, the equipment user shall advise other parties of the service conditions.

To complicate the issue even more, another NACE MR0103 was issued in April 2003 – yet another new standard – 'Materials Resistant to Sulphide Stress Cracking in Corrosive Petroleum Refining Environments'.

Refinery applications have always been outside the scope of the NACE MR0175. However, this standard has frequently been used as a reference – NACE MR0175 has always been a little over the top for refinery use. Sulphide corrosion cracking is not such a concern downstream (refineries have reduced chloride levels). Therefore, NACE decided to formulate a new refinery-specific standard inclusive of sulphide corrosion cracking to meet the specific needs of the oil refining industry.

The major differences between the old MR0175 and the new MR0103 for refinery use are as follows:

- The specific guidelines on whether environment is 'sour' differ from definitions in previous MR0175 versions.
- The new standard does not include environmental restrictions on materials.
- Materials and/or material conditions are included that are not listed in previous MR0175 versions (and vice versa).
- There is extra emphasis on welding controls due to welding being more prevalent in a refinery environment.

For most SRV manufacturers, the main difference between MR0175 and MR0103 is the fact that MR0103 does not include environmental limits on materials and they state that therefore usually their 'old' NACE bill of material can be used.

4.4 PED 97/23/EC (PRESSURE EQUIPMENT DIRECTIVE) – CEN

In Europe, since May 2002, compliance with PED for all member states became compulsory. The directive is somewhat different to what was known before (mainly based on ASME and API). It was introduced by the EU to harmonize national legislations for the supply of pressure equipment across the EEA by specifying a minimum number of safety requirements relating to pressure.

The PED is a legal requirement – failure to comply is a criminal offence. Products manufactured as per the harmonized standards are presumed to conform to the 'Directives' and can carry a CE mark.

The CE mark shows the product compliance to all applicable European Directives and is compulsory within the limits of its application. The CE mark is, however, not a mark of quality and has nothing to do with the standards.

For all current applicable directives for the CE mark, the following website can be consulted:

http://ec.europa.eu/enterprise/pressure_equipment/ped/index_en.html

As a definition for this directive, we can put that it applies to the design, manufacture and conformity assessment of pressure equipment and assemblies with a maximum allowable pressure (MAP or PS) greater than 0.5 barg.

The manufacturers have to comply with a mandatory set of 'essential safety requirements'.

Pressure equipment is defined as:

- Vessels — Safety accessories
- Piping — Pressure accessories

Pressure accessories are defined to have an operational function having pressure-bearing housings, therefore all valves and fluid-pressurized actuators are pressure accessories.

All equipment covered by the PED must meet the essential safety requirements (ESR) defined in Appendix I. They are to be addressed by the manufacturer for all 'operating and reasonably foreseeable working conditions'.

The manufacturer has to prepare a technical documentation file, demonstrating the conformity of the pressure equipment with the ESR. All applicable ESR have to be met, regardless of the risk level, regarding:

- Design
- Manufacturing
- Materials
- Specific pressure equipment requirements

The pressure equipment classification can be summarized as shown in Figure 4.16.

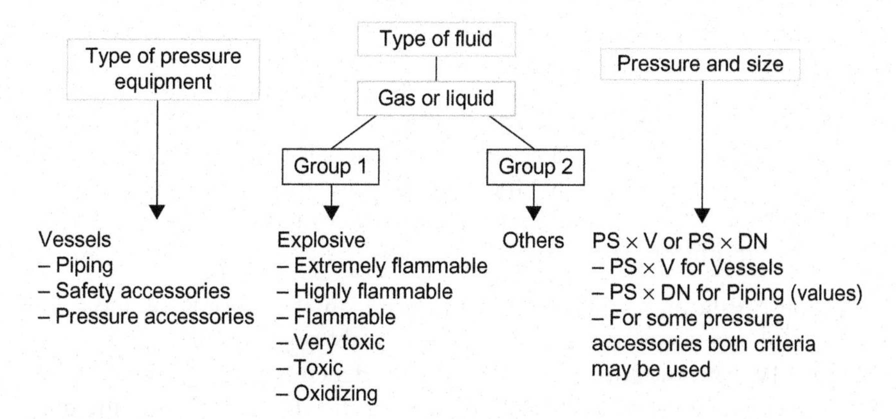

FIGURE 4.16
Pressure equipment classification

The Directive classifies the pressure equipment into four categories (I, II, III, IV) dependent on the level of risk presented by the pressure (Figure 4.17). Category IV represents the highest risk. The category of the pressure equipment is established from one of the nine tables in Annex II of the Directive. The categories determine the Conformity Assessment Procedures to be used.

All European valves professionals and organizations are recommended to use tables 6–9 (in Annex II of the Directive), which relate to pressure and size DN (maximum Category III). Safety accessories, however (such as SRVs, CSPRS, bursting disc safety devices, pressure/temperature switches, SRMCR devices, etc.), are classified in Category IV.

The PED is much more general than ASME. It consists of more broad guidelines with essential requirements ensuring safe use, to be approved by a NOBO (notified body approved by the EU).

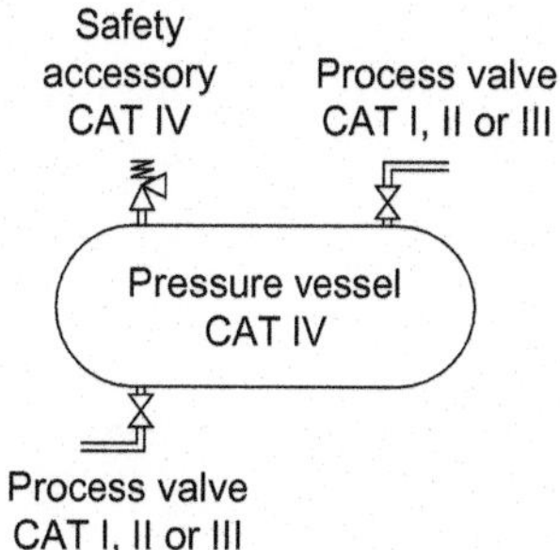

FIGURE 4.17
PED categories

What the PED does say in its ESR related to SRVs is that pressure-limiting devices must prevent the maximum allowable pressure being permanently exceeded. The safety valve (SV) should be set at PS (= MAP) or lower, with exception of momentary pressure surge (accumulation) which is 10% of PS in all cases (multiple valves also) and except for fire, this can be higher if proven safe. This is much more general but definitely differs with ASME as described in detail in Section 3.6.

About the materials, PED defines the pressure-retaining parts for the valves in the technical file of each valve type. Typically, for an SRV, this would be body, bonnet, disc and nozzle but could be different depending on the type of SRV. These parts must satisfy material requirements and have an EN10204 3.1B (now 3.1) material certificate, and are therefore traceable.

Materials for pressure-containing parts must come from 'recognized' suppliers, typically ISO 9000 certified by a recognized agency. Also they must either:

- Comply with a harmonized standard (EN).
- Be subject to a European approval for materials by a notified body (only if not covered by the harmonized standard).
- Be subject to a particular material appraisal by a notified body, included in the technical file.

Regarding certification, the categories depend on PS, size and volume, but SRVs are always Category IV except if sold only on equipment lower than Category IV. The module(s) of certification depends on the category, and there are several options possible to obtain certification, which is at the manufacturer's discretion (Figure 4.18).

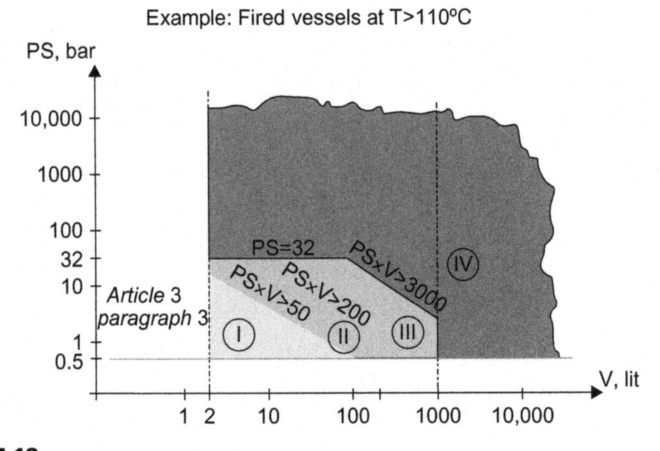

FIGURE 4.18

Example of PED categories for a fired vessel at temperatures above 110°C

Contrary to ASME what PED does *not* do is:

- Size SRVs
- Certify flow coefficients
- Fix quantity of SRVs
- Establish different accumulations for different situations (always 10%)

For a comparison against ASME VIII, refer to Section 3.6.

4.5 ATEX

ATEX

In Europe, at a certain moment in the early twenty-first century, all valves had also to comply with ATEX. However, concerning SRVs, there were also some changes which led to a lot of confusion. Most SRV suppliers conform to ATEX anyway, and the tag plates usually carry the sign on the left but they are actually doing this beyond the normal legal requirements.

The attached is based on a write up which was issued by the BVAA (British Valve and Actuator Association) concerning the important changes that were made to the directive.

Note that the attached is just a guideline. CEIR/PC3 and its committee of national experts have issued an ATEX guidelines draft, based upon a combination of the various guidelines already published by Valve Associations in the United Kingdom, France, Germany and Italy. After being reviewed and discussed at the CEIR meeting in Stockholm in early June 2005, the following guidelines were tentatively agreed by the national delegates.

4.5.1 European ATEX Guidelines

At a meeting of the ATEX Standing Committee in Brussels on 30 July 2005, it was unanimously agreed that the second revision of the European ATEX Guidelines should be formally approved.

These revised ATEX guidelines can now be found on the following website:

http://europa.eu.int/comm/enterprise/atex/guide/index.htm

The main issue here as regards to SRVs is the Guidelines on the Application of Directive 94/9/EC of 23 March 1994 on the approximation of the laws of the Member States concerning equipment and protective systems intended for use in potentially 'Explosive Atmospheres Second edition – July 2005'. The paragraph affecting SRVs is the section on what is called 'simple products' (including simple valves), which remains in the guidelines and reads as follows:

> *5.2.1. 'Simple' products*
>
> *In general, many simple mechanical products do not fall under the scope of Directive 94/9/EC as they do not have their own source of ignition (see chapter 3.7.2). Examples without own source of ignition are hand tools such as hammers, spanners, saws and ladders.*
>
> *Other examples that in most cases have no potential ignition source are given below. However, the manufacturer will need to consider each item in turn with respect to potential ignition hazard to consider whether Directive 94/9/EC applies (see also chapter 3.7.3):*

- *Clockwork timepieces; mechanical camera shutters (metallic)*
- *Pressure relief valves, self-closing doors*
- *Equipment moved only by human power, a hand operated pump, hand powered lifting equipment, hand-operated valves*

The issue of hand-operated valves has also been discussed. Given that these will move slowly, with no possibility of forming hot surfaces, as discussed in Section 3.7.3, they are not in the scope of the Directive. Some designs incorporate polymeric parts, which could become charged, but this is no different from plastic pipes.

Given that it is clear that the latter is outside of the scope of Directive 94/9/EC it has been accepted that such valves do not fall within scope.

Some manufacturers have argued that their valves are specially adapted for ATEX in that they have either selected more conductive polymers, or taken steps to ensure that no metal parts could become charged because they are unearthed.

Other manufacturers state that all their valves meet this requirement simply by the way they are constructed, and they see no distinction from valves used to process non-flammable materials. To avoid confusion between those who claim correctly that their valves have no source of ignition, and are out of scope, and those who claim that they have done some very simple design change and wish to claim that their valves are now category 2 or even 1, it has been agreed that simple valves are out of scope.

Nevertheless, as discussed in Section 3.7.3, where potentially flammable atmospheres exist, users must always consider the electrostatic ignition risks.

The statement above that 'some valves have polymeric parts which could become charged' is indicative that the committee have perhaps not understood the process of some valves, particularly a ball in a ball valve being electrically isolated between two polymeric seats and becoming electrically charged by the effects of the process media; the polymeric parts do not become charged, so it is not the same as a plastic pipe as they state.

There are two sentences in this section of the Guidelines that attempt to clear the committee of any responsibility for the problems that may occur following the adoption of the European Guidelines:

1. However, the manufacturer will need to consider each item in turn with respect to potential ignition hazard and consider whether Directive 94/9/EC applies.

2. Nevertheless, as discussed in Section 3.7.3, where potentially flammable atmospheres exist, users must always consider the electrostatic ignition risks.

Here, the responsibilities are in the camp of the manufacturer and end users without giving any further details. This is why most SRV manufacturers have chosen to just comply with ATEX and carry its sign.

In any case, the revised European Guidelines will allow manufacturers of SRVs to ignore the requirements of the ATEX Directive with regard to its directive for non-electrical equipment as they are classified under simple products and classified under the same category as clockwork timepieces, mechanical camera shutters, self-closing doors, hand-operated pumps, hand-powered lifting equipment and hand-operated valves. This is a retrograde step that could be a danger to users if the valve does not carry the ATEX approval sign, although it is not legally required.

CHAPTER 5

Design Fundamentals

With the existing safety relief valve (SRV) technology, codes and regulations, *NO* specific type of SRV is suitable for all overpressure protection conditions. Therefore, there are different types and designs suitable for different applications and process conditions.

Let's consider the objectives of the use of an SRV:

- Compliance with local, state, national and environmental regulations
- Protecting personnel against dangers caused by overpressure in the equipment when all other safety equipment has failed
- Minimizing material losses before, during and after an operational upset which caused the overpressure
- Reducing plant downtime during or after an overpressure in the system
- Preventing damage to capital investment
- Preventing damage to adjoining property
- Reducing insurance costs and premiums on capital investments
- Protecting the environment and minimizing hazardous areas within the plant

There are a lot of different SRV designs available on the market today because there is no such thing as one valve that fits all processes in this complicated industry. Everyone attempts to design the 'ideal valve', but that does not currently exist. We can, however, list the design considerations to make the optimal valve:

- Does not leak at system-operating pressure and remains preferably leak-free up to set pressure
- Opens at exactly the specified set pressure within determined limits
- Flows (minimum) a determined amount of product in a controlled way
- Recloses within determined limits
- Does not chatter or rapid cycle due to inlet/outlet piping losses
- Operation not influenced by backpressures

- Opening speed and blowdown can be set separately
- Easy to maintain, adjust and verify settings
- Complies with the applicable laws (ASME VIII, PED and others) worldwide

5.1 CONSTRUCTION MATERIALS

An important design consideration is, of course, the construction material. Compatibility with the process fluid is achieved by carefully selecting the construction material. Materials must be chosen with sufficient strength to withstand the pressure and temperature of the system fluid. Materials must also be resistant to chemical attack by the process fluid and, nowadays, the local environment, to ensure valve function is not impaired over long periods of exposure. Bearing properties are carefully evaluated for parts with guiding surfaces so that the risk of galling can be avoided at all times. A fine finish on the seating surfaces on disc and nozzle is required for tight shut off on a metal-to-metal valve and the correct choice of soft seats is important, because of their resistance to corrosion, temperature and pressure over a long period of time. Rates of expansion caused by temperature changes and the tolerances within the valve of the individual parts are other important design factors.

The variety of materials available on the market today is immense, and the better manufacturers have designs from carbon steel over stainless steel, to duplex, super duplex, up to Hastelloy® and even titanium. Usually body, bonnets and disc holders are made from castings, while a lot of the internal parts are made from bar, but almost any combination is possible today.

5.2 DIRECT SPRING-OPERATED SRVs

5.2.1 Introduction

An SRV is a safety device designed to protect a pressurized vessel or system during an overpressure event. An overpressure event refers to any condition which would cause pressure in a vessel or system to increase beyond the specified design pressure or maximum allowable working pressure (MAWP) (Section 3.6).

The purpose of this discussion is to familiarize one with the various parameters involved in the operation and design of an SRV.

Once a condition occurs that causes the pressure in a system or vessel to increase to a dangerous level, the safety relief valve (SRV) may be the only device able to prevent a catastrophic failure. Since reliability is inversely related to the complexity of the device, it is important that the design of the SRV be simple.

The SRV must open at a predetermined set pressure, flow a rated capacity at a specified overpressure and close when the system pressure has returned to a safe level. Safety relief valves (SRVs) must be designed with materials compatible with many process fluids, from simple air and water to the most corrosive media. They must also be designed to operate in a consistently smooth and stable manner on a variety of fluids and fluid phases. These design parameters lead to the wide array of products available on the market today and provide the challenge for future product development.

5.2.2 Functionality

The basic spring-loaded SRV has been developed to meet the need for a simple, reliable, system-actuated device to provide safe overpressure protection.

The opening and closing are maintained by a spring, as shown in Figure 5.1. The force of the spring determines when and how the valve opens and closes. With the adjustable spring, we can vary set pressure. Unfortunately as the seat assembly rises to open the valve, the spring is compressed more, resulting in a higher downforce restricting the valve opening.

The addition of a skirt, shown in Figure 5.2, creates a secondary area on the seat assembly. This provides a larger area for the inlet pressure to act on upwards as the valve begins to open, and redirects the flow downwards; both boost the valve to open more quickly with less simmer (seat leakage as the SRV approaches set pressure) and overpressure.

To achieve, however, the adjustable reseat pressure required by the codes, a nozzle ring is added in the design, as shown in Figure 5.3. The nozzle ring, the nozzle and disc holder form the so-called huddling chamber, the strength of which determines not only the opening action but also the reseat pressure.

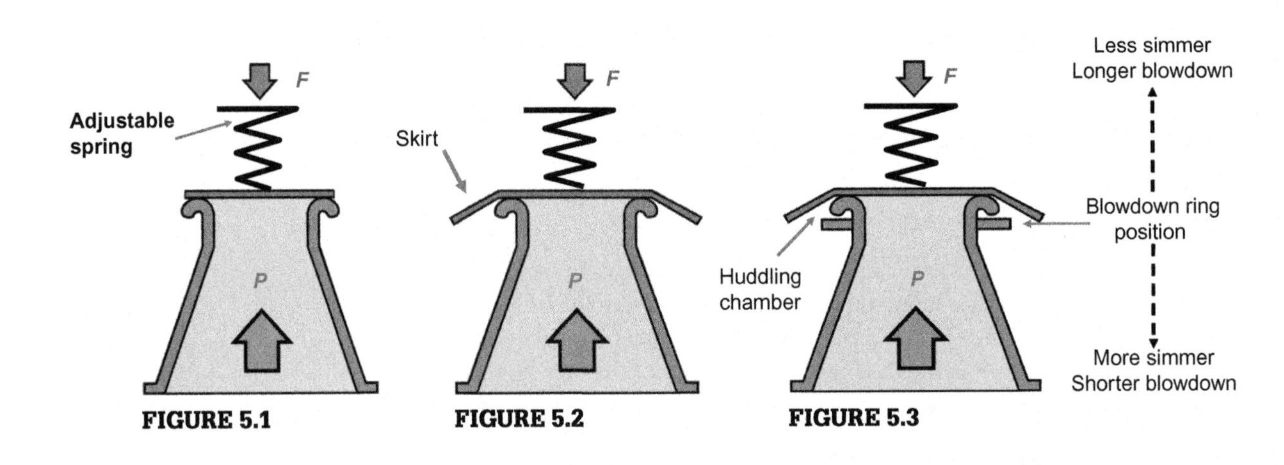

FIGURE 5.1

FIGURE 5.2

FIGURE 5.3

The huddling chamber is essentially a second orifice. A single-ring, metal-seated SRV cannot have both sharp opening action and short blowdown at the same time. As a consequence, the nozzle ring must be in a compromised position. If the nozzle ring is placed in a high position, the valve will snap open very quickly but will resist shutting and will have a longer blowdown. On the other hand, if the nozzle ring is set low, the release will be slower (because it needs more overpressure), but the blowdown will be shorter. So, the position of the nozzle ring and the volume in the huddling chamber determine release and shutoff. So overpressure and blowdown are linked to one another.

As per the General Gas Law, if pressure decreases (which occurs when the valve opens), the volume will increase proportionally with a constant temperature. Now, this is exactly what happens in the huddling chamber. When the valve opens, pressure is reduced and the enormous increase in volume is trapped in the huddling chamber. This volume acts on the secondary skirt (increased surface area), which forces the valve to snap open rapidly.

The operation of a spring-loaded SRV is shown step by step in Figure 5.4.

This operation results in the typical opening characteristics required by most codes, as set out in Figure 5.5.

As can be seen, the valve opening characteristics involve two steps, resulting in an expansive lift and a reactive lift (Figure 5.6).

It is important to mention that gas and vapours have different release characteristics than liquids, and that what is being discussed here is typical for gas, vapours and steam. Until 1985, the code allowed for liquid applications to have an overpressure of 25%. Until then, the same valve trim was used for both gases and liquids, resulting in the release characteristics shown in Figure 5.7.

However, since 1985, the code has also required a maximum overpressure of also 10% on liquid valves. This meant manufacturers had to redesign many trims so that the same valve could be fitted for both gas and liquid. So it is important to know the age of a liquid valve in order to determine whether it will flow full capacity at 10% or at 25%. Also, blowdown on liquid spring-operated valves was and is still rather unstable, and can be as high as 30% depending on the design. The code does not specify a blowdown requirement for liquid valves.

The liquid trim modifies the shape of the huddling chamber so that the full capacity at 10% overpressure can be achieved with a non-compressible fluid. Some manufacturers provide an 'all media trim' which fulfils the requirements for gas and liquid with one single trim; others have an interchangeable trim for each fluid.

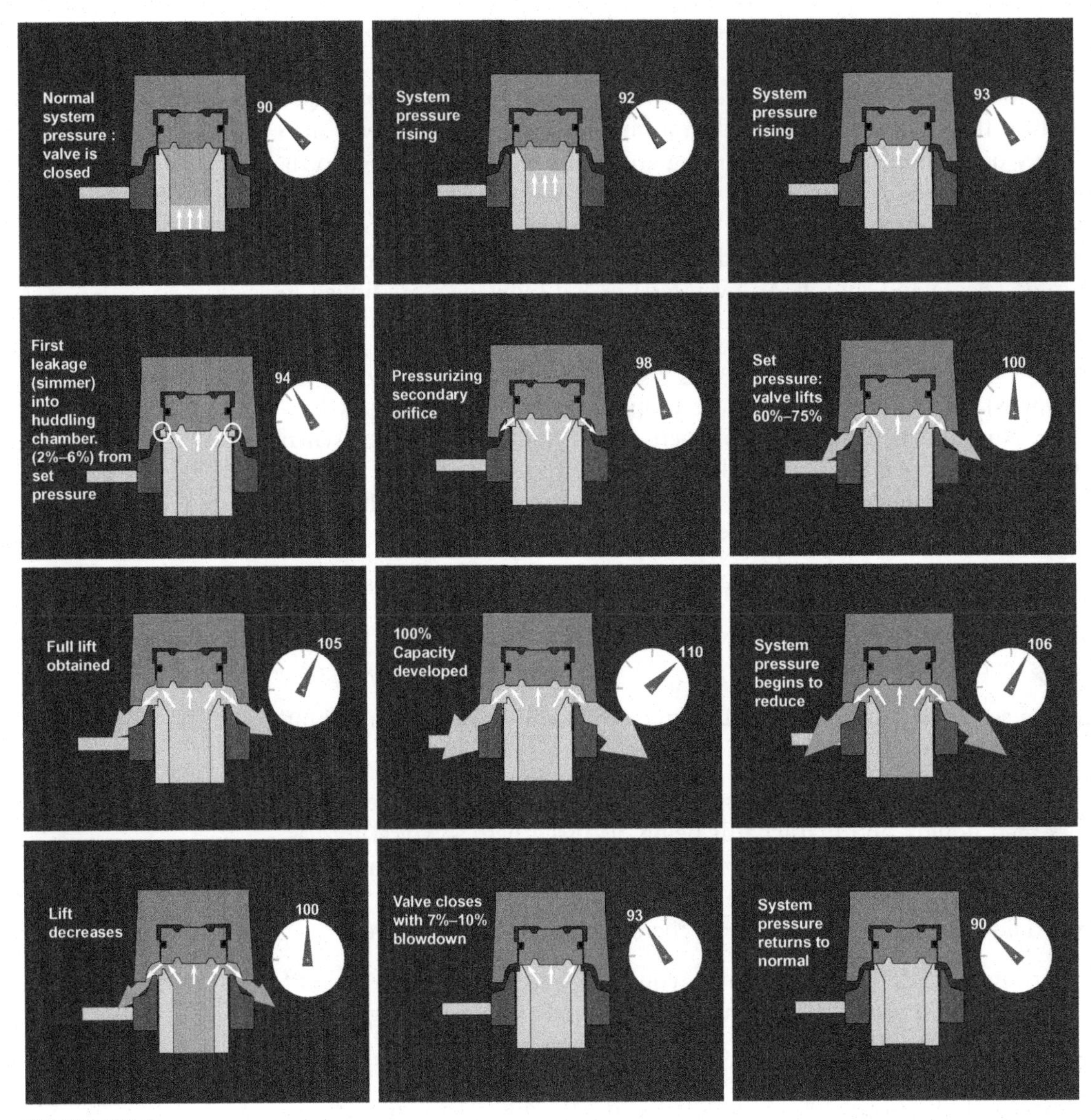

FIGURE 5.4
Valve operation

This results in the release characteristics for liquid shown in Figure 5.8.

It was a positive development when the code insisted on special liquid trim designs; besides the fact that the standard gas valve did not reach rated capacity at 10% for liquid, it was also demonstrated that the valve became very unstable during the opening cycle, as can be seen in Figure 5.9.

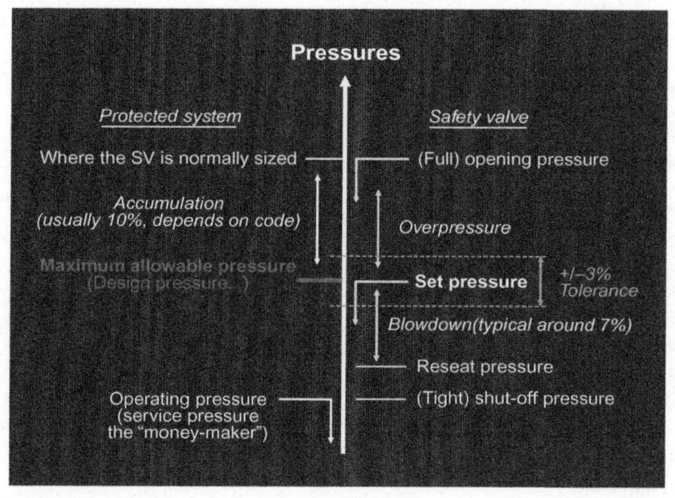

FIGURE 5.5

Typical opening characteristics of a spring operated SRV

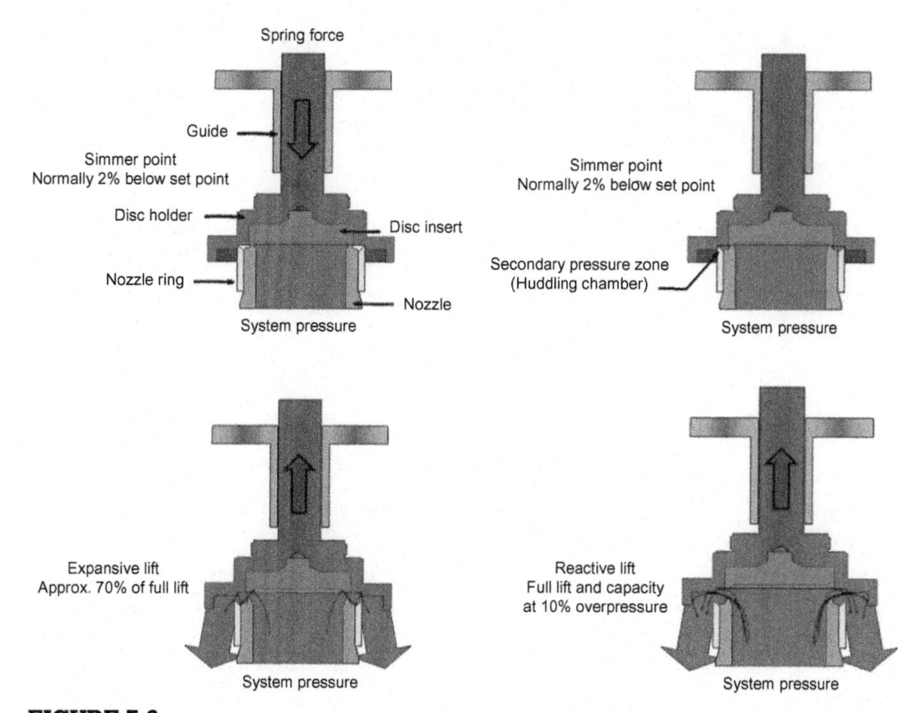

FIGURE 5.6

Expansive and reactive lift

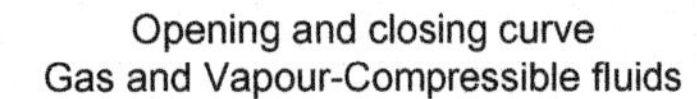

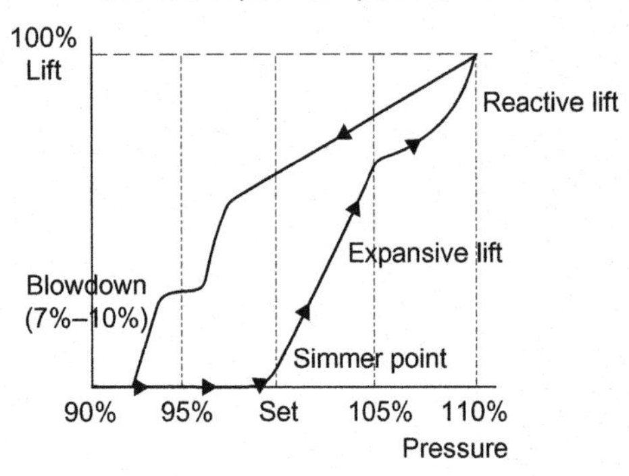

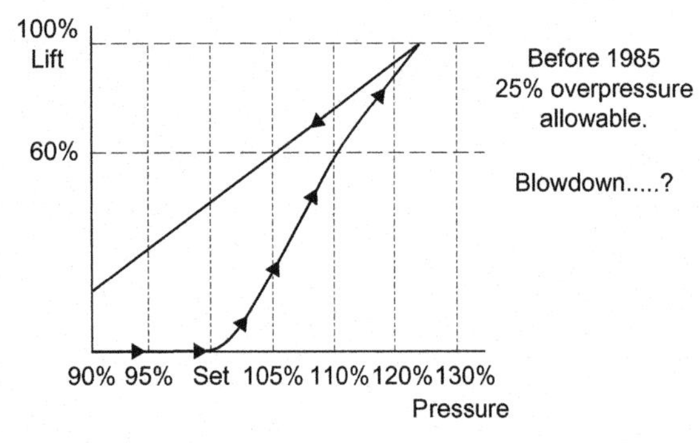

FIGURE 5.7

Typical opening and closing characteristics

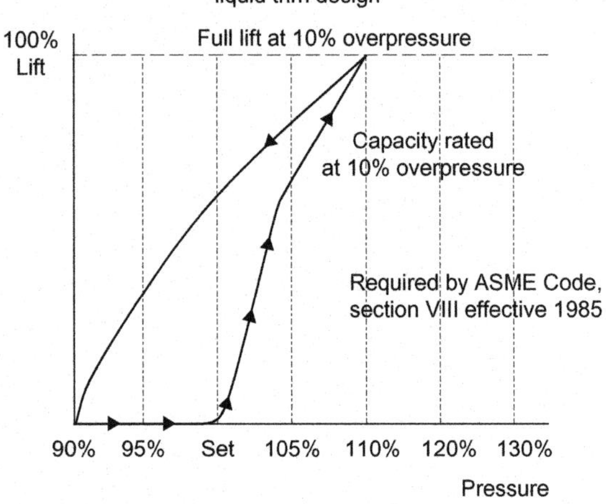

FIGURE 5.8

Typical opening characteristics for a liquid design pot 1985

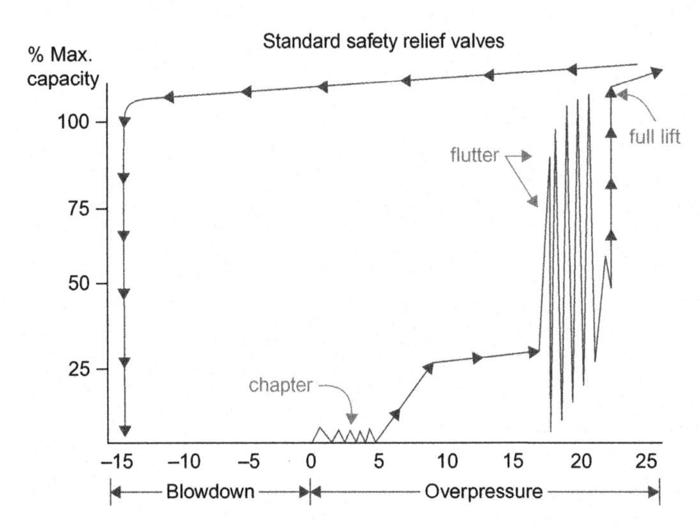

FIGURE 5.9

Unstable operation on liquid with a gas trim

5.2.3 General design

Figure 5.10 shows the construction and main components of a spring-loaded SRV.

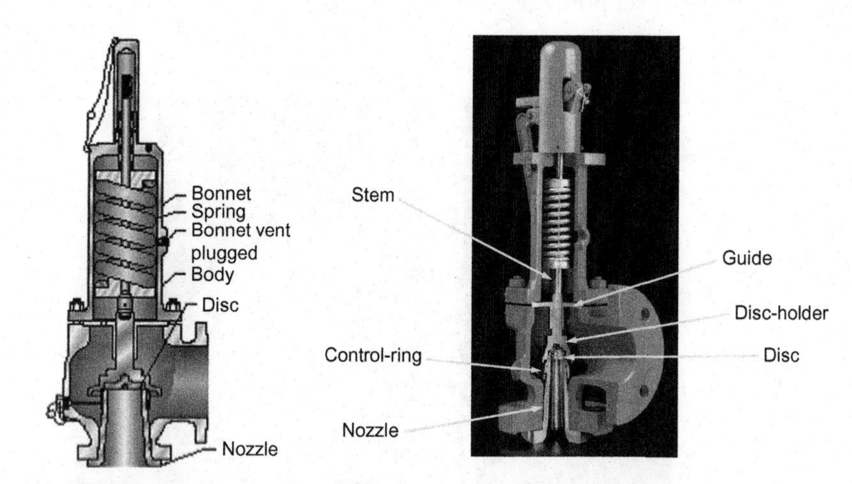

FIGURE 5.10
General design of a spring-operated SRV

The valve consists of a valve inlet or nozzle mounted on the pressurized system, a disc held against the nozzle to prevent flow under normal system-operating conditions, a spring to hold the disc closed and a body/bonnet assembly to contain the operating elements. The spring load is adjustable to vary the pressure at which the valve opens.

Let's take a closer look at some important components for the operation of a spring-operated SRV.

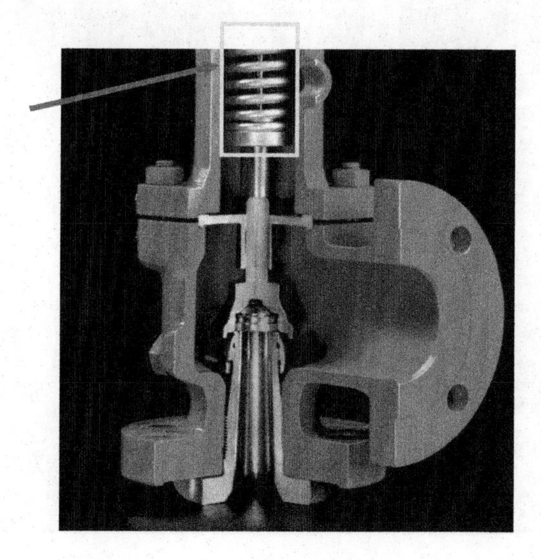

Spring: Comes in different spring ranges which can be found in the manufacturer's specific spring tables. Its compression is applied by the set screw which determines the exact set point at which the valve starts to open.

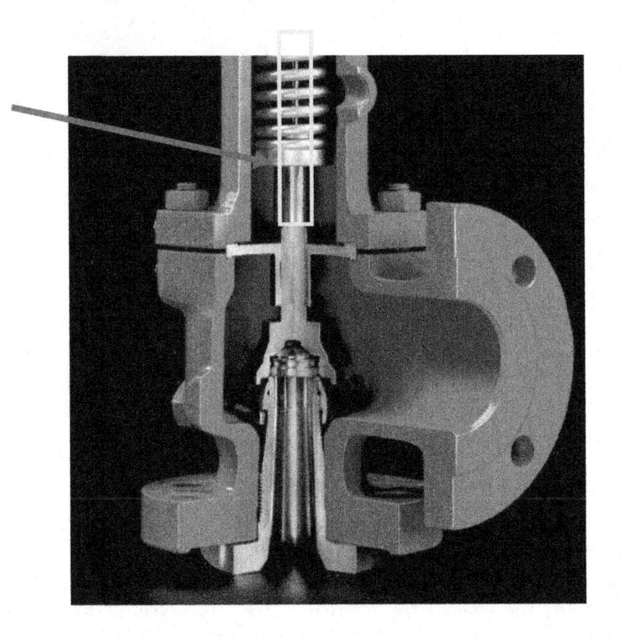

Stem: Transmits the spring force onto the disc in a uniform manner.

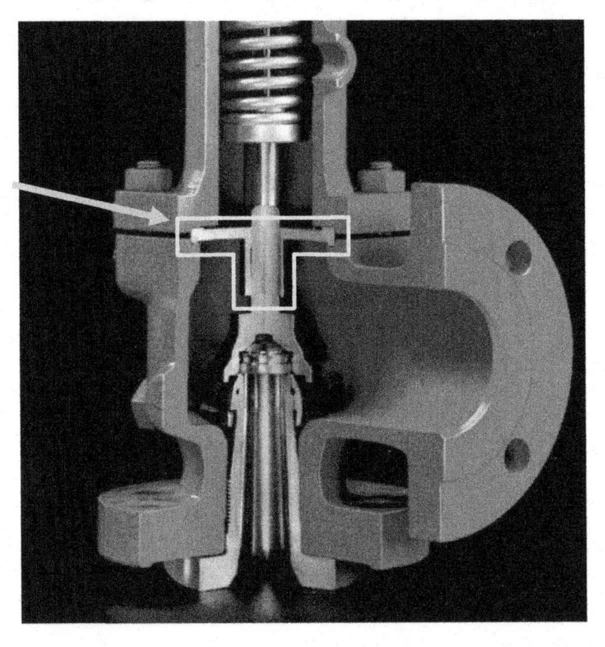

Guide: Guides the stem in order to have a perfect and uniform alignment to transmit the spring force on the disc. It is a very critical component for the correct operation of the SRV and also a vulnerable one because it is easily subject to galling, corrosion and so forth.

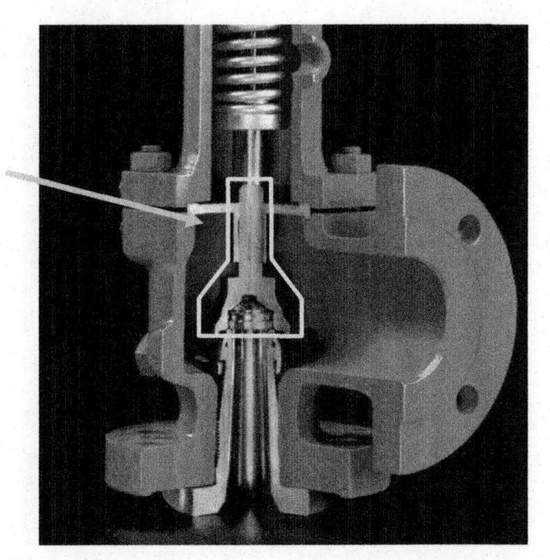

Disc holder: Has a dual function: First, it holds the disc, and second, its 'skirt' or 'hood' shape determines the opening characteristics of the valve as it forms the huddling chamber. It also determines the flow path and hence can have an influence on the flow coefficient. The design of disc holders is very different from manufacturer to manufacturer.

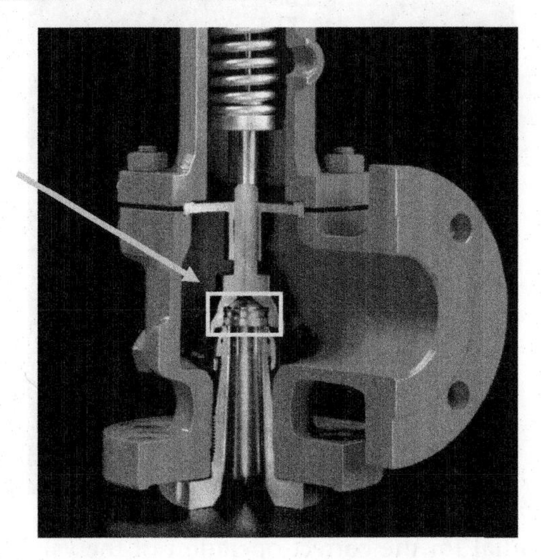

Disc (disc insert): A metal disc with (in the case of a metal-to-metal, spring-operated relief valve) a smoothly lapped surface which ensures tightness in contact with the nozzle. The disc is the main part which ensures a leak-free operation, and the part most subject to wear. It can be re-lapped or overhauled only a few times during maintenance. Removing too much material adversely affects tolerances in the valve and hence its operation.

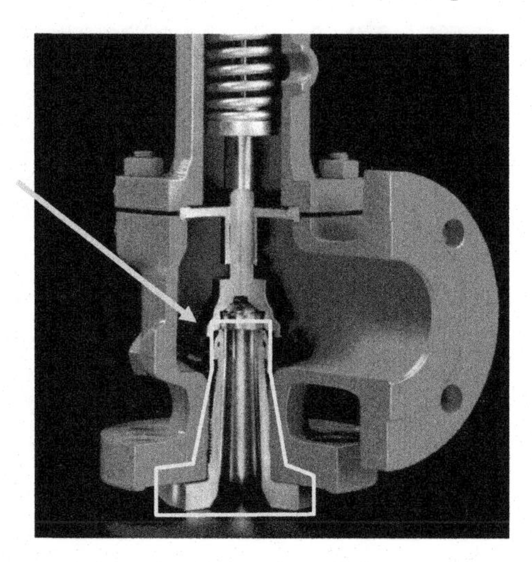

Nozzle: The second part of the seat ensuring the tightness of the valve. For full nozzle-type valves, it is usually integral, is screwed into the body and protects the body from the medium. It also forms the mating face of the flange.

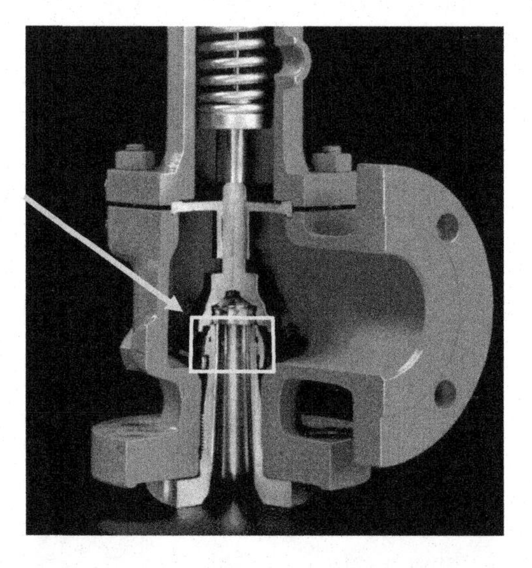

Control ring: Essential for the correct operation of the valve as it determines the overpressure and blowdown values. It is usually screwed on the nozzle and blocked by a pin.

5.2.4 Design of main assemblies

While most manufacturers provide a 3.1 (EN 10204) material certificate for traceability per PED on the body, bonnet, nozzle and disc only as being the pressure retaining parts, one could argue that this should be extended to all parts which truly contain the pressure, including bolts, studs and nuts. The code is not explicit on the matter and it is up to the user to evaluate the issue.

While the code describes pretty much how the valve needs to operate, manufacturers' specific designs for main assemblies may somewhat differ. Although their functionality will always be the same, the design can, for instance, facilitate maintenance or reduce the cost of ownership of the equipment.

Let's look in detail at the different parts and assemblies in a spring-operated SRV and at some of the most common options, and then compare some designs.

5.2.4.1 Cap design and styles

The primary function of the cap is to protect the set screw and the end of the spindle. Since an SRV is subject to regular maintenance and eventually needs resetting, this cap can be easily removed. Its secondary function is to house the mechanism of the lift levers in case they are required. PED code requires lift levers for steam, and ASME code requires them on steam, air and hot water.

The function of the lift lever is to allow manual opening of the valve. Usually, when no extra-lever mechanism is provided, at least 75% of the system's set pressure is required to overcome the spring force by opening the valve manually with a lift lever. While it is required by code, it is judged unsafe to manually open a valve under pressure, especially at higher pressures and higher temperatures, as the person operating the lever would be too close to the valve.

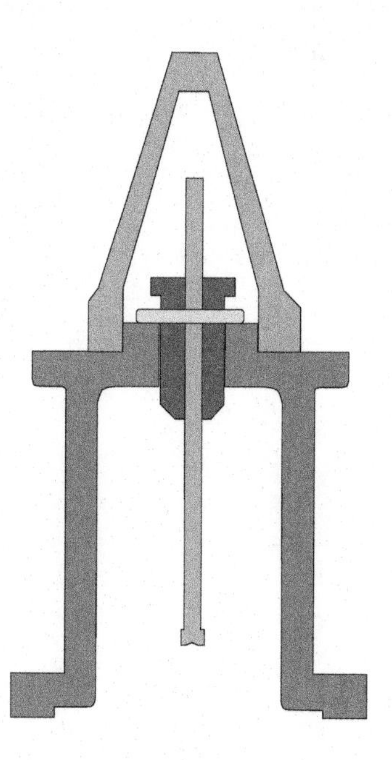

The screwed cap

The screwed cap is the most common, economic and simple design. It usually has a soft iron or other gasket for sealing the cap in case the valve opens, as vapours might be present in the cap during the opening of a conventional SRV.

The disadvantage is that it can come loose due to vibrations on the system. It could also become difficult to remove due to corrosion of the threads, especially in corrosive environments.

The bolted cap

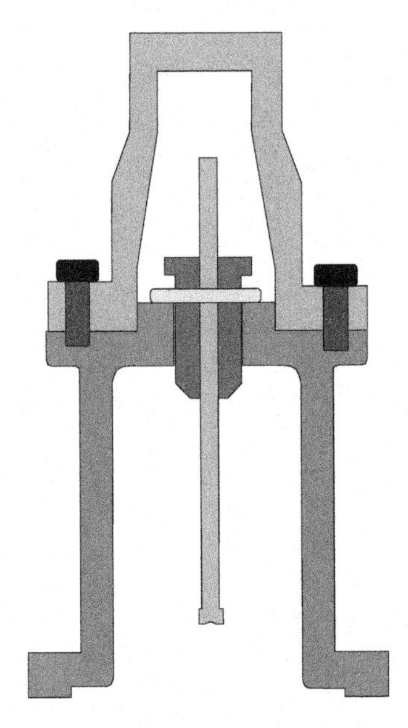

The bolted cap is a little more expensive design used for higher pressures and corrosive environments. Some designs have a metal-to-metal L-shaped seal, while others have a gasket seal.

Plain or open lever

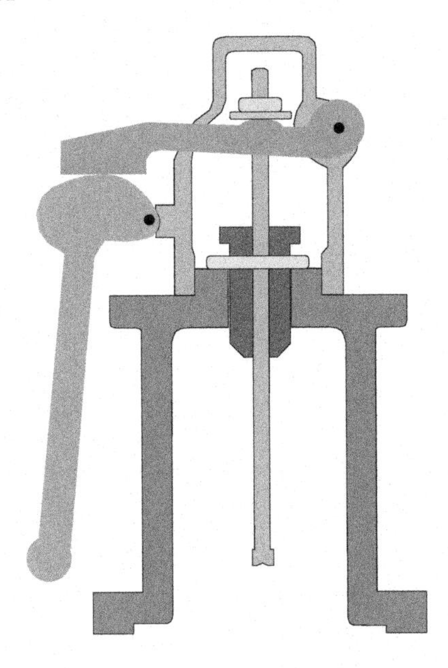

The plain or open lever is a simple construction which does not provide any seal to the atmosphere, so it can only be used when the system is allowed to vent to atmosphere. It should not be used on corrosive, inflammable or toxic products. It is also not recommended when the valve is used on a system with backpressure.

It requires a minimum 75% set pressure before it can be operated, as manual force must overcome the spring force.

It is usually combined with a screwed cap design.

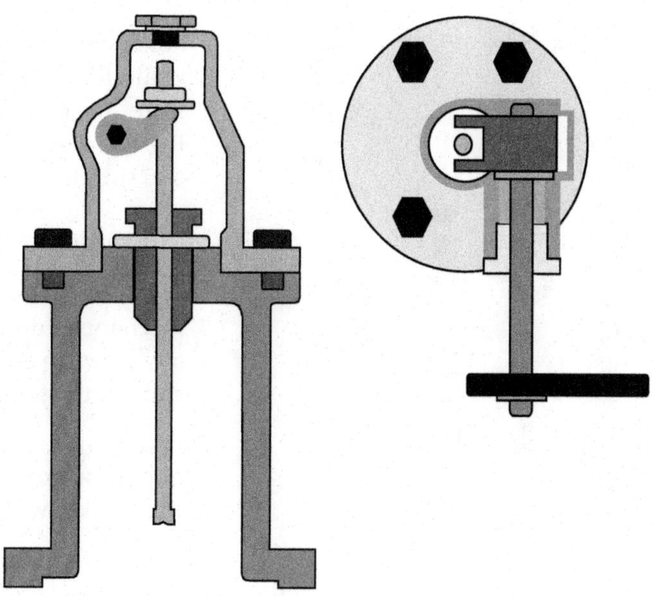

Packed lift lever

The packed lift lever design usually requires a little less lever force to operate.

It is packed and provides a tight seal to the atmosphere; therefore it can be used for corrosive or toxic products. It can also be used if there is backpressure because the fluids are contained in the cap.

Usually, it has a bolted design with a graphoil, soft iron, TFE or other type of gasket, depending on the product it needs to contain.

Test gag

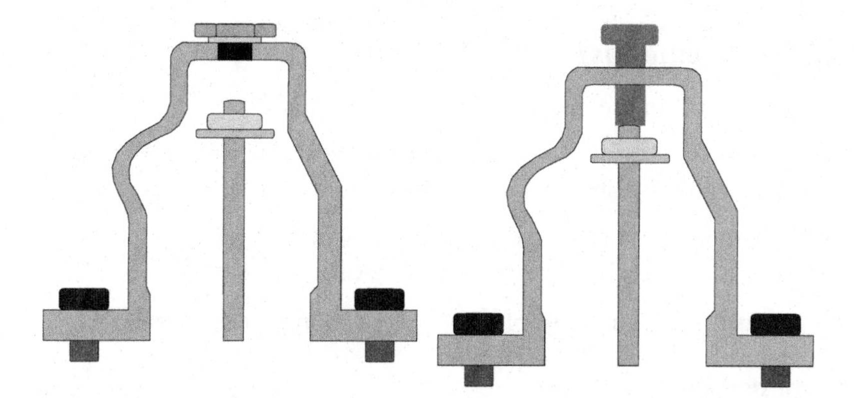

The test gag is intended to block the opening of the valve while a hydraulic test on the system is performed. It physically blocks the spindle from moving upwards, preventing the valve from opening.

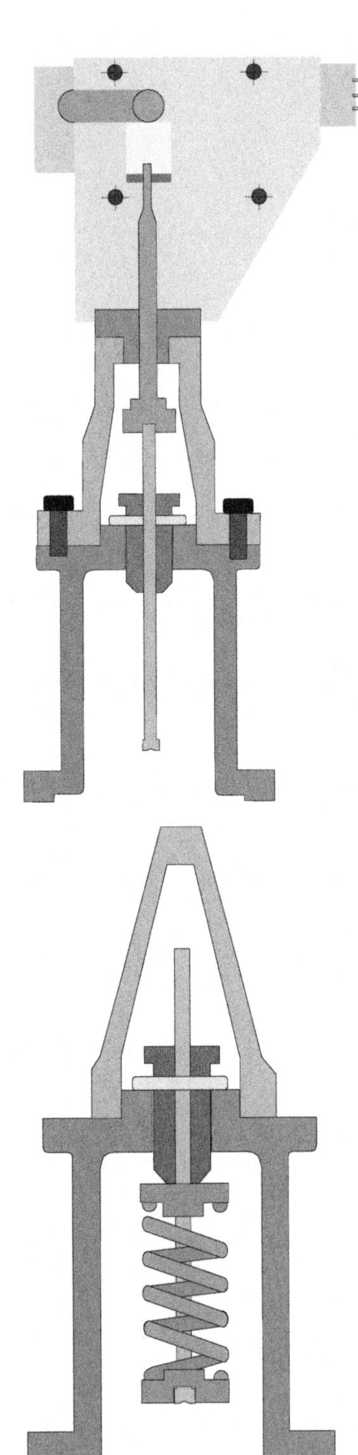

This device, however, is frequently misused and dangerous if not removed after hydraulic tests or before a new installation.

Many times it is also specified for holding the valve closed during transportation so that vibrations do not damage the seat or disc. This, however, is not the purpose of this device.

Some manufacturers supply valves standard with test gags while for others it is an option, which adds to the confusion. Always check that the gag is removed before installing the valve, or the system could be endangered due to this simple but hazardous little device.

Lift indicator switches

Some suppliers offer lift indicator switch mechanisms. The mechanism differs from supplier to supplier, but it is usually a switch which is mechanically activated by an extension on the stem.

Because a metal-to-metal SRV can be damaged after only one or two openings and require maintenance, some users want to know when a valve has opened.

The lift indicator switch will send an indication of the lift of the valve to a remote location.

The lift indicator can also work in combination with a lift lever.

5.2.4.2 Bonnet parts

The upper stem in the cap extends into the bonnet.

The bonnet is closed for a conventional valve and is vented for a balanced bellows valve, and it is usually bolted onto the body.

Its main function is to contain the spring, which is always supported by spring washers, which are usually unique for each spring.

It protects the surroundings of the valve if a spring should break.

Open bonnets

If emission of the fluid into the atmosphere is acceptable, the spring housing may be vented to the atmosphere in an open bonnet design, as shown in Figure 5.11. This is usually advantageous when the SRV is used on high-temperature fluids or for boiler applications as otherwise, high temperatures can relax the spring, altering the set pressure of the valve. However, using an open bonnet exposes the valve spring and internal mechanism to environmental conditions, which can lead to damage and corrosion of the spring. The type of the bonnet depends on the design; some bonnets look like yokes.

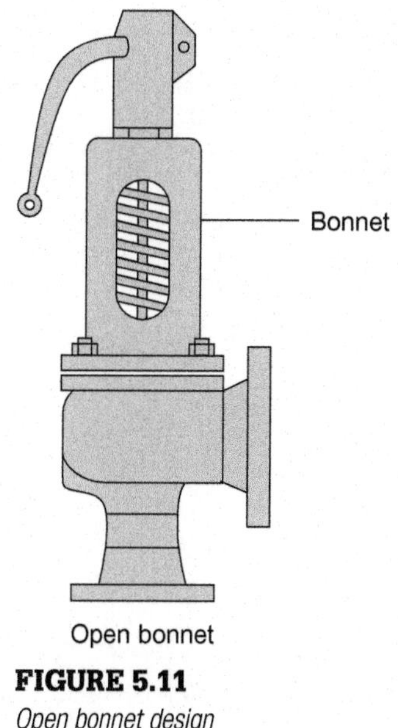

FIGURE 5.11
Open bonnet design

5.2.4.3 Non-wetted parts in the body

The stem and spring force is transmitted onto the disc holder via a guide which has vent holes in the bonnet. These vent holes are blocked when a bellows is installed as the bonnet needs to be at atmospheric pressure (vented bonnet) (Figure 5.12).

The way the stem is connected to the disc holder is different in each SRV design, depending on the manufacturer. It is of course very important that the force be transmitted equally; therefore, a lot of designs have a swivelling joint connection so that the disc is correctly aligned on the nozzle.

The way the disc is fitted in the disc holder also depends on the manufacturer's design. Since the disc is a critical part of the valve, it is important that it can be easily removed for rework, lapping or replacement during maintenance. For instance, the disc holder can be provided with a hole so the disc can be easily snapped out. Otherwise, it could become difficult to remove the disc (Figure 5.13).

Most discs can also be provided with soft seats for more frequent and trouble-free operation, but this design is limited by temperature. How the soft seat is fixed to the disc differs from manufacturer to manufacturer, but here two factors are important: First, the seat must be easy to replace, but the design must also be such that the soft seat cannot be blown out during a relief cycle (Figure 5.14).

The nozzle ring can be screwed up and down the nozzle to adjust the valve's operational characteristics. It is held in place by a lock pin which is accessible outside of the valve's body.

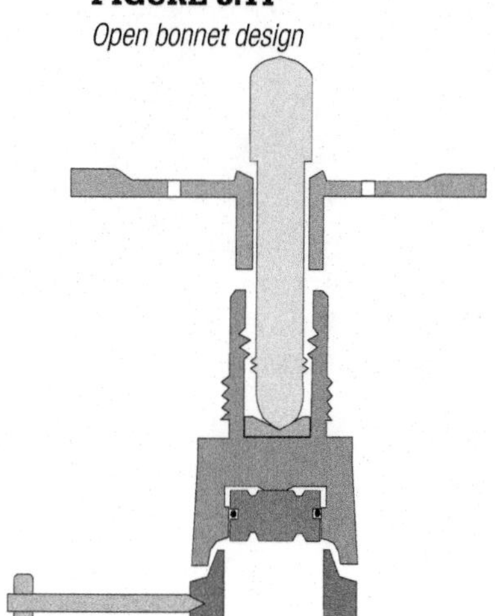

FIGURE 5.12
Trim assembly

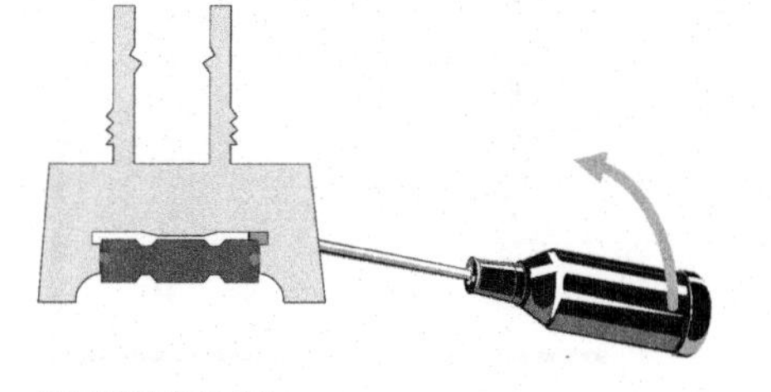

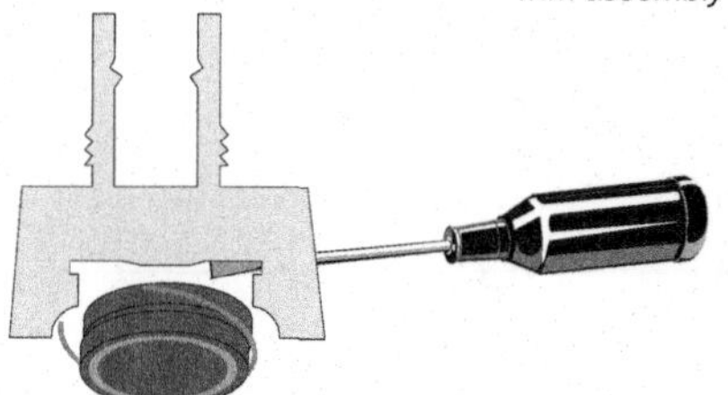

FIGURE 5.13
Disc in disc holder

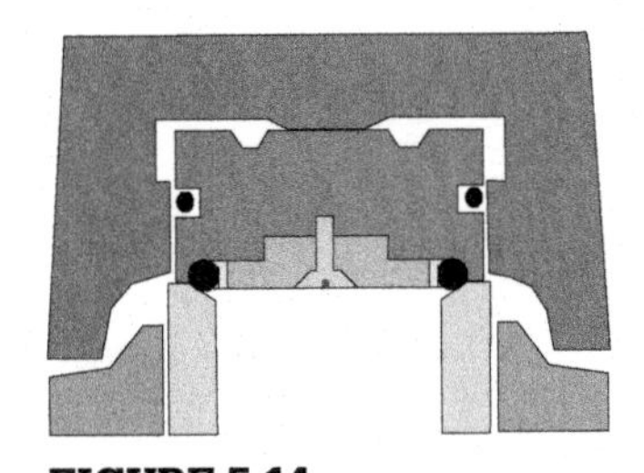

FIGURE 5.14
Example of a soft-seated design and fixture

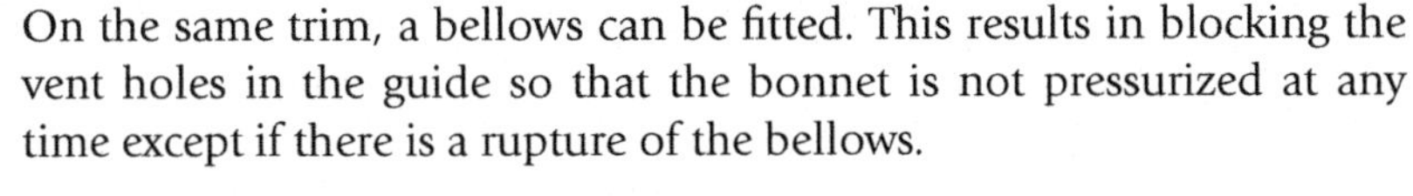

On the same trim, a bellows can be fitted. This results in blocking the vent holes in the guide so that the bonnet is not pressurized at any time except if there is a rupture of the bellows.

The bellows are fitted to cover the top side of the disc holder, in the same area as the nozzle. The way the bellows are fixed depends also on manufacturer design (Figure 5.15).

5.2.4.4 Bellows

To compensate for backpressure effects, the effective area of the bellows must be the same as the nozzle seat area. This prevents backpressure from acting on the top area of the disc, which is not pressure-balanced, and cancels the effects of backpressure on the disc. This results in a stable set pressure (Figure 5.16).

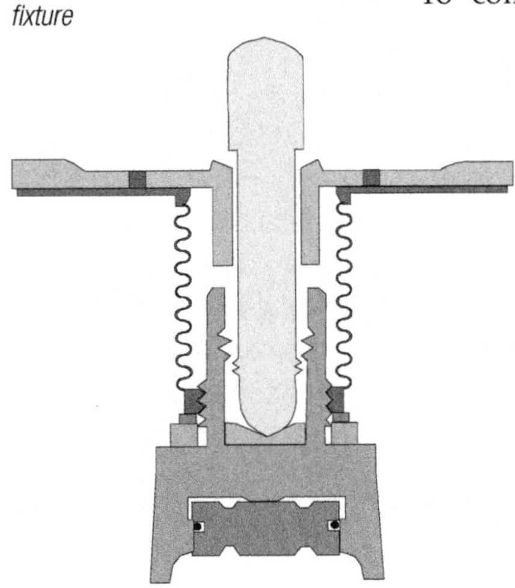

FIGURE 5.15
Bellows on trim assembly

As can be seen in Figure 5.17, the bellows effective area is equivalent to the seat area. The effective bellows area is exposed to the atmospheric pressure in the vented bonnet.

In this sectioned view of a balanced bellows valve (Figure 5.18), we can see that the bellows also protects spindle and guiding surfaces from corrosive fluids. It isolates the spring chamber or bonnet from the process fluid as well. Therefore, bellows provide good corrosive protection.

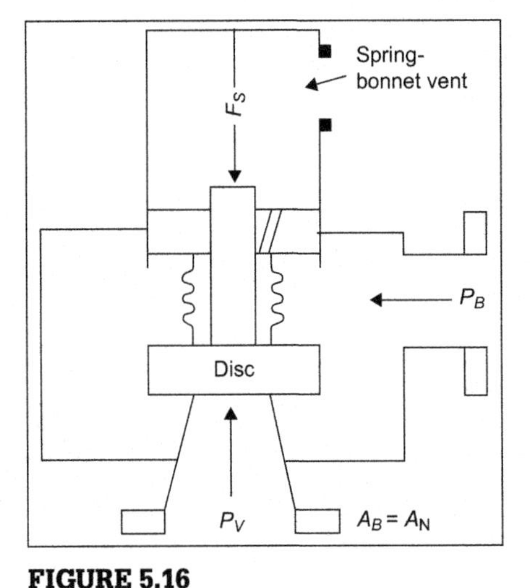

FIGURE 5.16
Bellows principle

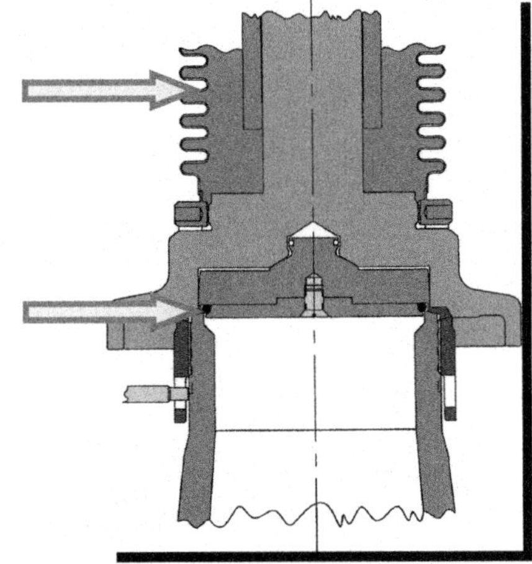

FIGURE 5.17
Bellow design

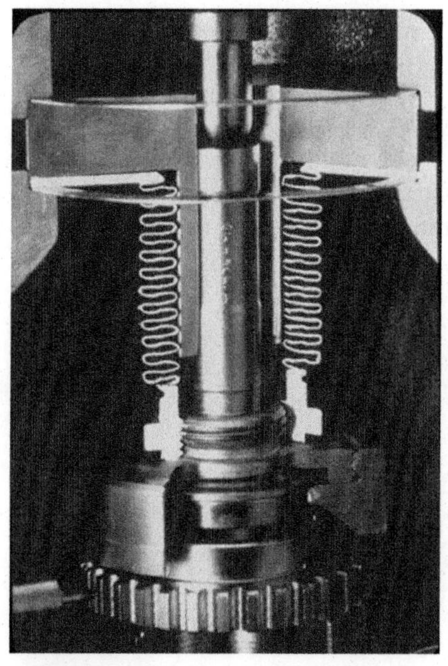

FIGURE 5.18
Sectioned view of a balanced bellows valve

FIGURE 5.19
Full nozzle and disc

5.2.4.5 Wetted parts

Many SRVs are full nozzle design, in which case, only the nozzle and the disc are in permanent contact with the fluid during normal operation (Figure 5.19).

The inside of the nozzle must have a very smooth finish. This improves the flow through the nozzle and prevents particles from getting trapped in some cavities. In such events, the particles would be ejected during opening of the valve and could be trapped between the nozzle and the disc, causing leakage.

In the full nozzle design, the nozzle sits on the mating flange, as shown in Figure 5.20.

In case of a semi-nozzle design, the body flange is in contact with the mating flange and the fluid is in permanent contact with the valve body.

Castings have cavities in which particles can be trapped (see above).

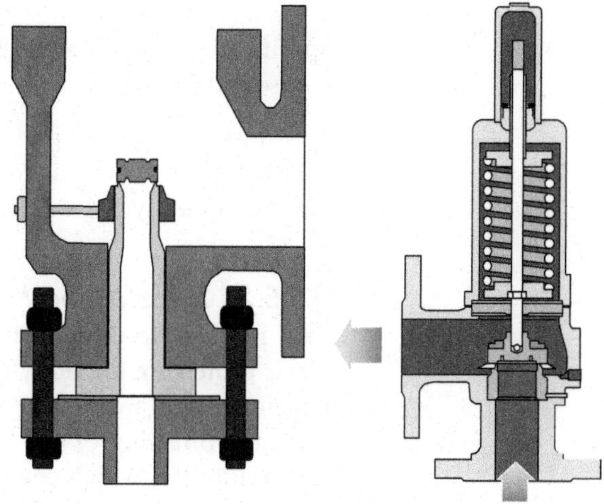

FIGURE 5.20
Full and semi-nozzle design

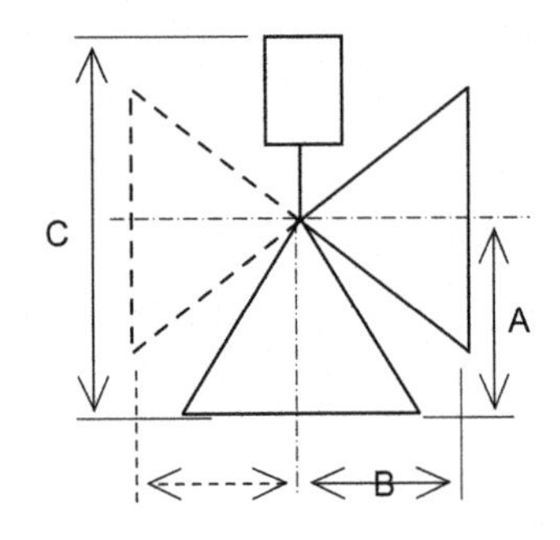

5.2.5 Design differences

We will now look at manufacturers' main component design differences.

Because the body A/B dimensions of the valve itself are fixed by API, all API 526 valves are interchangeable as a complete unit. However, as we will see, the individual components are never interchangeable. It is very important that original manufacturers' parts are used for replacement and maintenance purposes.

5.2.5.1 Disc designs

The disc designs from different manufacturers are very different. The different designs affect ease of maintenance and/or overhaul (Figure 5.21).

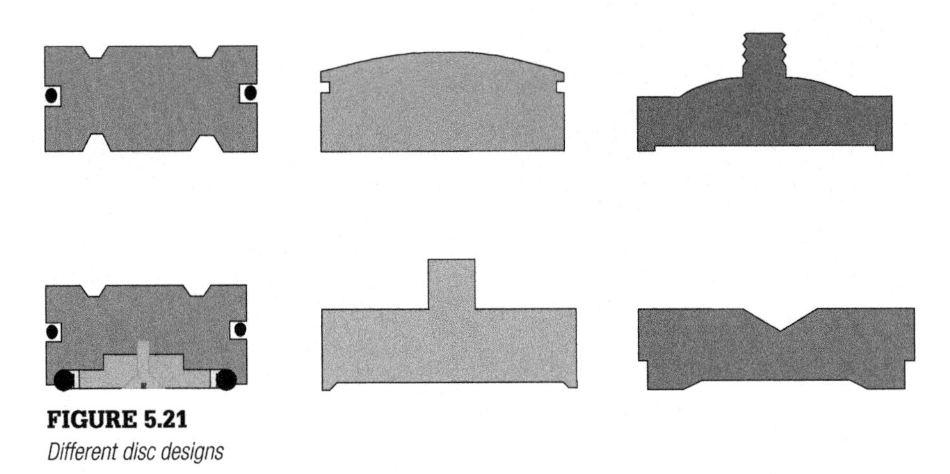

FIGURE 5.21
Different disc designs

Specific designs also affect how easily discs can be converted from metal- to soft-seated designs.

It is important that a perfect alignment between nozzle and disc is guaranteed. This can be obtained by a swivelling design of either the stem on the disc holder or by the disc design itself.

5.2.5.2 Nozzle/body design

In a full nozzle design, the entire wetted inlet parts are formed from one piece, protecting the body from being in constant contact with the fluid. Full nozzle designs are usually used for higher pressures and on corrosive fluids. Also, the finishing of the full nozzle is very smooth so no dirt or corrosion can accumulate which could damage the valve during opening.

Conversely, the semi-nozzle design consists of a seating ring fitted into the body bowl. The top of the nozzle forms the seat of the valve. The advantage here is that the nozzle can sometimes be easily removed for maintenance, but this depends entirely on how the nozzle is fitted into the body. It can be screwed, crimped or pressed, depending on the manufacturer.

While both designs have a huddling chamber arrangement for quick opening of the valve, it is much easier to fit a blowdown ring on a full nozzle design. Contrary to the ASME design, semi-nozzle designs are frequently used on the DIN design used in Germanic markets (Figure 5.22).

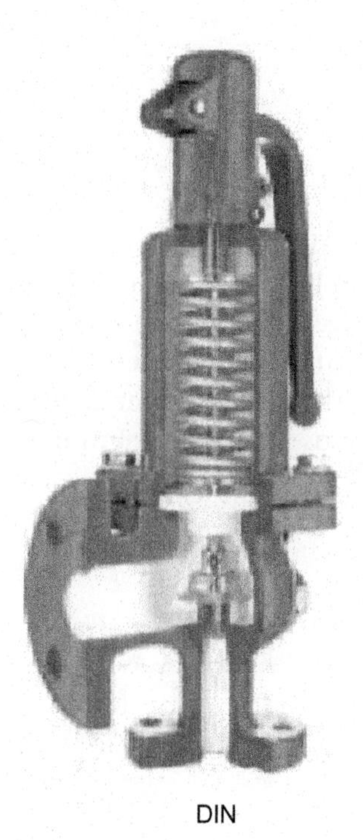

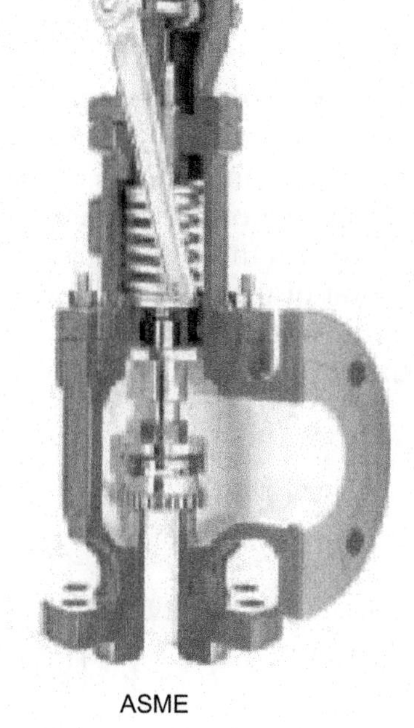

FIGURE 5.22
Full and semi-nozzle design

The way a full nozzle is fitted in the body can be very different from manufacturer to manufacturer – top, middle or bottom threads can be applied.

Most manufacturers use the bottom threads design (middle in Figure 5.23). This has its advantages and disadvantages: Fluid can easily be trapped between body and nozzle threads; on the other hand the nozzle is easier to remove.

Perfect alignment of the valve is critical for its operation, and the closer the threads are to the seat, the better the alignment.

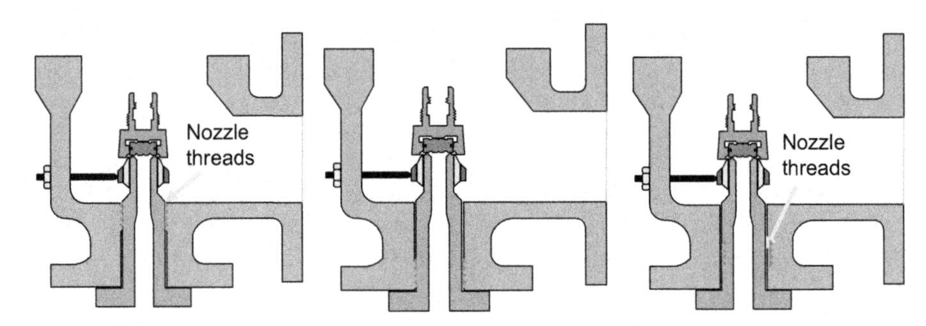

FIGURE 5.23
Full nozzle arrangements

5.2.5.3 Disc holder designs

The two main differences between the various designs are, first, the way the disc insert is fitted into the disc holder. This can be a threaded or snap design. Again, depending on the application (corrosive or not), one could be preferred above the other. In any case, all SRVs require regular maintenance, and easy removal of the disc insert is important. The second main difference in design is the way the stem is fitted to the disc insert as this determines again the alignment of the disc on the nozzle (Figure 5.24).

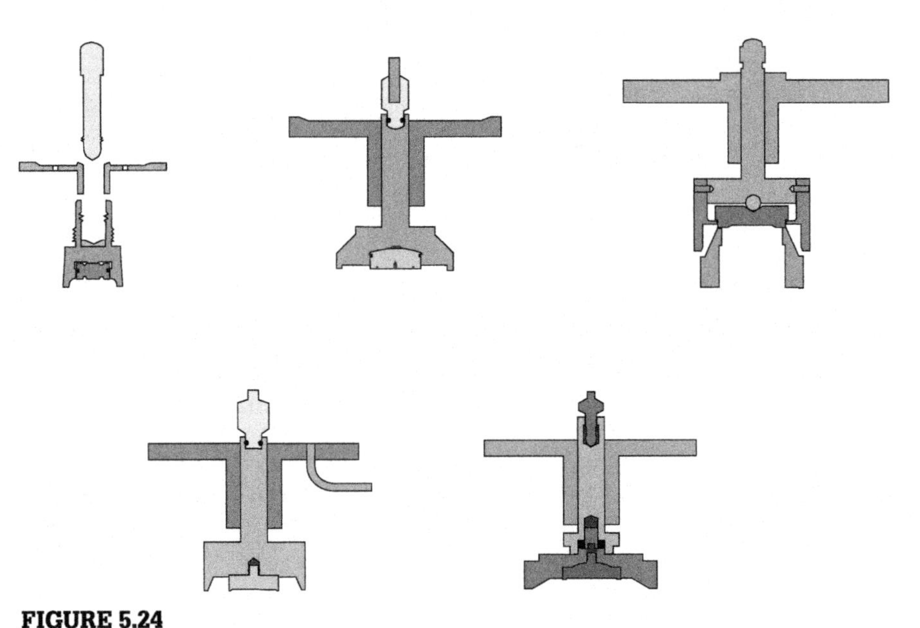

FIGURE 5.24
Disc holder designs

While we have different designs, one important constant is that the bottom of the guide must limit disc lift. This determines the valve lift or the curtain area and hence the flow capacity of the valve.

As already discussed, rated capacity is controlled by the nozzle diameter or bore.

The minimum seat plate lift to achieve rated capacity must be such that the annular curtain area around the periphery of the nozzle equals the nozzle area.

$$\text{(Curtain area) } \pi DL = \pi D^2/4 \text{ (Nozzle area)}$$

where D = nozzle bore diameter

L = Lift of seat plate

Dividing both sides by πD gives:

$$L = D/4$$

Theoretically: Capacity is achieved when lift = 25% of the nozzle diameter.

In practice: A lift of 40% is required for pressures below 1 barg because of flow losses.

5.2.6 Types of spring-operated SRVs

5.2.6.1 *Thermal relief valves*

Thermal relief valves are small, usually liquid relief valves designed for very small flows on incompressible fluids. They open in some proportion of the overpressure. Thermal expansion during the process only produces very small flows, and the array of orifices in thermal relief valves is usually under the API-lettered orifices, with a maximum orifice D or E. It is, however, recommended to use a standard thermal relief orifice (e.g. 0.049 in^2). Oversizing SRVs is never recommended since they will flow too much too short, which in turn will make them close too fast without evacuating the pressure. This will result in chattering of the oversized valve and possible water hammer in liquid applications.

Thermal relief valves are usually of rather simple design. A few are designed to resist backpressures. However, there are so many designs on the market, it is impossible to treat them all in this book.

As a general rule, they should not be used in steam, air, gas or vapour applications or in the presence of variable backpressure. Nevertheless, there are high-performance small spring-loaded valves which are designed specifically for specific services on gas, with backpressure or under cryogenic conditions.

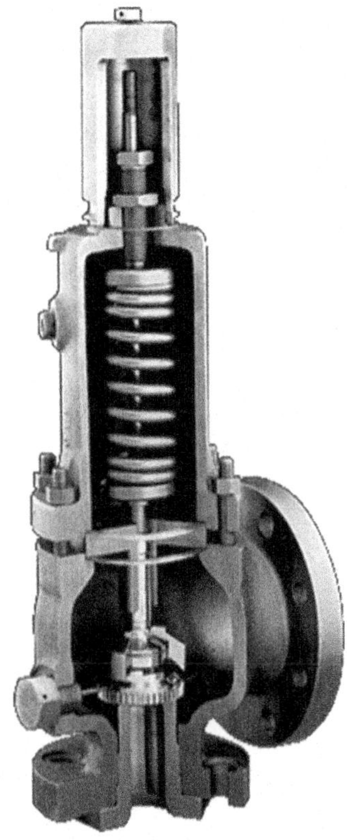

5.2.6.2 Conventional spring-operated SRV

These are the most commonly used SRVs in the process industry.

They can be used on any fluid: compressible or non-compressible.

They should preferably not be used on steam service, steam boiler drums or superheaters, where an open bonnet-type SRV is generally preferred because of the temperature of the spring, which is required to be cooled off in order for it to keep its characteristics and maintain stability in the set pressure.

Conventional SRVs are normally used in any services where the superimposed backpressure is constant and/or the built-up backpressure does not exceed 10% of the set pressure.

They have a pop or snap action opening, which means they open very rapidly. The maximum seating force occurs at the lowest system pressure. However, the closer to set pressure, the lower the seating force, as there tends to be equilibrium between spring force and system pressure, shown in the left graph in Figure 5.25.

This type of valve has a typical relief cycle, as shown on the right of Figure 5.25.

The SRV reseat pressure should always be above the normal operating pressure, or too much process fluid will be unnecessarily wasted.

Opening and closing characteristics can be adjusted via the blowdown ring (see Section 5.2.2).

Limitation

Conventional SRVs should not be used:

- On steam boiler drums or superheaters
- In cases of superimposed variable backpressure

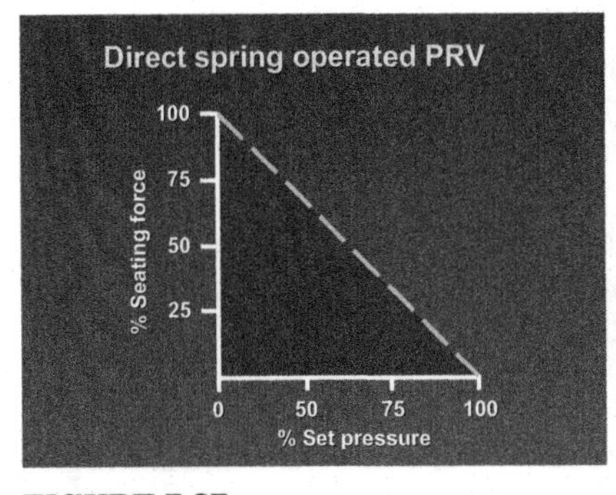

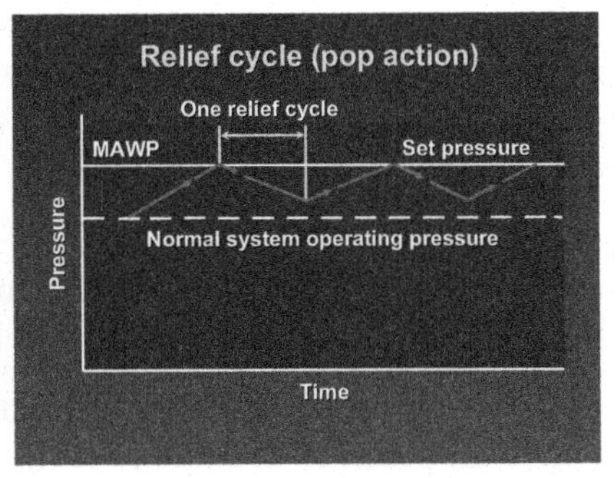

FIGURE 5.25

Seating force and relief cycle of a conventional SRV

- In cases where the built-up backpressure exceeds 10% of the set pressure
- As pressure control or bypass valves

Advantages	Disadvantages
Wide range of materials available	Prone to leakage (if metal seated)
Wide range of fluid compatibilities	Long simmer or long blowdown
Wide range of service temperatures	Risk of chattering on liquids, unless special trims are used
Rugged design	Very sensitive to inlet pressure losses
Compatible with fouling and dirty service	Limitation in pressure/sizes Highly affected by backpressure (set pressure, capacity, stability) Not easily testable in the field

5.2.6.3 Open bonnet spring-operated SRV

Open bonnet valves are used where the operating temperatures are high and the spring is required to be cooled off in order to retain its characteristics.

They are subject to an ASME I approval if used on the primary circuit of power boilers and superheaters. If the valve is designed according to ASME I, then the valve has a dual-ring control.

Open bonnet valves are characterized by a rapid opening or 'pop' action and small blowdown.

They can also be used for compressible media, such as air, steam and gases, with normal ASME VIII characteristics. In this case, the design is the same as that of a normal conventional valve.

Open bonnet valves are not recommended for liquid or toxic applications.

5.2.6.4 Balanced bellows spring-operated SRV

The basic design of a balanced bellows spring valve is the same as a conventional valve, but a bellows is added to compensate for variable backpressures and the bonnet is vented to atmosphere. A leak at the bonnet vent indicates a failed bellows. Because balanced bellows valves must have their bonnets vented to the atmosphere, a safe location for piping the vent must be determined. The balanced bellows valve, with or without a supplementary balancing piston, only discharges process medium from the bonnet vent in the case of failure of the bellows (Figure 5.26).

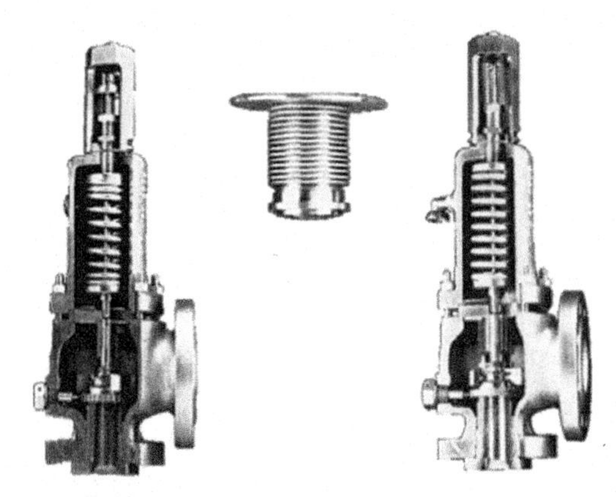

FIGURE 5.26
Balanced bellows design

The eventual supplementary balancing piston is actually fitted as a backup device; if the bellows fail in service, this device will ensure that the valve still relieves at the correct set pressure. Balancing piston valves are uncommon and expensive, but without the balancing piston fitted, the variable backpressure has an adverse effect on the set point of the SRV. This may result in the valve not relieving at its full capacity or not achieving full lift within 10% overpressure.

Theoretically, the balanced bellows valve can handle any backpressure but is usually limited up to 50% of set pressure. Balanced SRVs may be used anywhere the backpressure is either constant or variable.

The balanced bellows SRV is very effective in corrosive or dirty services because it seals the corrosive or dirty process fluid from contact with the guiding surfaces of the valve, thus preventing sticking as a result of corrosion or ingress of dirt at this contact point.

Limitation

Balanced type SRVs should not be used:

- On steam boiler drums or superheaters
- As pressure control or bypass valves

Advantages	Disadvantages
Guiding surfaces protected	Prone to leakage (if metal seated)
Set point unaffected by back pressure	Long simmer or long blowdown
Capacity reduced only at high levels	Risk of chattering on liquids, unless special trims are used
Wide range of materials available	Very sensitive to inlet pressure losses
Wide range of fluid compatibilities	Limitation in pressure/sizes
Wide range of service temperatures	Limited bellows life
Rugged design	High maintenance costs
Compatible with fouling and dirty service	Not easily testable in the field

5.2.6.5 High-performance resilient-seated SV

Some manufacturers have special designs for particular applications, but high-performance valves extend the characteristics of the normal API/ASME/PED and EN requirements. Many are resilient-seated design.

- Possibility for independent opening and blowdown settings
- Optimum tightness up to set pressure with soft-seated designs

- Snap opening at set pressure without requiring overpressure
- Blowdown settings between 3% and 25%
- Balancing against backpressures without the use of vulnerable bellows

Such resilient-seated valves allow for much higher system-operating pressures, which results in enhanced profitability and minimized emissions. They minimize unnecessary product losses.

Soft-seated, leak-free and snap-acting valves are also recommended for cryogenic applications as the typical leakages allowed on metal-to-metal API valves could cause icing of the seat area, which could cause the valve to fail.

Application

Resilient valve seats are frequently used when a greater degree of seat tightness is required than is likely with metal-to-metal seats.

- Where the service fluid is corrosive or hard to hold with metal seats. Slight leakage of a corrosive gas, vapour or liquid could deteriorate or foul the moving parts of a valve.
- When small, hard foreign particles are carried in the flowing fluids, they can easily scratch or mark metal seats once the valve discharges, which results in probable leakages. The resilient seal can absorb the impact of the particles, shield the mating metal-seating surface and reduce the probable incidence of leakage to a certain extent.
- To prevent loss of expensive fluids and to minimize the escape of explosive, toxic or irritant fluids into the environment.
- Where operating pressures may be too close to the set pressure. As the operating pressure approaches the set pressure, the net differential forces on the disc are reduced. Resilient seats provide a better degree of tightness than metal ones.
- When an SRV is subject to a minor pressure relief situation, the disc may only lift enough to cause a slightly audible escape of fluid or visible drip (if liquid). This may relieve the system pressure, but the valve does not significantly pop or lift open. Under this condition, with metal seats the disc may not reseat properly and the valve may continue to leak below the system normal operating pressure. A resilient seat provides tight shutoff when the system pressure falls after a minor relief.
- Vibrations and pulsating pressures tend to reduce the effect of the spring load on the disc, causing a rubbing movement of the disc on to the nozzle seat. This results in seat leakage. Where SRVs are subject to this condition, a greater degree of tightness can be maintained with resilient seats than with metal ones.

Limitation

- A large variety of elastomers and plastics are currently available for seals in valves. At present, there is no single material suitably resilient to all pressures, temperatures and chemicals. Therefore, each resilient seat application should be selected after considering the specific fluid and service conditions. Where certain materials may be excellent with respect to chemical resistance, they may not be suitable for the intended service temperatures, and vice versa.
- Explosive decompression can occur on some gases at some very high pressures (Figure 5.27). This can blister and split O-rings when the pressure drops suddenly (Figure 5.28). This is of great concern in high-pressure applications. In particular, elastomers (which normally provide the best tightness) are rather porous. Gas can dissolve/diffuse into these microscopic pores under high pressure. When the valve relieves, there is a sudden enormous pressure decrease. If pressure decreases, volume proportionally increases and these gases trapped in the elastomers need to expand very fast, which can make the soft seat explode. Special attention to the porosity of the elastomer is always required for high-pressure applications as damages may not always be visible at the outside of the seal.

FIGURE 5.27
Explosive decompression in O-rings

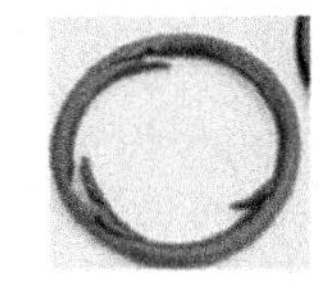
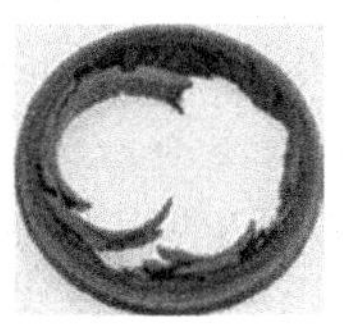

FIGURE 5.28
Exploded O-rings

Past plant experience and pressure/temperature limitations are probably the best guides in choosing an elastomer or a plastic.

Some basic guidance:

Plastic:

- Poor resilience, hard
- Poor memory
- Not good for dynamic friction (unless with energizing spring)
- Usually a wide chemical compatibility

Elastomer:

- Good resilience
- Good memory
- Soft
- Limited chemical compatibility
- Temperature limitations

Some frequently used resilient seats and their characteristics within SRVs:

Plastics:

- Urethane
  - Inert
  - Rather standard available
  - Large chemical compatibility
  - Acrylic polymer
  - Temperature range: −54°C to +150°C
- PTFE (Teflon), Polytetrafluoroethylene
  - Inert
  - Rather standard available
  - Large chemical compatibility
  - Somewhat softer than urethane, therefore suitable for lower pressures
  - Very sensitive to erosion, scratches
  - Thermoplastic, mouldable,
  - Temperature range: −267°C to +260°C
- SS-filled PTFE
  - Same as PTFE but can resist higher pressures
  - Problems on liquid services
- PCTFE (Kel-F)
  - Inert, large compatibility
  - Suitable for high pressures (very hard)
  - Not too sensitive to erosion, scratches
  - Thermoplastic, mouldable
  - Temperature range: −240°C to +204°C
- PEEK (Polyetheretherketone)
  - Inert
  - Large chemical compatibility
  - Suitable for high pressures
  - Not too sensitive to erosion, scratches

- Thermoplastic, difficult to mould
- Temperature range: −62°C to +288°C (melts at 340°C), +200 on steam and hot water
- Not suitable on chlorides and highly concentrated acids

- VESPEL (SP-1)
  - Polyamide resin
  - Suitable for high pressures
  - Temperature range: −252°C to +315°C
  - Not recommended for water, steam, ammonia

Elastomers:

- Buna-N
  - Nitrile
  - Mechanically the best
  - Excellent for abrasive applications
  - Excellent on sweet oil and gas, dry ammonia, butane, propane...
  - Beware of H_2S contents in the fluids (maximum 10 ppm), 'Chloro-', acids ...
  - Temperature range: −54°C to +135°C
- Viton (A), FKM
  - Fluoroelastomer
  - Good for oil and gas, H_2S (maximum 2000 ppm)
  - Can resist small percentage of methanol
  - Not good on pure methanol
  - Not suitable for high percentage of CO_2, amines, ammonia
  - Temperature range: −54°C to +204°C
- EPR (Ethylene propylene)
  - Excellent for abrasive applications
  - Good for water, steam, H_2S, hydraulic fluids...
  - Not good for oils and lubricants...
  - Temperature range: −54°C to 162°C
- Kalrez (FFKM, Perfluoroelastomer)
  - Mechanically not very good
  - Chemical compatibility excellent, depending on the compound: acids, H_2S, chlorine, steam
  - Very dependent on the compound
  - Temperature range: −29°C to +315°C

5.2.6.6 Balanced piston SRV

The difference between a balanced bellows valve and a balanced piston valve is that additional assurance for safe operation is built into the latter by adding a balanced piston on top of the guide (Figure 5.29).

A balanced piston SRV can handle all applications mentioned for balanced bellows SRVs, but should the vulnerable bellows fail, the piston on top of the guide maintains the proper performance of the valve, with no change in opening pressure and no reduction in valve capacity. The balanced piston ensures a stable valve performance.

5.2.6.7 Designs out of forged blocks – Block design

With modern-day process systems requiring greater throughputs at ever increasing pressures, it has become an engineer's dilemma as to what SRVs to specify when all existing design possiblities have been exhausted and none suffice (Figure 5.30).

For those applications involving very high set pressure and backpressure out of the range of normal standards, some manufacturers offer a custom-made block design. The entire internal function of the valve is enveloped within a forged block maintaining code design requirements, safety and ease of maintenance.

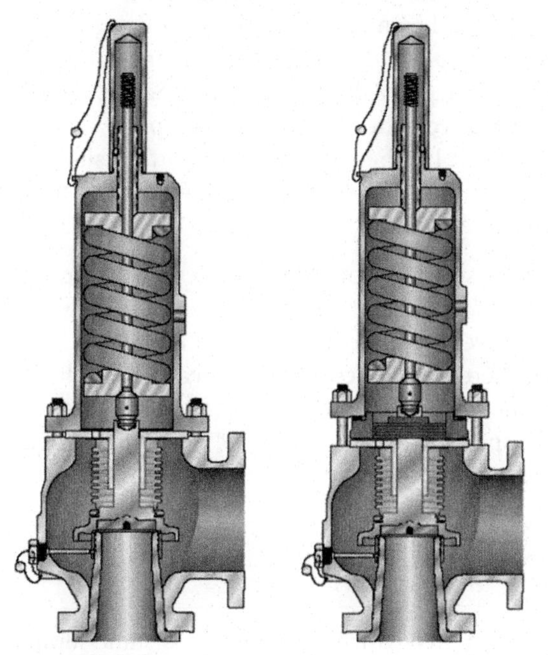

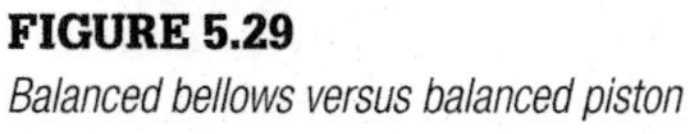

FIGURE 5.29
Balanced bellows versus balanced piston

FIGURE 5.30
Forged block design

The required performance characteristics in terms of lift, reseat and high flow coefficient are ensured, but the whole valve is customized for a specific application.

Usually the size of the body and bonnet are determined by the design of the special spring (pressure and material).

If possible, the customized centre-to-face dimensions are compliant with API in order to maintain existing piping.

The design almost always exceeds the limits of API 526 but is, of course, fully conforming to the ASME VIII code and can be stamped per the code. Since they can handle extremely high temperatures and pressures, one valve can handle many times the capacity of multiple valves.

Forged block designs are completely custom engineered to perfectly conform to the application but are extremely expensive.

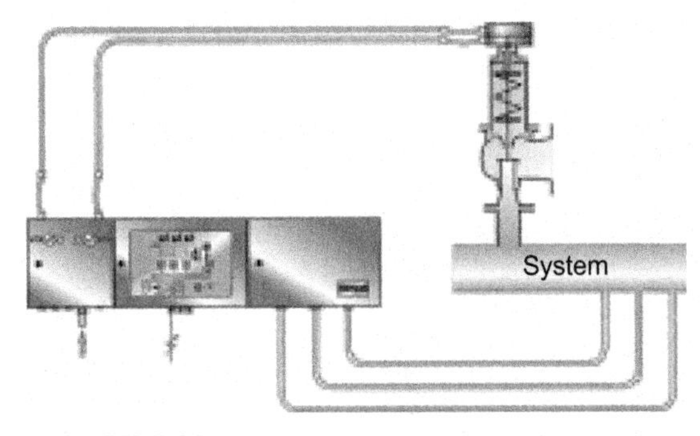

FIGURE 5.31
Typical CSPRS system

5.2.6.8 Controlled safety pressure relief system

The controlled safety pressure relief system (CSPRS) (Figure 5.31) is a special type of actuated SRV that has mainly been used in Europe and in power applications, primarily in the German power industry, for the last 40 years. So far, it has not been used extensively in other parts of the world, but it has become more important since it has its own EN/ISO 4126 Part 5 code in the new European regulation.

It is interesting to note that the German wording 'gesteuerte sicherheitsventile' has no true English equivalent; it is 'controlled safety valves' in the technical sense. Translated into English, the German words 'gesteuert' and 'geregelt' both mean 'controlled'. Therefore, the English expression 'controlled safety pressure relief system' (CSPRS) was created for the European Standard EN ISO 4126-5 to prevent any confusion between 'controlled safety valves' and 'control valves'.

For the past 40 years, CSPRS have been successfully used in Germany as safety devices for protection against excessive pressure, especially in power plants, both conventional and nuclear.

A CSPRS is, in fact, a spring-operated SRV which can operate as stand alone but in normal operations is controlled by an actuator (usually pneumatic or hydraulic) and which opens upon the signal of an instrumentation loop. Due to the additional force on the spring, more accurate set pressure can be obtained, therefore allowing the user to operate at higher process pressures. Raising the valve's pressure below set pressure is also possible by reversing the force from the actuator. This inevitably results in a high degree of safety as it is a redundant system in itself. If the control system fails, the stand-alone valve will normally still function as a normal SRV. It should in no way be confused with a pilot-operated safety relief valve (POSRV) which is solely powered by fluid.

There are many possible configurations; therefore, we will limit ourselves to one example where this type of configuration is composed of an SV body assembly with a pneumatic actuator which acts as additional force to the valve closure (Figure 5.32).

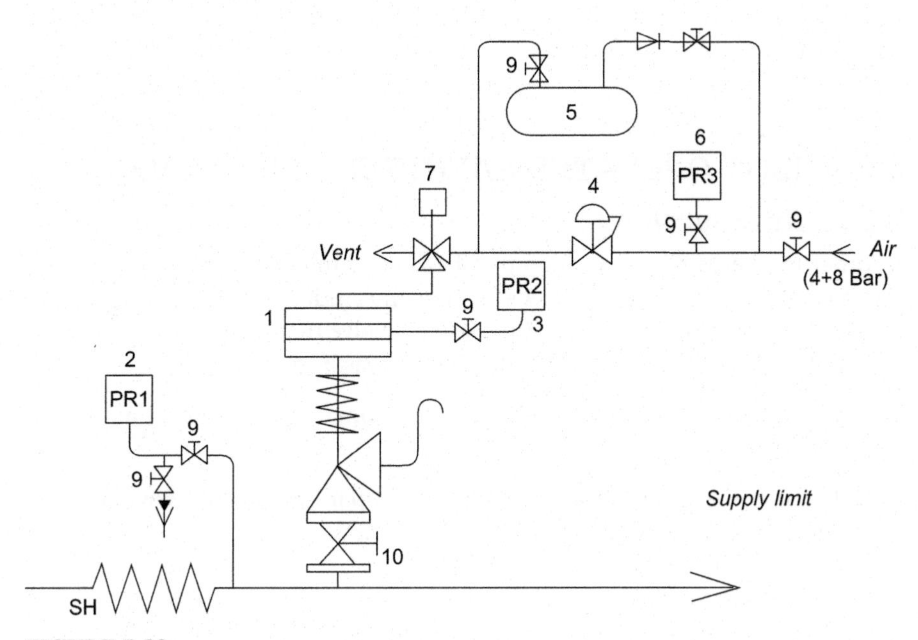

FIGURE 5.32
Possible typical CSPRS set-up

The air is fed to the actuator through a solenoid which is controlled by a pressure switch measuring the pressure at which the actuator is triggered or in this case simply by a push button for the manual test of the valve.

When the solenoid is de-energized, the valve closes.

Energizing causes the air to be discharged from the actuator, opening the valve instantaneously, and forces it to reclose when the pressure switch, sensing the controlled pressure, de-energizes the solenoid.

With these types of valve, an electric and a pneumatic board are usually supplied, complete with all the accessories necessary for operation, signalling and manual testing, in order to guarantee the safety and reliability of the service.

In this particular case, the actuator has two diaphragms as a redundancy measure. The existence of two diaphragms on the actuator assures continuity of valve operation even if the diaphragm fails under pressure, which, in any case, should be signalled by means of an alarm.

The air lock is another additional security in case pneumatic supply should fail.

Due to its performance, this type of valve is in many cases installed in addition to a spring-loaded SRV and is set to open at a lower pressure, avoiding possible interventions of the spring-loaded SRV in case of, for example, light unexpected plant load reductions.

5.3 PILOT-OPERATED SAFETY RELIEF VALVES

5.3.1 Introduction

In order to overcome the various problems encountered when using spring-operated SRVs and also at NASA's request, the industry started looking for ideal SRV characteristics and tried to design a valve that came as close as possible to these characteristics. Thus, the POSRV was born.

The main objectives achieved with some pilot-operated valves on the market were:

- Compensation for variable backpressure up to 90% to 100% without the use of vulnerable bellows
- Tightness up to set pressure (98%)
- Rated capacity at set pressure, not requiring any overpressure
- Opening and blowdown independently adjustable
- Large range of blowdown adjustment to overcome inlet piping losses
- Possibility of installing the valve further away from the process
- Easy maintenance
- In-line functionality testing possibilities
- Reducing the unnecessary loss of material during opening (cost and environmental considerations)
- Reducing the weight of the valves in order to reduce mechanical supports and effects of reaction forces on the valve during opening
- Reducing the noise during the opening cycle
- Stable and reliable operation in case of dual flow (flashing) applications

Unfortunately, a lot of the operational objectives for the POSRV described above could not be achieved as economically as with a spring valve; the use of soft seats was imperative to obtain all advantages, which limits the use of most POSRVs in high temperatures (typically up to 300°C maximum). There are now POSRVs in the market with metal-to-metal seats, but here tightness, especially after a few operations, decreases much more than with resilient-seated valves and even traditional spring-operated metal-to-metal valves.

Due to its somewhat more complicated design, in the early days API did not recommend its use on dirty service or polymerizing fluids. However, since then some renowned manufacturers have found solutions to overcome these problems.

The advantage of soft seats is also that they are very suitable for cryogenic applications.

5.3.2 Functionality

The POSRV consists of two basic components: a main valve, which provides the capacity, and a pilot, which controls the main valve. In normal operation, the pilot allows the system pressure (P) to act upon the top of the piston. The unbalanced piston assembly is increasingly forced down onto the nozzle with increasing system pressure due to the piston seal area (A_D) being larger than the seat seal area (A_N). The surface (A_D) on top of the piston on which system pressure acts is approximately 30% greater than the seat area. So the closing force ($F = PA_D$) increases as pressure rises (Figure 5.33).

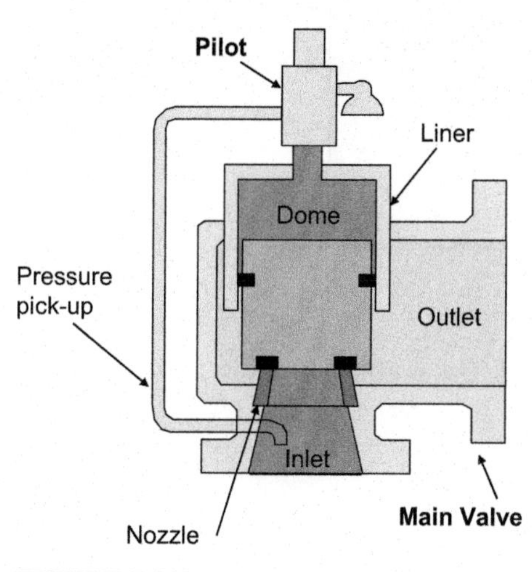

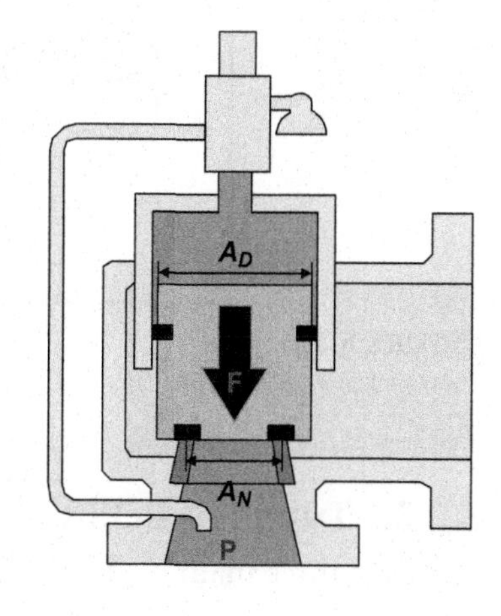

FIGURE 5.33
POSRV principle

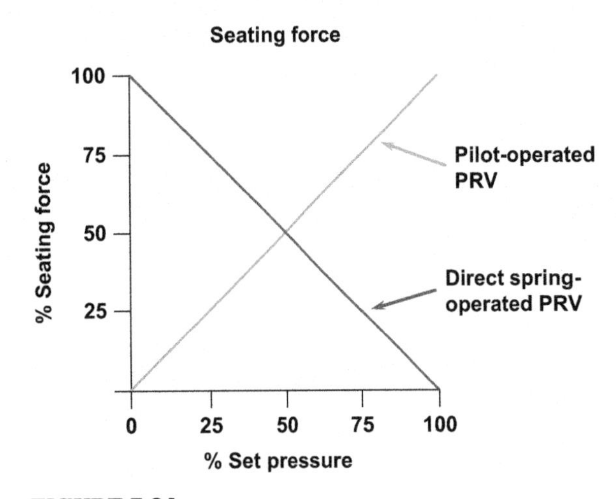

FIGURE 5.34
Seating forces

As opposed to the spring-operated process, the more system pressure (P) increases towards set pressure, the greater the closing force (Figure 5.34).

When the pilot senses that set pressure is reached, it vents the pressure above the piston sufficiently for it to be forced open by inlet pressure, reversing the unbalanced direction. During a relief cycle, when the reseat pressure is reached, the pilot shifts internally, admitting system pressure again into the volume (dome area) above the piston, and again closing the main valve.

The inherent ability of a POSRV is to maintain premium tightness close to set pressure, allowing optimization of the process output, thus allowing a higher normal system-operating pressure than with direct spring SRVs (Figure 5.35).

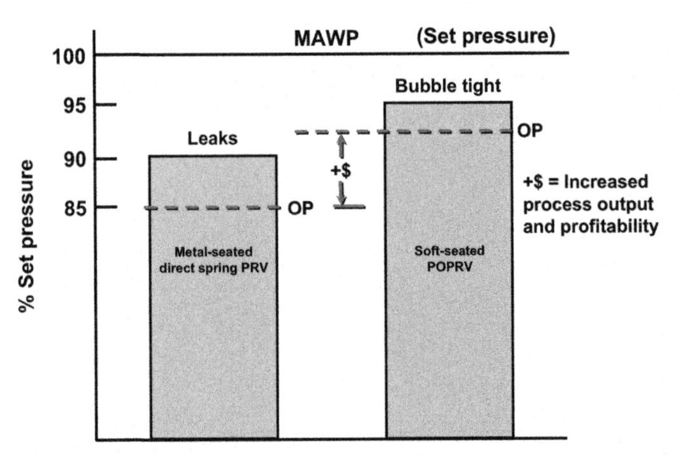

FIGURE 5.35
Premium tightness of POSRVs

5.3.3 Types of POSRV

There are three main types of POSRV:

- Low-pressure diaphragm types
- Snap-acting high-pressure types
- Modulating high-pressure types

5.3.3.1 Low pressure diaphragm type

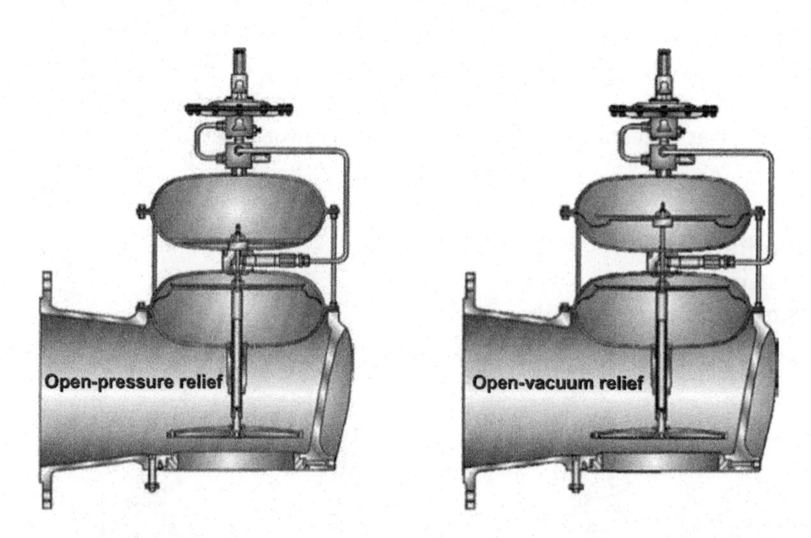

These are valves which are generally not subject to any code and handle pressures less than 1 barg; therefore, we will not go into detail. They can be used as breather valves for a combination of vacuum and pressure relief. Due to the low pressures they handle, diaphragms are used.

5.3.3.2 Snap (pop) action high-pressure POSRVs

Snap action POSRVs are usually of the piston-type design. It is important that the pilot be a non-flowing design. This means that there is no constant flow through the pilot after it vents the dome pressure to atmosphere. A pilot is a delicate part of the valve as it controls the complete valve operation, and fluid flowing through the pilot could have particles in the flow which could cause galling, blockage and malfunction of the pilot.

The main valve is basically a simple valve body design with a free-moving piston on a nozzle (semi or full). The main valve represents the capacity of the valve and is designed to flow the required rated flow. The pilot represents the quality of the valve and makes sure the valve works correctly. A pilot is usually the same for all sizes and pressures for a certain range of pilot-operated valves and is relatively small. The pilot is the intelligence of the valve and is where the set pressure is adjusted.

To demonstrate the operation of a typical non-flowing type, snap-action POSRV, we have taken the schematic example of the Anderson, Greenwood & Co. model. While there are subtle differences between manufacturers the basic operation is pretty much the same.

When the process pressure is below set pressure, as in the first schematic (Figure 5.36), the process fluid flows freely on top of the unbalanced piston via the open blowdown seat. This allows the process fluid to fill the dome, forcing the valve to close because the surface on top of the dome is approximately 30% larger than the nozzle area, as explained earlier. Since $F = PA$, the closing force increases as pressure increases towards set pressure.

where

F = Closing force
P = System pressure
A = Surface on top of the piston

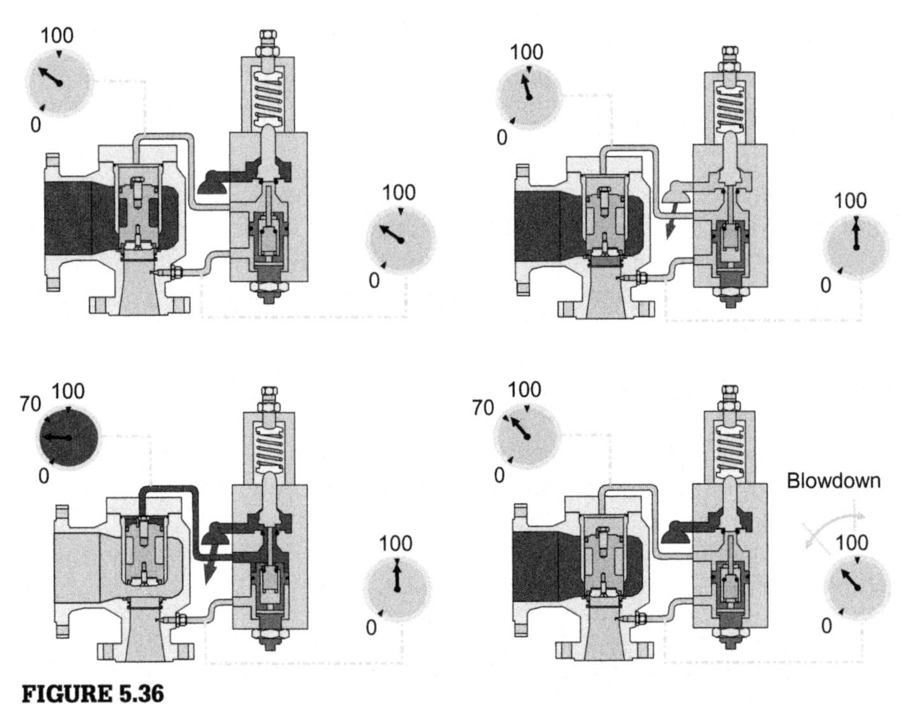

FIGURE 5.36
Function of Anderson, Greenwood & Co. pop action pilot-operated valve

When system pressure approaches set pressure, as shown in the second schematic in Figure 5.36, the pilot starts to operate. The top spindle overcomes the spring force and starts allowing the piston dome volume to vent to atmosphere. At the same time, the blowdown seat moves upwards to seal the process pressure from flowing through the pilot. This way the pilot is nonflowing while the valve is open.

In the third schematic, the main valve is now fully open: When the dome is vented up to approximately 70% of set pressure, the process pressure lifts the piston, and the valve can flow its rated capacity. Seventy per cent is dependent on the proportion of the unbalanced piston and varies from manufacturer to manufacturer. In this case, the top dome area of the piston is approximately 30% larger than the seat area. The dome volume to be vented is dependent on the valve size. For a 2J3 valve, this volume is only around 40–60 cm^2.

In this type of valve, full lift on gas is typically achieved in 0.1–1.50 seconds.

During the full-lift relief cycle, the pilot spring (compressed further when the spindle snapped open) downforce is opposed by the decreasing system pressure pushing up on the blowdown seat seal. When the valve reseat pressure is reached, the pilot spring recloses the relief seat and simultaneously opens the blowdown seat, allowing the main valve dome to be re-pressurized to force the piston down to the closed position. Blowdown is adjusted by varying the additional pilot spring compression when the pilot snaps open by varying the degree to which the blowdown seat travels upwards. The blowdown adjustment screw is at the bottom of the pilot.

5.3.3.3 Modulating action high-pressure POSRVs

The modulating acting pilot is designed to flow only the required amount of fluid in order to regain a safe situation. In general, this pilot regulates the venting of the dome in a much more controlled manner than does a snap action pilot. This causes the piston to open gradually, although very few pilot valves have a full proportional opening. It is important to carefully check with the manufacturer as to how the valve operates and what its opening characteristics are. In particular, check whether the pilot valve is truly modulating around set point and not pulsating. Pulsating valves can have a damaging effect on both the valve and piping, and cause serious system vibrations, especially when working close to set pressure.

Here again, there are different designs on the market, but it is important to use non-flowing pilots so that during a relief cycle fluid does not flow through the pilot, causing impurities to damage it.

The use of non-flowing pilots is highly recommended for liquids and especially flashing dual-flow applications where the amount of mixture (vapour/liquid) is not exactly known.

The modulating pilot valves are getting more popular because they are environmentally friendly and waste less product, and therefore can be more economic in the long term.

The graph in Figure 5.37 shows the product loss comparison for a spring valve, a pop/snap action pilot valve and a modulating valve. It is obvious that the product gain of a modulating valve is important (Figure 5.38).

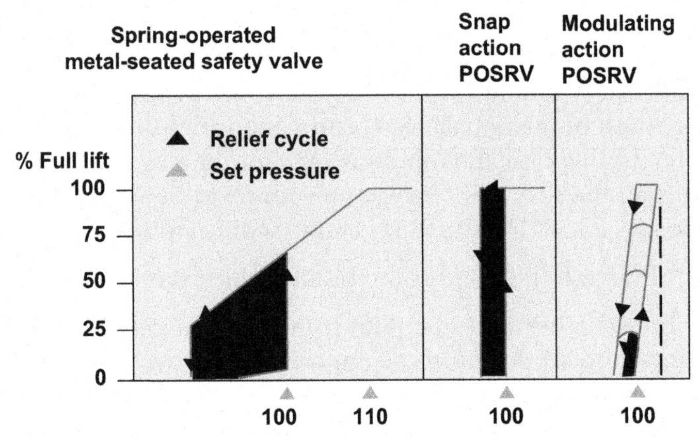

FIGURE 5.37

Table with product losses per type of valve

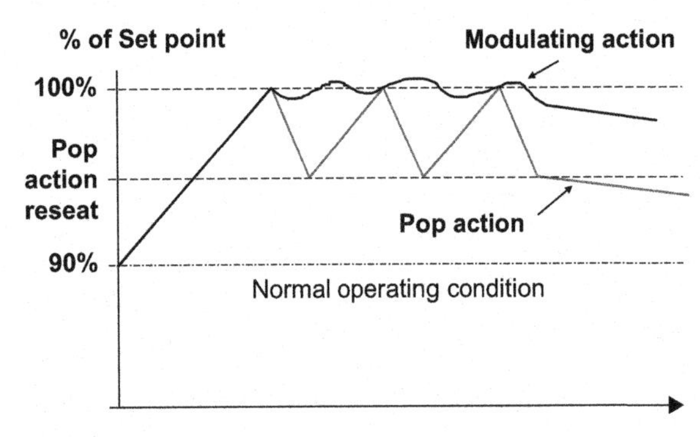

FIGURE 5.38

Modulating versus pop action pilot valve operation

5.3.3.4 *Special features of POSRVs*

Backpressure

One of the main reasons why POSRVs are used in the process industry is because they can resist much higher backpressures than any other valves. This is without the use of vulnerable bellows, which can easily rupture. Worse, it is difficult to detect whether a bellow inside a valve has failed; if the bellows have failed, the system is no longer protected against overpressure.

The graph in Figure 5.39 shows the effect backpressure has on the lift of the three main types of SRVs discussed here.

However, never confuse the lift of the valve with its capacity, as even a perfect convergent/divergent nozzle's flow rate is reduced beyond the medium's critical pressure ratio, as shown in the graph in Figure 5.40. In principle, a perfect nozzle has a K_D (flow factor) = 1.

Therefore, an SRV has a backpressure correction factor which is applied in the calculation and which is different for each manufacturer depending on the design of the valve and, in particular, the nozzle, huddling chamber arrangement and the shape of the body bowl.

Per EN 4126 (but not per API), it is required that manufacturers demonstrate the correctness of its backpressure correction factors on the basis of effective

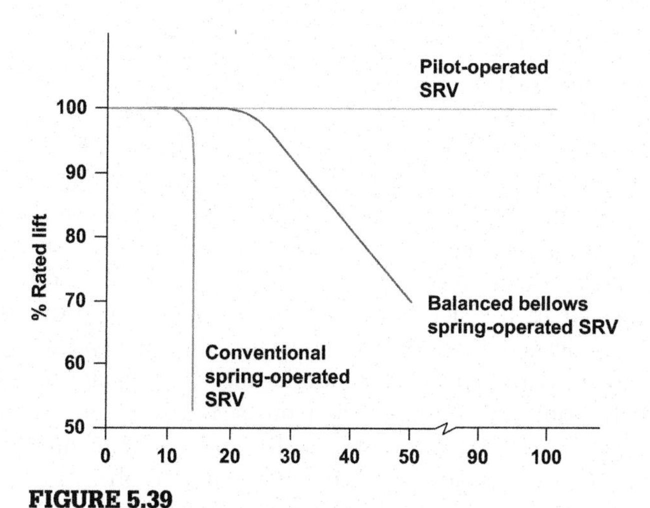

FIGURE 5.39
Effect of backpressure on the lift of an SRV

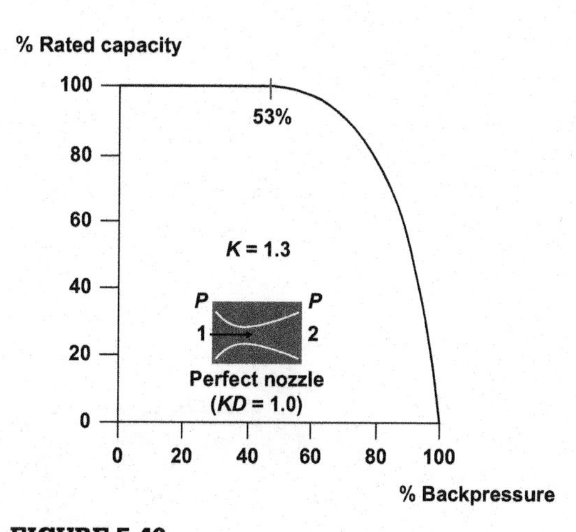

FIGURE 5.40
Backpressure effects on perfect nozzle

flow tests. Keep in mind that EN/ISO is not law but recommendation, and not all manufacturers have complied. Because of this, it is impossible to provide a typical curve, but a typical graph could look as per Figure 5.40.

Full, customized and restricted lift

Another advantage in some pilot-operated designs is the possibility of having a customized orifice so that the valve fits the required flow exactly. By adjusting the bolt on top of the piston, the curtain area of the valve can be adjusted to accommodate for the perfect flow.

We have seen that it is dangerous to oversize the valve; the ability to adjust flow both compensates for this and limits unnecessary product loss (Figure 5.41).

The full lift position affords the biggest possible API orifice, where the nozzle area gives the capacity. In this case, when the bolt is fully screwed down, a lower size valve provides more capacity than another standard API valve, which could make it more economic, lighter, and so forth.

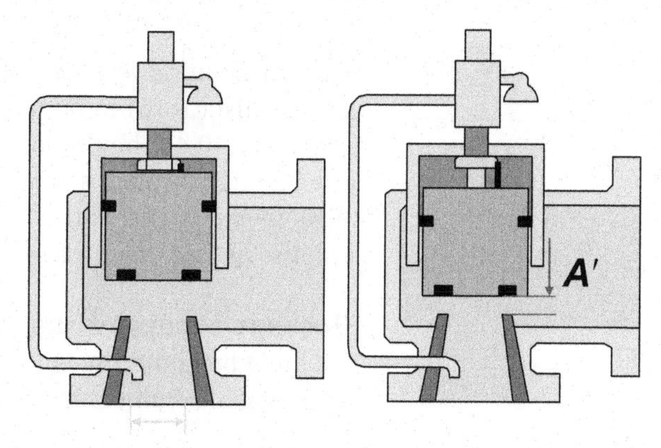

FIGURE 5.41
Full customized or restricted lift

The possibility of restricted lift, on the other hand, provides the opportunity to achieve smaller or customized API orifices. Here, the curtain area determines the capacity. Later, if capacity needs to be increased, only one bolt need be adjusted to get a larger orifice valve. It is recommended to check with the valve manufacturer for details, as each has its own design.

In-line maintenance and testing

Pilot valves usually can be easily and accurately tested in-line by means of simple bottle-compressed air or nitrogen, a pressure gauge and a regulator. A simple field-test connection must be provided with the valve (this is an option with most manufacturers) and is basically a check valve on a T-connection on the valve. Further, just a little bottle of nitrogen or compressed air (like a scuba-diving bottle) is needed, with a vent valve, a manometer and a hose. The hose is connected to the field-test connection. When the pressure delivered by the bottle is higher than the system pressure, the check valve admits the test pressure to the pilot, which pops when the actual set point is reached. The check valve on the inlet pressure line is to avoid sending all the nitrogen into the process. The volume in the dome and the pilot (if non-flowing design) is usually so small that a bottle of compressed air can test a large number of valves. Also, because of the small volume of nitrogen or air required to pop the pilot, the test is not only accurate but also fully opens the main valve, which is not possible with a similar test on a spring-operated valve. Other *in situ* tests are described in Section 10.4.

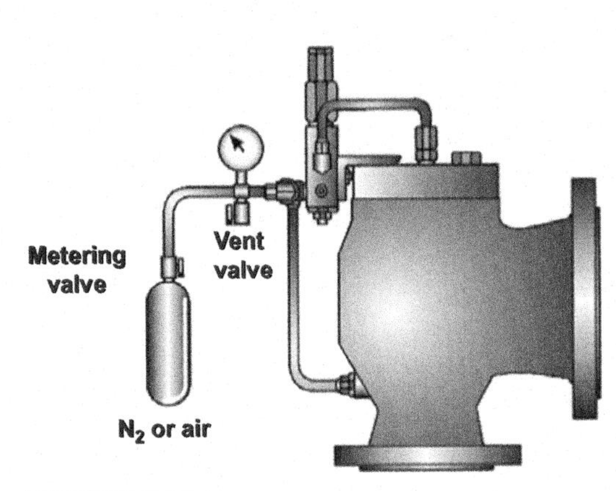

FIGURE 5.42
In situ *testing of a pilot valve set-up*

Because set pressure can be tested on the pilot, the main valve can remain on the system and be dismantled and maintained without having to be removed with cranes and other auxiliary equipment. The system must, of course, always be depressurized.

Some mechanical tests on spring valves are not always accurate; some systems can be satisfactory but remain rather cumbersome and expensive, as explained in Section 10.4. They require a very accurate knowledge of the system pressure, the nozzle area, spring compression, and so on. During some tests, the valve always opens, which is not recommended as it can damage the nozzle and disc in case of a metal-to-metal valve. It is also very time-consuming.

Blockage of supply lines in pilot-operated valves

In a lot of literature, the fail-safe operation of a pilot valve is questionable due to the potential accumulation of dirt, hydrates, and so on, in the pilot supply lines.

In fact the real question here is: What is the maximum pressure at which the POSRV opens if there is a blockage in the pilot or in the sensing line?

In the past, certain manufacturers have run tests which showed that, typically, a high quality, non-flowing POSRV would start to open at about 30% to 35% overpressure above set pressure.

However, caution needs to be taken as this overpressure fully depends on:

1. The pressure at which the blockage occurs (as this pressure is on top of the dome, preventing the piston to lift)
2. The severity of the blockage (full, partial…)
3. The type of fluid (compressible or not)
4. The temperature of the fluid
5. The size of the valve

Therefore, it is virtually impossible to predict the exact overpressure at which the main valve will start to open if this famous blockage in the supply lines should occur.

We could ask whether it is therefore acceptable to install an SRV that could start to open at a pressure higher than the maximum overpressure allowed to protect the equipment and allowed by the codes.

Almost all the pressure vessel codes in the world, particularly the European PED, require that the SRV start to open at a pressure no higher than the MAWP of the protected equipment, and that it be fully opened at no more than 10% above this pressure. So unless the SRV is set at a much lower pressure than the MAWP, this is not a viable option.

The point, however, is that an SRV should never be installed if there is a doubt about its suitability for the service. If the customer's HAZOP (hazard and operability) review shows some potential for 'blockage' (caused by dirt, polymerization, hydrates formation, etc.), then the valve selected for installation should be such that it can still operate properly under these conditions.

If there is a risk of blockage, this risk should be eliminated and that applies to a spring-operated valve as well as for a pilot-operated valve. Some manufacturers have acted on this and can propose many different accessories and configurations to achieve this which have been proven to work. Pilot valves can be protected against dirt with a variety of options, and most spring valve suppliers can supply such items as steam or electrical jackets, to avoid, for instance, polymerization or formation of hydrates.

A few possible tips to overcome possible pilot blockage

Primarily, it is important to always make sure to use a non-flowing pilot–type as otherwise the particles continue to flow through the pilot and will block it.

Solid dirt, scale, sand, and so forth

Use a remote sense line and an auxiliary filter. In that case, the sense line is not mounted integrally to the valve body but can be fitted directly onto the vessel to be protected or at a place which is less contaminated. The length of the remote sense should be limited to approximately 30 m and the pick-up

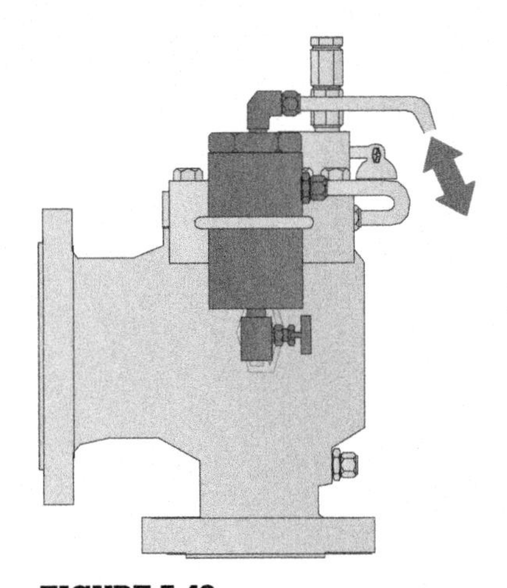

FIGURE 5.43

Pilot valve equipped with remote sense and auxiliary filter

tube should have an ID of 10 mm minimum. Take care when using valves with remote sense lines; the most frequent mistake made on site by people not familiar with the valves is that they forget to connect the remote sense. The valve is not a valve if the supply line is not connected.

As an additional feature, an auxiliary filter can be fitted with a drain valve to regularly clean the filter and prevent clogging. This drain can also, on demand, be piped to the main valve outlet or it can be piped on site to a purge collector. The purge valve can be programmed to open at pre-set intervals. The mini-valve for the drain, shown in Figure 5.43, can also be an automated ball valve. I would recommend not lower than a 55 μ filter element with a minimum area of 80 cm^2 and an approximate volume of 50 to 75 cc.

Polymerization

Polymerization is a more serious problem for which only a few manufacturers offer a suitable solution. The best solution here is to install a purge system to keep the pilot clean. In fact, this is the same sort of protection that is used to isolate the instrumentation from polymerization effects. Figure 5.44 depicts only one of the possibilities from one specific supplier, but the principles remain the same among different types. A clean gas source is applied to leak a positive pressure away from the controlling (or measuring) component.

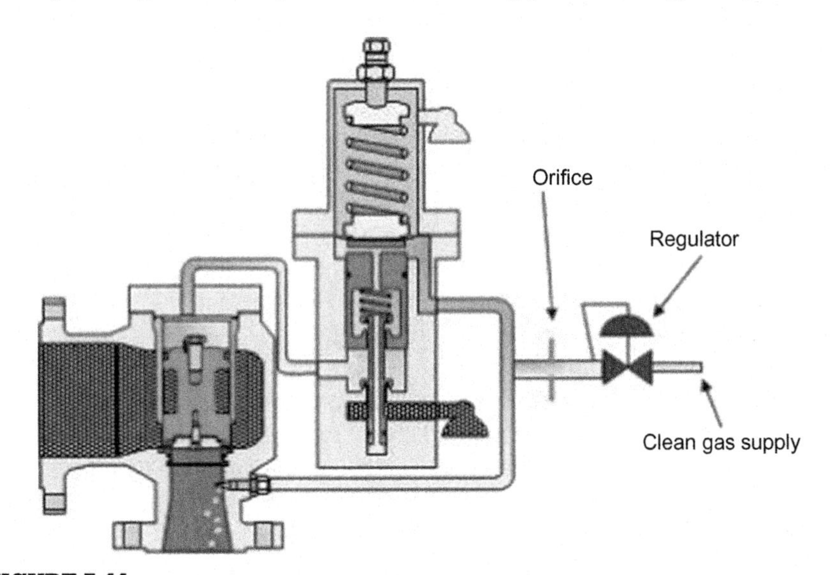

FIGURE 5.44

Purge system on a pilot-operated valve

Hydrates formation

Here, by far, heat-tracing is the best solution and has been proven to work perfectly for both spring and pilot valves. But some suppliers offer a sort of clean medium barrier (for instance glycol). In this case, the pilot is only in contact with the clean medium, which is vented each time the valve opens until the buffer tank is empty (Figure 5.45).

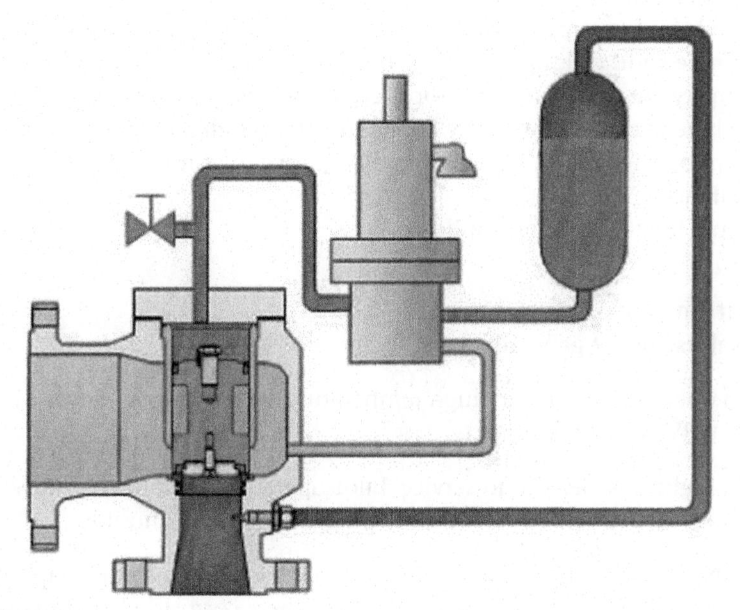

FIGURE 5.45
Pilot valve with clean medium buffer tank

Wax, paraffin in crude oil

There are a very limited number of good solutions available on the market, and what there are should be carefully discussed with the manufacturer. Here, the recommendation would be to use a spring-operated valve whenever possible.

So contrary to some literature, dirt does not have to be a reason to turn away from, for instance, the pilot technology if this would be best for the application. However, this situation must be handled with caution and preferably discussed with specialists. If there is concern about the overpressure at which a blocked pilot valve will open, just take away the potential cause and take the necessary, proven precautions available on the market.

5.3.4 Summary

Application

Pilot-operated valves are used primarily in the following services:

- Where large relief areas at high set pressure are required. Many POSRVs can be set to the full rating of the inlet flange
- Where the differential between normal operating pressure and set pressure of the valve is small
- On large low-pressure storage tanks to prevent icing and sticking
- Where short or long adjustable blowdown is required
- On cryogenic applications for preventing icing (no simmer)
- Where backpressure is very high and balanced design is required. A pilot-operated valve with the pilot vented to atmosphere is fully balanced
- Dual flow or flashing applications.

Limitation

These valves are not generally used:

- In operations with very high temperatures (especially in the case of POSRVs with soft goods).
- In highly viscous liquid service. Pilot-operated valves have relatively small orifices which can become plugged by viscous liquids.
- Where chemical compatibility of the process fluid with the diaphragms or seals of the valve is questionable or where corrosion build-up can impede the actuation of the pilot.

Advantages	Disadvantages
Good and repeatable seat tightness, before and after cycle	Limited on dirty or fouling service, unless with special configuration
Smaller and lighter valves, larger orifice sizes	Requires special configuration on polymerizing fluids
Pop or modulating action available	O-ring and soft seals limiting chemical and temperature compatibility
Easily testable in the field	
Not affected by backpressure	Liquid service limitation, requiring modulating pilot
Operation not affected by inlet pressure losses with remote sense In-line maintenance (for some designs) High flexibility of designs and configurations to suit applications	

5.4 DIN DESIGN

The DIN (Deutsche Industrie Normen) design had its roots in Germany. It is less utilized globally but has its place in less critical processes in Germany, Eastern Europe and some Western European countries where they also use hybrids between the DIN and API designs in less critical industries (Figure 5.46).

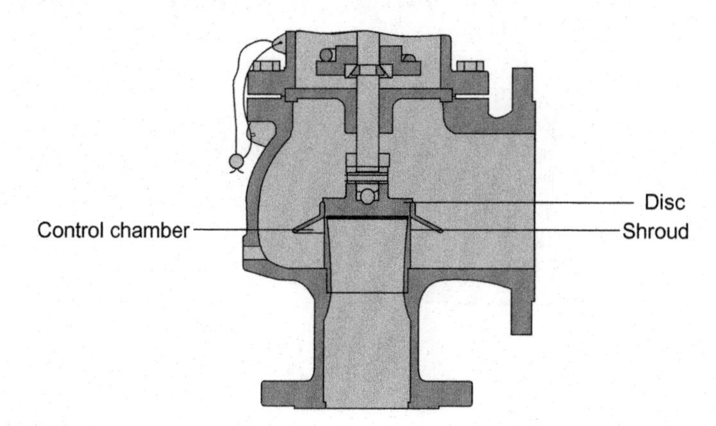

FIGURE 5.46
Typical DIN valve construction

Although the main elements of a conventional API SRV and a DIN valve are somewhat similar, in order to obtain a more or less similar operation, the design details can vary considerably. DIN valves have different inlet and outlet connections and flange drillings than the API valves.

In general, the DIN-style valves tend to use a somewhat simpler construction, often with a fixed skirt (or hood) arrangement. The ASME-style valves have a design which includes one or two adjustable blowdown rings to form the huddling chamber which controls the opening characteristics of the valve. The position of these rings can, as we have seen in Section 5.2, indeed be used to fine-tune the overpressure and blowdown values of the valve. Therefore, typical DIN valves offer less control. Typically, they open around 10% overpressure on compressible fluids and 25% on liquids, and have a non-adjustable blowdown around 10% for compressible fluids and 25% for liquids. Different designs might give slightly different, but fixed, values. This usually also makes these valves a little more economic than the API-type valves.

For a given orifice area, there may be a number of different inlet and outlet connection sizes, as well as body dimensions such as centreline-to-face dimensions. Many competing products, particularly those of European origin, have different dimensions and capacities for the same nominal size, unlike the API valves which have strict dimensional and capacity standardizations. DIN valves are, therefore, not necessarily interchangeable.

The operating principle is similar to that of API valves with the exception that the blowdown ring and resulting adjustable huddling chamber are usually semi-nozzle design (Figure 5.47).

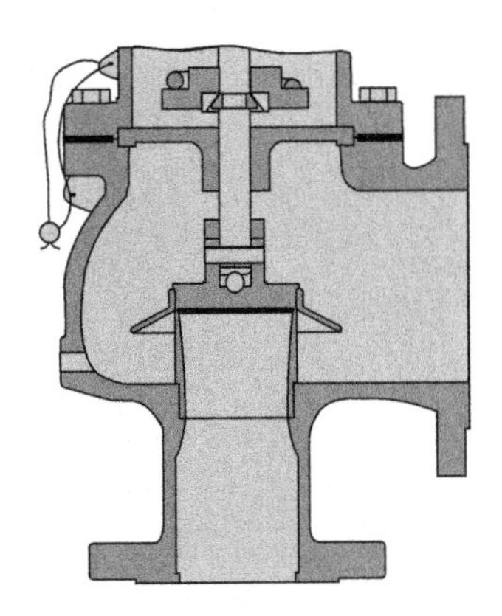

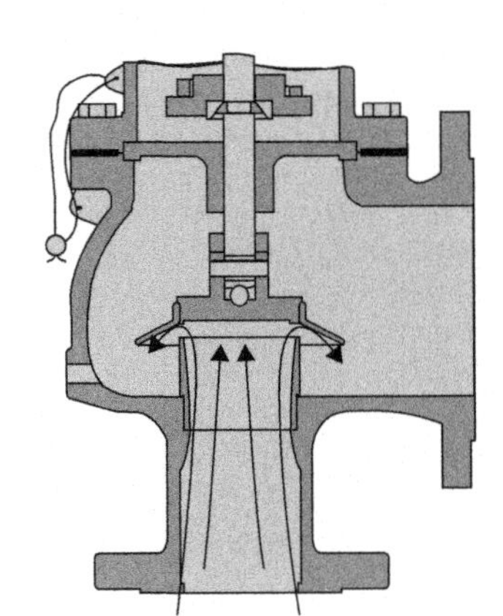

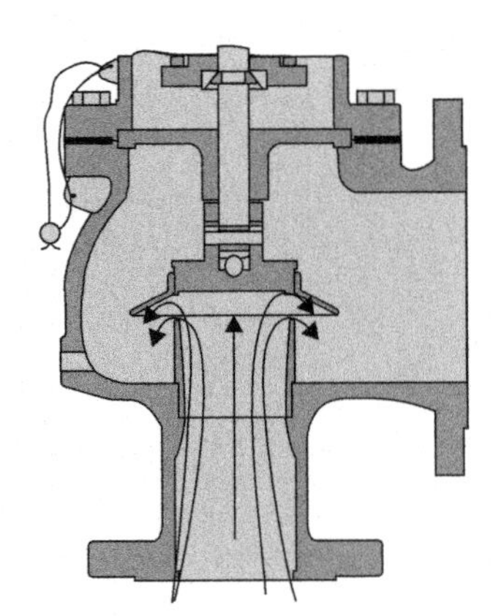

FIGURE 5.47
Operation DIN type SRV

5.5 NON-RECLOSING PRESSURE RELIEF DEVICES

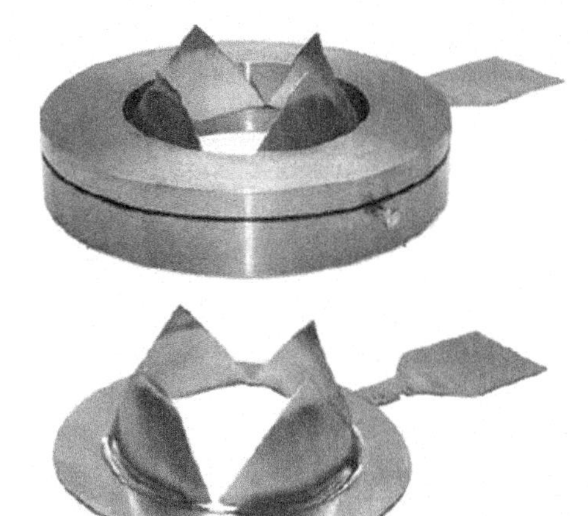

FIGURE 5.48
Open bursting or rupture discs

A non-reclosing pressure relief device is a pressure relief device designed to remain open after operation. A manual reset may be provided on certain valves.

While the bursting discs are by far the most commonly used non-reclosing pressure safety devices, '(buckling) pin valves' can also be found in applications where the operation is closer to the device's opening pressure (found on rupture discs) (Figure 5.48).

A rupture or bursting disc is a non-reclosing pressure relief device also actuated by inlet static pressure and designed to function by bursting a fragmenting or non-fragmenting bursting device.

Provided that the appropriate regulations allow this, a single rupture disc can be used on its own for system protection. Rupture discs can be used for both primary relief and as an additional secondary relief in combination with another pressure relief device.

There are two basic types of rupture discs: the conventional type, which has its concave surface towards the process pressure, and the so-called reverse buckling type, which has its convex surface towards the process pressure.

In the conventional type, the metal is always under tension. As pressure increases, the metal becomes thinner and eventually bursts when the ultimate tensile strength of the metal is reached. This type of disc if not scored usually ruptures in a random pattern, and it is possible for pieces of the disc to break off. This is also called a *fragmenting disc*. Therefore, this type is not really recommended if it is used upstream of an SRV, as these pieces or fragments may damage the valve seating area. The main advantage of these conventional discs is that they may be used at lower bursting pressures than the reverse buckling type (typically 0.2 barg versus 0.8 barg for buckling types). The maximum operating pressure for conventional discs is normally between 70% and 80% of the burst pressure.

The reverse buckling disc is under compression and is, therefore, less affected by creep or metal fatigue. It may be used under vacuum conditions without vacuum supports and the maximum operating pressure is at 90% of the bursting pressure. Most reverse buckling discs have four blades (knives) in the upper disc holder, which cut the disc when it bursts, ensuring that it nicely folds back against the wall of the holder instead of fragmenting. Therefore, they are more suitable for usage upstream of an SRV. Since there is a smooth convex surface oriented towards the process, there is also less risk for product building up on the disc, which could of course affect its bursting pressure.

ASME VIII specifies the following requirements for rupture discs:

- Every disc must carry a stamp indicating the bursting pressure at a specified temperature and must be guaranteed to burst within a 5% tolerance of the stamped values shown on the disc.
- A rupture or bursting disc may be installed between a SRV and the pressure vessel provided that:
  a. The combination of the SRV and the rupture disc has sufficient capacity to meet the requirements of the code and that the calculations of the SRV have taken into account the use of a rupture disc upstream.
  b. The stamped capacity of the SRV may also be de-rated with 10%, or alternatively the capacity of the valve and disc combination shall be established by test in accordance with the code and the combination is approved.

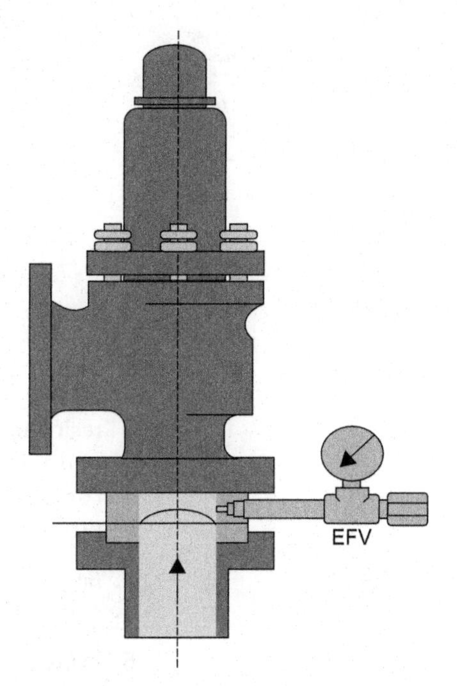

FIGURE 5.49
Rupture disc of an SRV combination with burst indicator

Rupture discs can also be used in pairs or in parallel with an SRV for additional safety, operational and installation cost benefits and system integrity.

It is equally recommended that a pressure gauge between rupture disc and SRV be installed, functioning as a burst indicator

(Figure 5.49). If the pressure gauge is indicating operating pressure, then it is highly possible that the rupture disc needs replacing. The advantage here is that the process is still protected by the SRV.

Advantages	Disadvantages
Instantaneous full opening	Non-reclosing (vent until inlet and outlet pressures equalize)
Zero leakage	Requires high margin between operating and opening pressures
Very large sizes easily and relatively economically available	Can fail by fatigue due to pulsations of pressure
Wide range of materials easily available	Burst pressure highly sensitive to temperature
Economical when exotic materials are imposed for the process	No possibility to check the burst pressure in the field
Virtually no maintenance	Requires depressurizing equipment for replacement after bursting
Full pipe bore (almost) Low pressure drop Low cost	Tolerance usually +/ − 5%

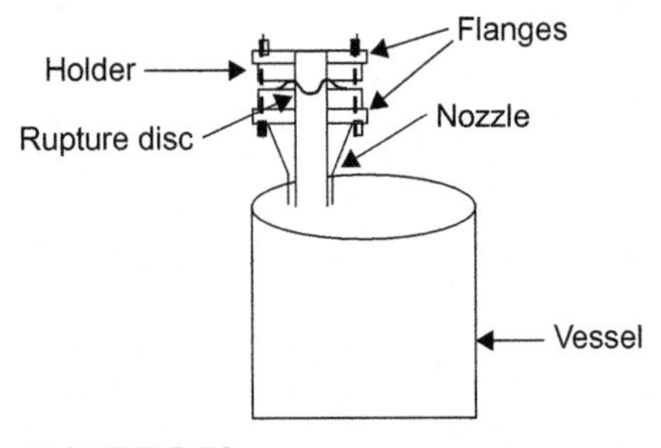

FIGURE 5.50
Typical stand-alone rupture disc installation

In which cases would we use a stand-alone rupture disc?

The reasons for using a stand-alone unit given here are probably not exhaustive but are in practice the most common reasons used within the process industry (Figure 5.50).

1. *Capital savings:* Rupture discs simply cost less than SRVs. The costs are especially less when exotic materials are required.
2. *Maintenance savings*: They generally require little to no maintenance – only replacement.
3. *No risk if product is lost after a pressure upset:* Rupture discs are non-reclosing devices. Whatever is in the system will get out and continue to do so until it is stopped by some form of intervention. If loss of contents is not an issue and it is installed as such, that action can be prompted.
4. *Benign service:* If the relieving contents are non-toxic, non-hazardous, and so forth.
5. *Extreme fast-acting pressure relief is required:* Rupture discs can be considered as process pressure protection when there is a potential risk for chemical runaway reactions. Depending on the choice of the SRV (for instance, a non-coded spring-operated valve), there could be a risk that they would

not react fast enough to prevent a catastrophic failure. However, an SRV set at a lower pressure may still be installed on the vessel in parallel to protect against other relieving scenarios. High-performance SRVs that open at set pressure can also be used for this type of application, but they are far more expensive. So this case is in fact dependent on economic considerations and on the selection of the correct SRV.

6. *Tube rupture on heat exchangers:* A lot of engineers prefer to use rupture discs for heat exchanger tube rupture scenarios rather than SRVs. They are concerned that SRVs won't respond fast enough to pressure spikes that may be experienced if gas/vapour is the driving force or if liquid flashing occurs. Again, this depends completely on the SRV that is selected as pilot-operated valves are ideal for flashing fluids and they can be equipped with spike equalizers.
7. *The fluid can plug the SRV during relief:* There are some liquids that may actually freeze or polymerize when undergoing rapid depressurization. This may cause blockage within a traditional SRV that would make it useless. Again, the selection of the correct SRV for the correct application is key as some high-performance valves are ideal even for cryogenic applications. Also, if the vessel contains solids, there is a danger of plugging the SRV during relief.
8. *High-viscosity liquids:* If the system is filled with highly viscous liquids such as polymers, the rupture disc can be considered because the SRV will have to be equipped with steam jackets, which will increase the price of the device considerably.

As a summary, if the correct SRV is selected for the right application, the only, but important, remaining major factor of selecting only a rupture disc is price, because the loss of product after opening is not an issue. Therefore, a short cost comparison between comparable stand-alone rupture discs and SRVs is presented here.

Rupture disc manufacturers burst at least two discs per lot before shipping them to a customer. As a consequence, even if you want just one rupture disc, you will be paying for three. Therefore, the first usable rupture disc is comparatively expensive.

Also, for new installations each installed rupture disc must be purchased along with a disc holder. This holder basically consists of two flanges between which the rupture disc is held. Disc holders are typical for specific manufacturers and are not interchangeable. The same holder may only be used for replacement purchases as long as you buy the exact same rupture disc from the same manufacturer. So the disc type and holder are pretty much a set.

Below is a capital cost comparison between a rupture disc manufacturer and SRV manufacturer based on 2008 pricing, which is not expected to vary a lot in the future.

Rupture disc: 3 in. Hastelloy C discs = € 2.500 for first usable disc, then €850 per disc plus a 3 in. Hastelloy C holder = € 3.100 ea. Total for 3 pair : €13.500, versus a Hastelloy C 3 in. × 4 in. standard spring-operated valve for around € 40.000.

This cost comparison will of course vary considerably with size and material of construction and type of selected SRV, but you get the point. However, please note that everything has a value and the loss of contents should also be considered in the overall cost difference between a rupture disc and an SRV.

Therefore, in a lot of cases, we will find the rupture disc upstream and/or downstream of an SRV when:

1. One-hundred per cent positive seal of the system needs to be ensured; the system contains a toxic, extremely corrosive or flammable fluid, and there is a concern that the SRV may leak.
2. The system contains large solids or a fluid that may plug the SRV over time under normal operating conditions.
3. *Economic considerations:* If the system is a corrosive environment, a rupture disc with the more exotic and corrosion-resistant material can be selected upstream and downstream of the SRV. It acts as the barrier between the corrosive system and the relief valve.

5.5.1 Summary

A stand-alone rupture disc is used when:

1. You are looking for capital and maintenance savings.
2. You can afford to lose the system contents.
3. The system contents are benign.
4. You need a pressure relief device that is fast acting.

A rupture disc/relief valve combination is used when:

1. You need to ensure a positive seal of the system.
2. The system contains large solids that may damage the SRV over time.
3. If the system is a corrosive environment, the rupture disc is specified with the more exotic and corrosion-resistant material to protect the SRV.

CHAPTER 6

Installation

Correct installation of safety relief valves (SRVs) is as important, if not more so, than the correct selection of the SRV for the correct application. Incorrect installation is the most frequent cause for SRV malfunctioning. This chapter cautions the piping designer, process engineer and user that the performance of a properly sized and selected SRV can be severely compromised when used in conjunction with improper companion piping or incorrect handling and installation. SRV installation guidelines and their rationale, as well as some precautions, are offered to ensure optimum performance and safety. About 75% of all the problems encountered with pressure relief valves (PRVs) are due to improper installation or handling.

6.1 INLET AND OUTLET PIPING

SRVs are extremely sensitive to the effects of improper companion piping both on inlet and outlet sides. In this section, we consider some basic reasons for this.

One of the main malfunctions on primarily spring-operated SRVs is incorrectly designed inlet piping and the resulting excessively high pressure drops.

American Petroleum Institute (API) recommends a maximum pressure drop of 3%. This should be taken into account when sizing valves; higher pressure drops not only cause reduced flow but also cause quick damage to the valve due to chatter, especially when the blowdown is set below the effective pressure drop. Therefore, always try to obtain short and simple inlet piping. Complex inlet piping is only effective if the piping is at least one or two pipe sizes larger than the SRV inlet and its size is reduced just before the SRV inlet.

Looking at piping technology, it is obvious why a long radius elbow is better underneath an SRV as the pressure drop through it is about 50% less. If the inlet piping needs turns and elbows, try to use long runs and turns (Figure 6.1).

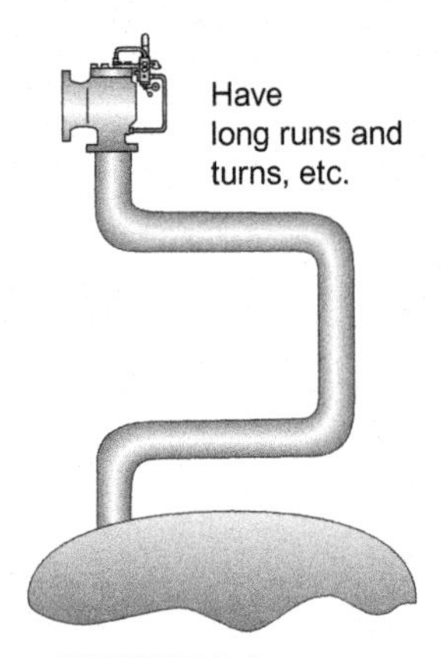

FIGURE 6.1
Long runs and turns are required on inlet piping

An equal-legged tee, shown in the middle of Figure 6.2, has very severe pressure drops due to extreme turbulence and the radial right-angle bend. This also creates a lot of vortices, increasing the pressure drop further. It can also make the SRV unstable during opening.

A globe valve (which has a typically high pressure drop) mounted under an SRV can in fact only be suitable for liquid thermal relief due to the very low lift of the SRV and the very small amount of product discharged per relief cycle. Other valves may be used under a PRV as long as they are full bore and can be 'locked open'. In Figure 6.2, some *L/D* values are given for some traditional valve inlet configurations.

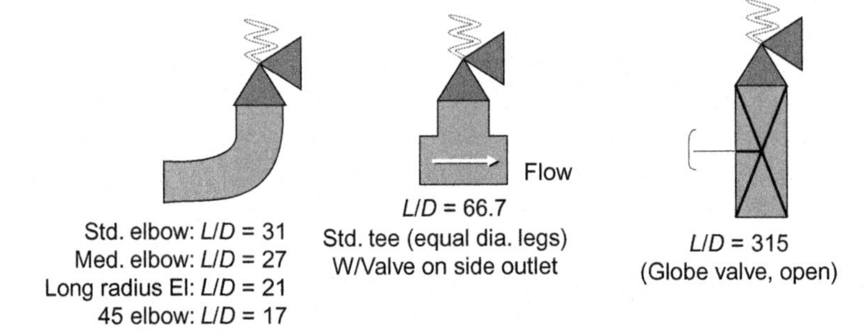

FIGURE 6.2
Some L/D values for typical inlet piping configurations

Let's have a closer look at what excessive pressure drops do to the SRV operation. In normal closed position, the SRV sees static pressure and there is no influence on the opening of the valve. However, once the system pressure reaches the set pressure range and the valve starts to open, the pressure becomes dynamic and the inlet pressure drop takes effect. When the SRV pops into high lift, the 3% inlet pressure drop shown in Figure 6.3 is established. Even though the SRV blowdown may have been carefully set for typically 7% pressure drop, the installed valve blowdown is reduced to 4% by the amount of existing inlet pressure loss while in a dynamic flowing condition. There is no problem with this particular installation as long as we have set the valve to reclose below the pressure drop value.

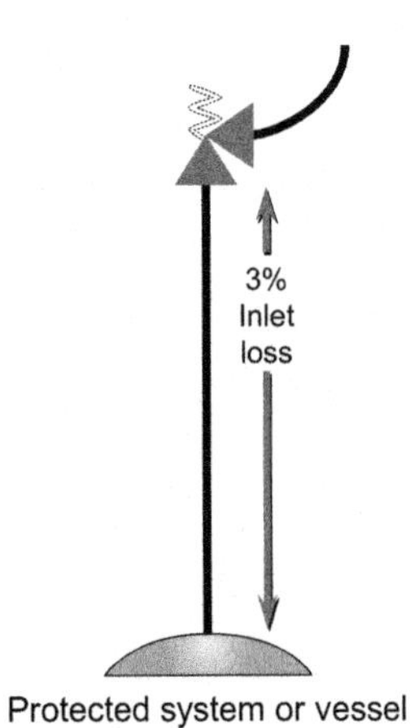

FIGURE 6.3
System with 3% pressure drop per API recommendations

Example per above conditions in Figure 6.3:

Valve set: 100 barg
Blowdown: 7%
Inlet loss: 3%
Valve closes when inlet pressure is 93 barg
System pressure at closing of valve: 93 + 3 = 96 barg
Actual system blowdown: 4%

For pop-action, spring-operated SRVs or pop-action pilot-operated safety relief valves (POSRVs), the actual system blowdown must always be a positive number. When the net blowdown is a negative number, the SRV will be unstable unless a pilot valve with remote sensing is used or a modulating pilot valve. The remote sensing overcomes the pressure drop in the inlet piping as it is bypassed by the pilot sensing line (a modulating pilot is proportional anyway).

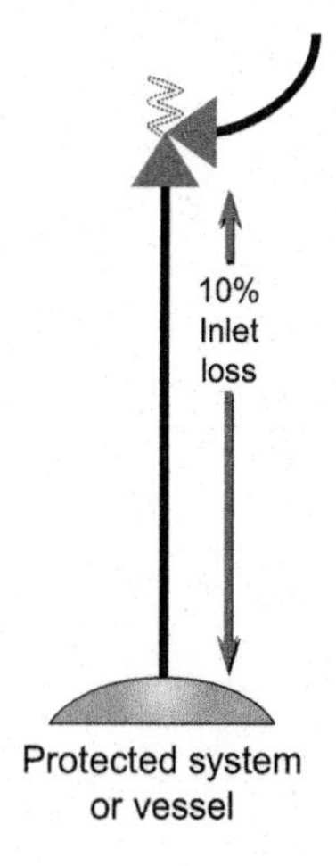

In the next example, an inlet pressure drop of 10% or greater, using snap-action integral POSRVs or spring-operated SRVs definitely creates a chatter problem. As in the previous example, the valve is set to close at 7% (blowdown). When the valve opens and the pressure becomes dynamic, the pressure at the inlet of the valve immediately becomes 3% lower than the 7% at which it has been set to close. Therefore, the valve is set to reclose before the process has been depressurized. The pressure builds up again because, when the valve is closed, pressure becomes static again and the cycle repeats over and over again, causing the valve to excessively chatter and destroy itself within minutes, without depressurizing the system.

Example:

Valve set: 100 barg
Blowdown: 7%
Inlet loss: 10%

The valve closes when inlet pressure is 93%, but this occurs immediately when the valve pops open and goes from a static to a dynamic phase. The inlet pressure at the valve instantly becomes 90 barg due to the inlet loss, causing severe chatter for the following reasons:

System pressure at closing of valve: $93 + 10 = 103$ barg
Actual system blowdown: $-3\% =$ chatter

When chatter due to excessive pressure drops occurs, the disc holder slams upwards against the lower end of the guide and the disc slams down again onto the seating surface of the nozzle. This occurs many times per second, making a sound like a machine gun. So each time the valve makes a full lift, travelling at very high speed, vibrations negatively affect the SRV causing vibrations and excessive noise, and jeopardizing the overall safety of the system:

- Internal parts of the valve are damaged as the SRV violently travels/slams from fully closed to fully open position, many times per second.
- The valve flows less than half its rated capacity as it is closed half the time.

- It causes damage to the companion piping.
- SRV internal stress and inertia reversals are caused by violent opening and closing cycles.
- Pressure surges occur in liquid service as the violent, multiple closing cycles cause water hammer and place undue stress on companion piping, piping supports, connections, internal components in pressure vessels, instrumentation, and so on.

Modifying the piping to eliminate chatter due to excessive pressure loss is recommendable but is also usually the least desirable solution due to extended process downtime, modification cost and the required involvement of various tradesmen – pipe fitters, welders, inspectors and so forth. It can also simply be impossible due to the isometrics of the complete system.

A more economic solution is to replace the valves with POSRVs with remote pressure sensing. Instead of having a pressure pick-up to the pilot at the inlet of the main valve, pressure can be measured on the system, bypassing the inlet piping and its pressure losses. This solution is also easily field convertible if the snap-action pilot has integral pressure sensing. A typical remote sense point could be a tee into a gauge tap.

The installed SRV capacity should, however, always be calculated to ensure its flow is sufficient, with the correct inlet pressure losses being considered. For inlet piping loss calculations, see Section 6.1.1.

One can also consider replacing the chattering valve with a pilot-operated safety valve with modulating action. This option would require a new capacity calculation.

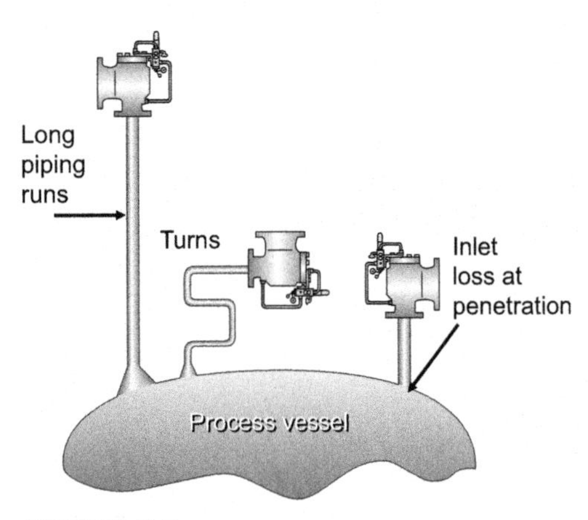

FIGURE 6.4
Causes for excessive inlet pressure losses

If the installed chattering valve can handle it, simply lengthening the blowdown settings beyond the pressure loss values is an option, but only if there is sufficient blowdown adjustment available on the valve.

Using complicated inlet piping leading to the inlet of the SRV should be avoided at all times during construction and design engineering.

Several of these configurations in Figure 6.4 may look ridiculous, but they do occur much more often than one might think!

The left one, where excessive long inlet piping is used, is typical when pressure vessels are mounted inside a building and the valve is installed for venting outside the building.

The middle one is typical in revamped installations where the isometrics become difficult or complicated due to added piping.

The right one is less obvious, but the method of penetration and inlet configuration within the vessel can cause significant inlet pressure losses, as is demonstrated in Figure 6.6.

It is also of utmost importance that the inlet piping leading to the SRV is always at a minimum, the size of the SRV inlet. It is preferred that it be one size larger than the diameter leading to the SRV inlet (Figure 6.5).

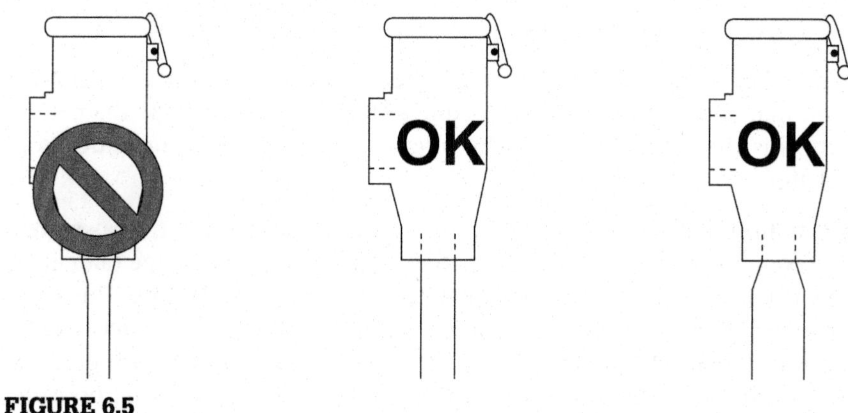

FIGURE 6.5
Inlet piping configurations leading to the SRV

In the first case, the diameter of inlet piping leading to the valve is smaller than the valve inlet bore, which is a frequently made mistake and to be avoided at all times. Even on short runs, this leads to excessive pressure losses and high turbulences when opening, which in turn leads to unstable operation of the valve.

The second case, where there is a straight run of the same size diameter as the valve inlet diameter, is acceptable provided the inlet runs are not excessively long.

The third case is by far the preferred configuration and also allows longer inlet runs.

The way the inlet nozzle enters the vessel it is protecting can also have a serious impact on inlet pressure losses. Of course, the ones which give the biggest pressure drops are also the most economic vessel penetrations.

A concentric reducer, as shown on the left of Figure 6.6, is a near-ideal penetration configuration. The relatively gradual acceleration of the process into the SRV inlet creates a minimum of turbulence.

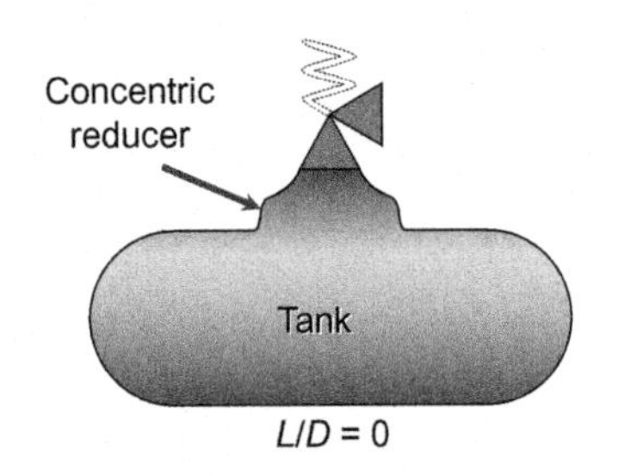

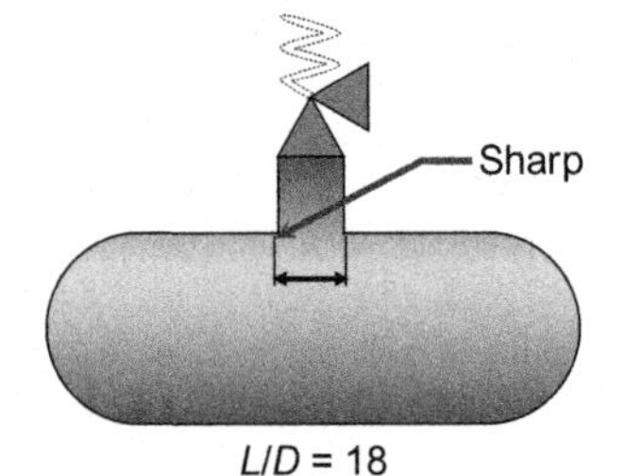

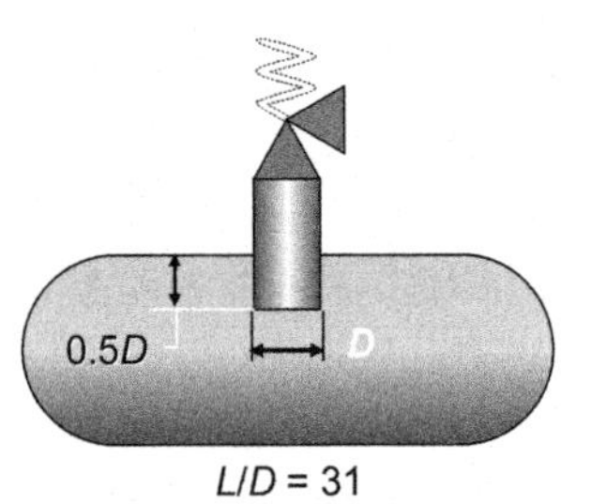

FIGURE 6.6
Different vessel penetrations

A sharp-edged penetration, as shown in the middle figure, results in a configuration that creates a pressure drop equivalent to 18 straight pipe diameters. Add to this the actual length of the inlet riser for total inlet pressure drop calculations.

A penetration that extends into the vessel, as shown on the right, creates even more severe pressure drops. This penetration adds 31 straight pipe diameters to the actual riser length for total inlet pressure drop calculations. This configuration, however, is often used to enable backwelding of the inlet riser to better resist the discharge reaction forces when the SRV relieves. While this is indeed positive for taking the reaction forces, it is very bad for inlet pressure losses, and a good compromise must be made and carefully evaluated when designing the system.

Beware of using three-way ball valves, plug valves or changeover valves to select either of two redundant SRVs. In these changeover systems, usually radical right-angle turns within the valve(s) or system cannot be avoided, which can inherently create very important pressure drops (Figure 6.7). Even if the switching system is designed to be the same size as the SRV inlets, the system itself, combined with the elbows, results in a very significant inlet pressure drop between the protected system and the SRV inlet. Many times SRVs mounted on changeover systems are more maintenance-intensive because of the chatter effect.

3-Way full-port
ball valve

FIGURE 6.7
Changeover system

For a typical purpose, a designed changeover valve, as shown in Figure 6.8, also known as a 'bull horn', easily reaches a pressure drop of 15% to 20%. This must be taken

FIGURE 6.8
Common changeover valve designs

into account in both the sizing calculations and in the blowdown setting. It is wise to know which changeover system is going to be used, if any (and its relative C_v values), before sizing the SRVs.

There are, however, specially designed switch-over systems on the market that respect the API recommendation of a maximum pressure drop of 3% and that are designed to have a smooth flow path. This needs careful checking with the manufacturer, as the 3% is considered over the valve only and does not take into account any additional pipe riser runs (Figure 6.9).

Long runs of piping before an SRV is never good practice. A good 'target' inlet length is five pipe diameters or less. To compensate for long piping lengths and/or elbows, make the piping large and reduce size only just before the SRV inlet. Always use long radius elbows (Figure 6.10).

A very good practice for avoiding pressure losses is the flared penetration, as shown in Figure 6.11. It is very common in LNG carriers and results in practically no pressure drop due to the smooth, non-turbulent acceleration of the product into the SRV inlet piping.

FIGURE 6.9
Special pressure drop-limiting changeover valve

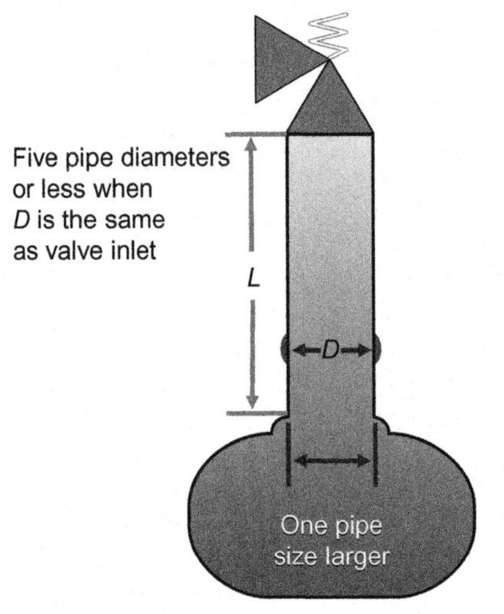

FIGURE 6.10
Recommended inlet piping

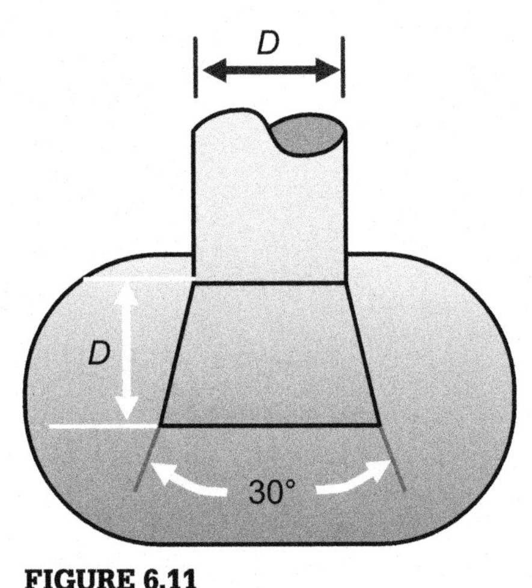

FIGURE 6.11
Flared vessel penetration

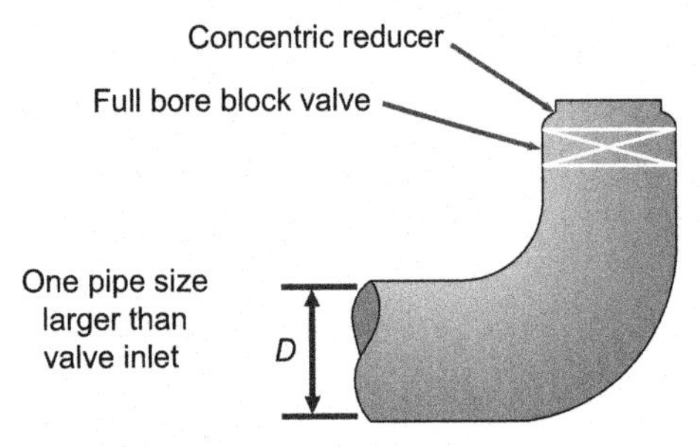

FIGURE 6.12
Use of block valves in inlet and outlet piping

One always has to be very careful with the use of isolation valves at the inlet of an SRV. If it is absolutely necessary to use them, always use full-bore valves and ensure that the incoming piping is one size larger in diameter than the SRV inlet (Figure 6.12). Always make sure the isolation valves can be blocked in open position and that they are provided with a clear position indicator. If isolation valves are used in both inlet and outlet, they need to be provided with an interlock system in order to be sure both valves are together in closed or open position.

Always avoid an SRV installation at the end of horizontal non-flowing lines as this configuration results in debris collecting at the entrance to the horizontal riser (Figure 6.13). When the SRV opens, it must then pass the bits of weld slag, pipe scale, and so forth – usually resulting in an SRV in need of immediate maintenance because of severe leakage. When (if) it closes, particles get trapped between disc and nozzle, resulting in constant leakage or small leaks and causing further erosion of the nozzle and disc.

As explicitly mentioned in ASME, never install an SRV horizontally, both for the above reason – debris collection – and also for valve function. Although this seems obvious, the strangest installations exist in the field, with end users wondering why the valves do not function correctly (Figure 6.14).

Always avoid installing process lateral piping into the SRV inlet piping. The SRV is normally sized and selected for the worst-case scenario that could

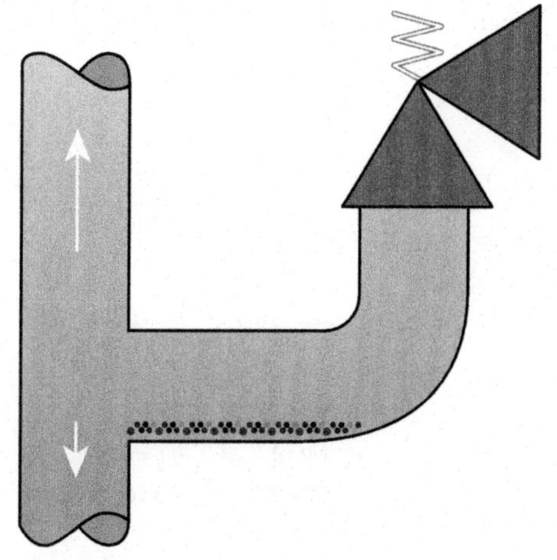

FIGURE 6.13
Installation on top of a horizontal riser is not recommended

FIGURE 6.14
Bad example of an SRV mounted horizontally

occur in the vessel. However, when the SRV opens, it not only flows fluid from the vessel but also from the lateral process connection, possibly causing an undersized SRV condition for the protected vessel (Figure 6.15).

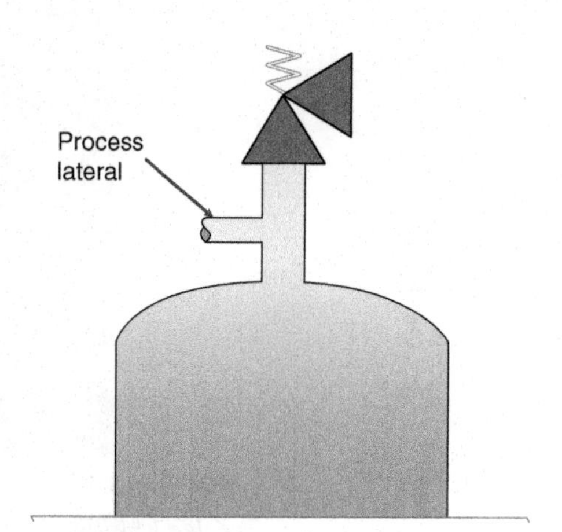

FIGURE 6.15
Lateral inlets in SRV supply lines are not recommended

For the outlet piping, the general rule is again to keep it as simple and direct as possible.

When piping the outlet to headers, flash drums, scrubbers, and so on, it is highly recommended to diverge at least one size up from the outlet piping to allow the fluids to expand; this avoids unnecessary built-up backpressure and turbulences in the outlet, which creates instability in the valve. This is basically the same reason the outlet of an SRV is always one size up from the inlet (Figure 6.16).

When discharging to atmosphere via a vertical-rising tail pipe, avoid using a flat-ended tail pipe, as shown in Figure 6.17. It is best to provide a 45° angle ending to the tail pipe to reduce built-up backpressure, reaction forces and noise while discharging.

As a general rule, the length of the tail pipe should never exceed 12 times the pipe diameter.

When a discharge goes straight to atmosphere, as shown in Figure 6.18, tail pipe drainage should be provided in the event that snow, rain or other substances enter the stack. If, for instance, water accumulates in the discharge piping – especially with a spring-operated valve, which is very vulnerable to the effects

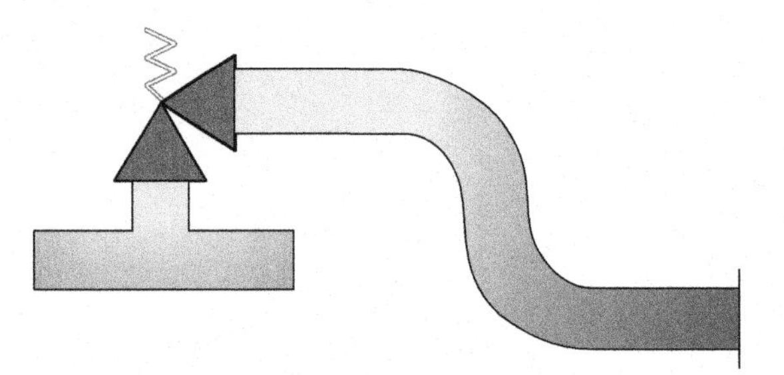

FIGURE 6.16
Outlet piping to flash tanks, scrubbers, and so forth: one size up recommended

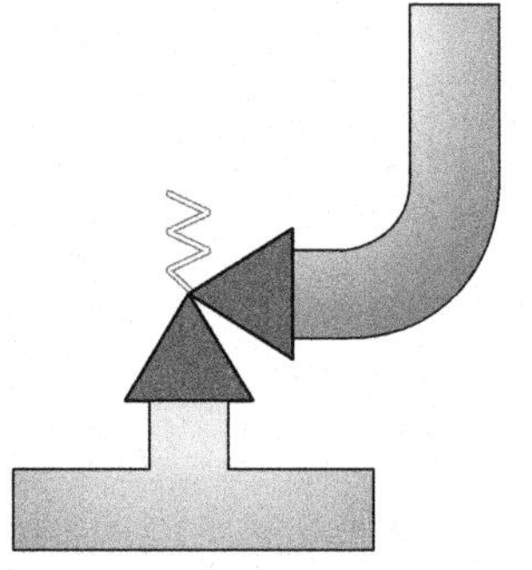

FIGURE 6.17
On vertical-rising tail pipes, flat endings are not recommended

FIGURE 6.18
Correct tail piping configuration

of built-up backpressure – it will probably start to chatter and not provide the desired overpressure protection because it will not flow its rated capacity.

Due to reaction forces created during the opening of the valve (see also Section 6.3), outlet piping must always be well supported. In particular, a direct spring-operated valve with metal seats can be affected by body distortion caused by stresses. These can be due to the valve's discharge reaction forces or because the valve was forced into place in existing piping. Correct alignment of an SRV is paramount for correct operation. Misalignment can result not only in valve leakage but also in binding of the guided diameters inside the valve, resulting in erratic set pressure and blowdown.

Formation of pockets in the inlet piping must be avoided at all times (Figure 6.19). Condensate can be formed, destabilizing the valve when it is opening. Also, as discussed before, the opening and sizing characteristics of a gas, steam or liquid are totally different, and the sizing of the valve will probably be incorrect.

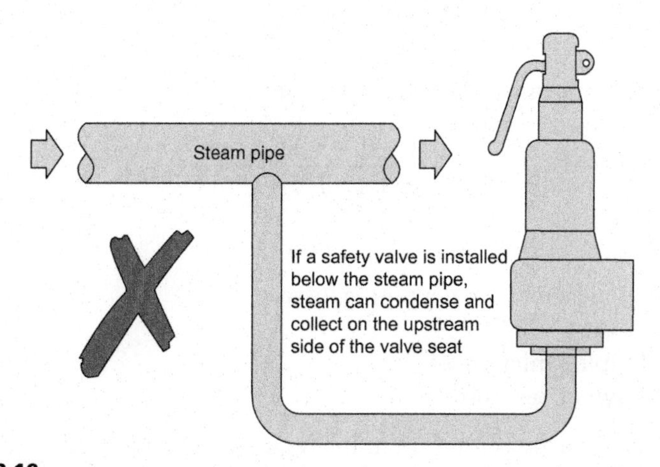

FIGURE 6.19
Avoid possible formation of pockets in the inlet piping

6.1.1 Calculating piping losses

All the recommendations on inlet and outlet piping are very useful, but in many cases engineers are faced with existing situations where it seems impossible to adhere to the above recommendations without going through major investments and possibly process shutdown. Of course, the potential investment needs to be carefully weighed against jeopardizing the safety of the plant or the possibility of having to purchase larger valves. In this chapter, we provide some guidance on how to deal with existing situations to make the correct evaluation.

The sequence of events for making the evaluation is as follows:

1. Estimate the total inlet pressure losses.
2. Calculate a revised capacity of the SRV.
3. Verify that the revised capacity is equal to or greater than the required or rated capacity necessary for a safe process condition.
4. Verify that the set pressure minus inlet loss is greater than the blowdown.
5. Size the vent pipe for the fully rated SRV capacity or select a different SRV (type or size).

Here we focus on the calculation of the piping inlet losses. The total inlet pressure loss is the sum of the different components explained hereafter:

$$\Delta_{Ptotal} = \Delta P_i + \Delta P_p + \Delta P_v$$

Once the total pressure loss is calculated, we can determine the de-rating factor to be multiplied by the effective relieving capacity of the SRV, which can be found on its tag plate.

The de-rating factor can be calculated as follows:

$$\alpha = \frac{1.1 \times P_{set} - \Delta P_{total} + P}{1.1 \times P_{set} + P}$$

where

α = De-rating factor for inlet piping losses
P_{set} = Set Pressure
P_{total} = Total piping inlet losses
P = Atmospheric pressure

If the de-rated capacity is lower than the required capacity for the system, it is highly recommended to look closely at either the inlet piping configuration or to increase the size of the SRV.

Check that the de-rated relief capacity is equal to or greater than the required capacity. If this is not the case, have a close look at either the inlet piping configuration or at increasing the size of the valve.

Always verify that the set pressure minus the inlet losses is greater than the blowdown. If the set blowdown is not known, check the valve manufacturer's standard blowdown procedures or check with the maintenance shop which last maintained the valve. Most manufacturers set the blowdown standard at 6% to 7%. Valve shops are sometimes less predictable. If maintenance is done in-house, it is wise to set procedures for blowdown setting or make sure the same procedures as the manufacturer's are used.

Use the normal (rated) capacity of the valve as the basis for (conservative) vent pipe sizing, not necessarily the flow rate on which the valve was sized. Always specify vessels with larger diverging relief connections. In this case, bigger is always better.

If changeover valves are required, choose the largest three-way valve or the one with the least pressure drop. There are changeover valves available which have proven pressure drops of maximum 3%. They are usually more expensive than the traditional types, but they are certainly recommended if piping losses run too high.

The calculations of the pressure drops are in English units only. The different possible pressure loss components are as follows.

6.1.1.1 Inlet losses due to entrance effects

These are the pressure losses created by the fluid prior to entry into a nozzle. They depend greatly on the type of nozzle (K_e coefficient). See Section 6.1 for typical types of nozzle entries. An empirical equation is presented for the computation of entrance pressure losses:

$$\Delta P_i = \frac{K_e \rho V_2^2}{288 g_c}$$

where

ΔP_i = Entrance pressure loss (psi)
ρ = Density of fluid (lb_m/ft^3)
K_e = Inlet loss coefficient. For a square edged entrance K_e is in the range 0.4 to 0.5.
V_2 = Gas velocity in the downstream pipe section (ft/s)
g_c = Gravitational constant, 32.2 ($lb_m ft/s^2 lbf$)

Some K_e coefficients can be found in the following table and most can be found in piping books.

Entrance loss coefficients for pipe or pipe arch culverts:

Type inlet design	Coefficient, K_e
General	
Square cut end	0.5
Socket end	0.2
Socket end (grooved end)	0.2
Rounded entrance (radius = 1/12 of diameter)	0.2
Mitred to conform to fill slope	0.7
End section conformed to fill slope	0.5
Bevelled edges, 33.7° or 45° bevels	0.2
Side slope tapered inlet	0.2
Corrugated metal pipe or pipe arch	
Projecting form fill (no headwall)	0.9
Mitred (bevelled to conform to fill slope)	0.7

Headwall or headwall with square edge wingwalls	0.5
End section conforming to fill slope	0.5
Bevelled ring	0.25
Headwall, rounded edge	0.2

There is also an empirical formula that can be used to determine the K_e factor, as follows:

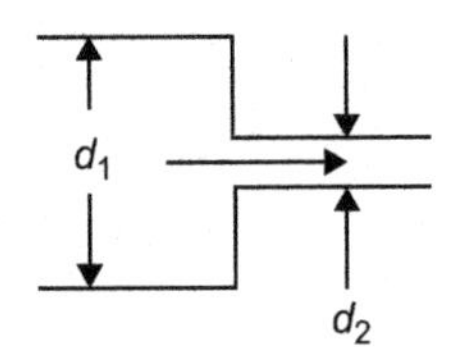

$$K_e \approx 0.42\left[1-\left(\frac{d_2}{d_1}\right)^2\right]^2$$

6.1.1.2 Inlet losses due to piping

These inlet losses are mainly dominated by frictional effects in the pipe. They are represented by the K_f factor and depend largely on the roughness of the pipe's internal wall and its diameter. They can be found in most piping books and are specific for the type of pipe:

$$\Delta P_p = \frac{K_f \rho V^2}{288 g_c}$$

where

ΔP_p = Piping pressure loss due to frictional effects (psi)
K_f = Loss coefficient due to frictional effects
ρ = Density of fluid at the upstream pressure and temperature (lb_m/ft^3)
V = Gas velocity in the downstream pipe section (ft/s)
g_c = Gravitational constant, 32.2 ($lb_m ft/s^2 lbf$)

Here, K_f is the determining factor and can also be calculated as follows:

$$K_f = \frac{12f}{dL_{eq}}$$

where

f = Friction factor
d = Inner diameter of the pipe (in.)
L_{eq} = Equivalent length of piping and fittings (ft)

The fully rough friction factor *f* can be determined as follows:

$$\frac{1}{\sqrt{f}} = -2\log\left(\frac{12\dfrac{\varepsilon}{d}}{3.7}\right)$$

where

ε = Pipe roughness (ft)
D = Inside pipe diameter (in)
f = Moody friction factor

Pipe roughness, or friction factor *f*, can be obtained with the pipe supplier or in piping books. As a reference, commercial steel piping roughness is assumed 0.00015 ft for schedule 40- and 80-type pipes. The following table provides some guidance but is not to be used as reference.

	Schedule 40		Schedule 80	
NPS (in.)	ID	*f*	ID	*f*
¾"	0.824	0.0240	0.742	0.0247
1"	1.049	0.0225	0.957	0.0230
1¼"	1.38	0.0210	1.278	0.0214
1½"	1.61	0.0202	1.5	0.0205
2"	2.067	0.0190	1.939	0.0193
2½"	2.469	0.0182	2.323	
3"	3.068	0.0173	2.9	
4"	4.026	0.0163	3.826	
5"	5.047	0.0155	4.813	
6"	6.065	0.0149	5.761	

6.1.1.3 Pressure drop effect due to upstream devices

Earlier in this book, we cautioned against using the type of changeover valves or inlet isolation valves related to inlet pressure drops. The main factor here is the C_v (valve flow coefficient) value of the changeover device or block valve used upstream of the SRV. If individual full-bore valves are used, the same principle applies as for changeover valves. Note once more, however, that if individual valves are used upstream of an SRV, they must be interlocked if used in a switch-over system or locked open if used in an individual isolation valve. In any case, it is important to know the C_v value of these isolation valves and take them into account in the pressure loss calculation.

The inlet loss is calculated as follows:

$$\Delta P_v = \left(\frac{7.481\mu}{C_v\rho}\right)^2 SG$$

where

ΔP_v = Pressure loss due to flow through the valve (psi)
C_v = Valve flow coefficient (gal/min/psi)
μ = Mass flow rate of fluid through the valve (lb_m/min)
ρ = Fluid density at the upstream pressure and temperature (lb_m/ft^3)
SG = Specific gravity of fluid evaluated at upstream pressure and temperature

6.1.2 Calculating outlet piping

There are two kinds of discharge systems: open and closed. Open systems discharge directly into the atmosphere, whereas closed systems discharge into a manifold or other fluid recuperation device, eventually, along with other SRVs. Both systems can create backpressures, which need to be taken into account at all times when sizing and selecting the correct SRV.

It is recommended that discharge piping for steam, vapour and gas systems should rise, whereas for liquids, the discharge piping should fall. Remember to provide drainage for rising discharge piping. Note that any drainage systems form an integral part of the overall discharge system and are therefore subject to the same precautions that apply to the discharge systems, notably that they must not affect the valve performance and any fluid must be discharged to a safe location.

Horizontal piping is, in fact, less recommended, but if it cannot be avoided, it should have a downward gradient of at least 1 in 100 away from the valve; this gradient ensures that the discharge pipe is draining away from the SRV.

Whatever the configuration, it is essential to ensure that fluid cannot collect on the downstream side of the SRV as this impairs the performance of the valve. In addition it can cause corrosion and/or damage of the spring and internal parts. Many SRVs, but by far not all, are provided with a built-in body drain connection; if this is not used or not provided, then a small bore drain should be fitted in the outlet piping close to the valve outlet.

One of the main concerns in closed systems is the built-up backpressure in the discharge system as this can drastically affect the performance of an SRV.

The EN ISO 4126 standard states that the pressure drop should be maintained below 10% of the set pressure. To achieve this, the discharge pipe can be sized using the following equation:

$$d = \sqrt[5]{\frac{L_e Q^2 \left(\frac{\rho + 2}{2}\right)}{0.08P}}$$

where

d = Outlet pipe diameter (mm)
L_e = Equivalent length of pipe (m)
Q = Discharge capacity (kg/h)
P = Valve set pressure (barg) × Required percentage pressure drop
ρ = Specific volume of fluid at pressure P (m^3/kg)

The pressure (P) should be taken as the maximum allowable pressure drop according to the relevant standard. In the case of EN ISO 4126, this would be 10% of the set pressure, and it is at this pressure we determine ρ.

■ Example

Calculate the discharge piping diameter for an SRV designed to discharge 1000 kg/h of saturated steam, given that the steam is to be discharged into a vented tank via a piping system, which has an equivalent length of 25 m. The set pressure of the SRV is 10 barg and the acceptable backpressure is 10% of the set pressure. (Assume there is no pressure drop along the tank vent.)

Solution

If the maximum 10% backpressure is allowed, then the gauge pressure at the SRV outlet is:

10/100 × 10 barg = 1 barg

Using the saturated steam tables, the corresponding specific volume at this pressure is $\rho = 0.88\,\text{m}^3/\text{kg}$.

Applying the formula:

$$d = \sqrt[5]{\frac{25 \times 1000^2 \left(\frac{0.88 + 2}{2}\right)}{0.08 \times 1}} = 54\,\text{mm}$$

This shows that the piping connected to the outlet of the SRV should have at minimum an internal diameter of 54 mm. With a schedule 40 pipe, this outlet pipe would require a DN65 (3 in.) pipe which is the next standard size up.

If it is not possible to reduce the backpressure to below 10% of the set pressure, a balanced bellows or pilot-operated SRV should be considered. ■

Ideally, SRVs installed outside a building with a vertical-rising discharge directly into the atmosphere should be covered with a hood or screen. This screen allows the discharge of the fluid but helps to prevent the build up of dirt and other debris in the discharge piping, which could affect the backpressure. Even birds' nests have been found in discharge tail pipes and, needless to say, this is not good for either the valve or the birds when the valve opens. The hood or screen should also be designed so that it too does not affect the backpressure.

There are many types of closed discharge systems and they can be very complex. They must be subject to careful, individual analysis of piping.

In general, manifolds must be sized so that in the worst case (i.e. when all the manifold valves are discharging at the same time), the downstream piping is large enough to cope without generating unacceptable levels of backpressure. The volume of the manifold should ideally be increased as each valve outlet enters it, and these connections should ideally enter the manifold at an angle no greater than 45° to the direction of flow, as shown in Figure 6.20. The manifold must also be properly secured, supported and drained where necessary.

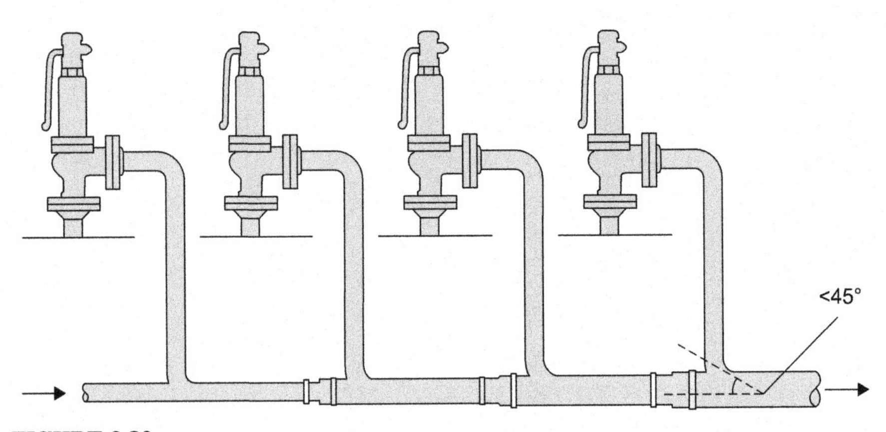

FIGURE 6.20
Manifold discharge system

For steam applications, it is generally neither recommended nor usual to use manifolds or headers, but they can be utilised if proper consideration is given to all aspects of the design and installation, and proper draining is provided via, for instance, automatic steam traps. On the other hand, considering current environmental or safety requirements, a lot of process vapours are manifolded into flare headers before being flared off or otherwise treated and disposed of.

6.2 LOCATION OF INSTALLED SRVs

An SRV is a very pressure-sensitive device; it could sense turbulence as varying pressure, act accordingly and possibly become unstable. Exercise caution when installing SRVs downstream of these:

- Pressure-reducing stations (turbulence)
- Orifice plates or flow nozzles (vortex turbulence)
- Positive-displacement compressors (pressure pulsation and/or mechanical vibrations)

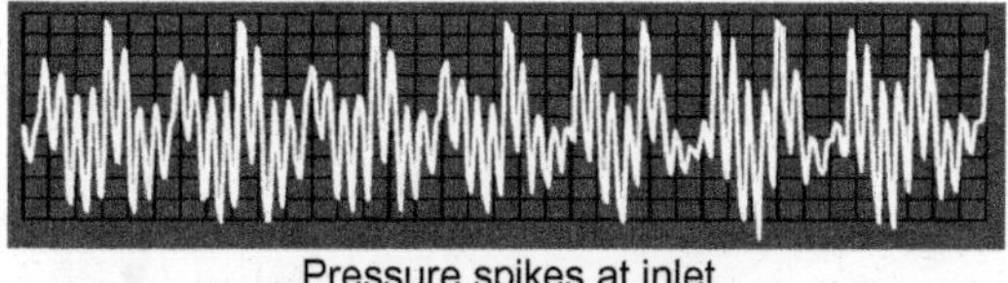

FIGURE 6.21
Pressure spikes measured before and after a pulsation dampener

Pressure pulsations can make an SRV relieve below set pressure when the pressure spikes perhaps reach set pressure. These spikes have such high frequency that they are unseen on charts or gauges. Nevertheless, they do act on the SRV, making it vibrate at high frequency.

In particular, vibrations and pulsations due to positive-displacement compressors can cause premature opening when the forces within the valve are anywhere near equilibrium and are acting as another upward force. For pilot-operated valves, some manufacturers provide pulsation dampeners in their pressure pick up lines so that the effect of these pressure spikes are compensated. As can be seen in Figure 6.21, for spring-operated SRVs, unfortunately, no provisions can be taken to compensate for this effect.

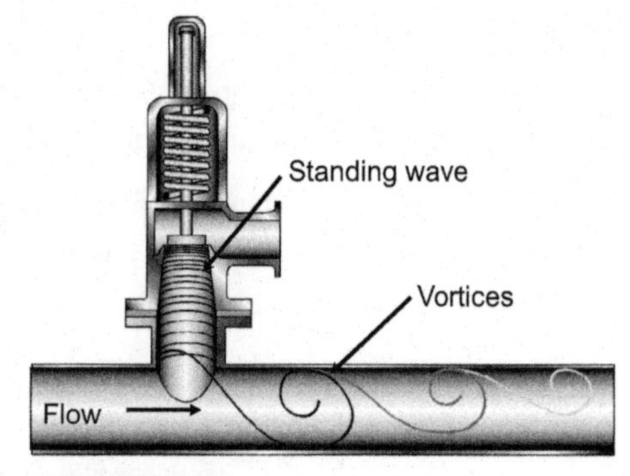

FIGURE 6.22
Pressure pulsations caused by sharp-edged nozzle entrances

Figure 6.22 shows how a sharp-edged SRV inlet entrance can induce pressure pulsations (vortices) that apparently make the SRV relieve low, leak severely and/or wear out prematurely.

In 1984, two engineers from the Southwest Research Institute presented an extensive ASME technical paper on this phenomenon, which can be obtained from ASME.

6.3 REACTION FORCES AND BRACING

When an SRV is opening, very high reaction forces can potentially occur, so correct bracing must be provided. The reaction forces are proportional with pressure and size and are highest on compressible fluids.

Besides using good bracing, another possible option is to use an SRV with a dual outlet as shown in Figure 6.23, so that the reaction forces lift themselves. This is

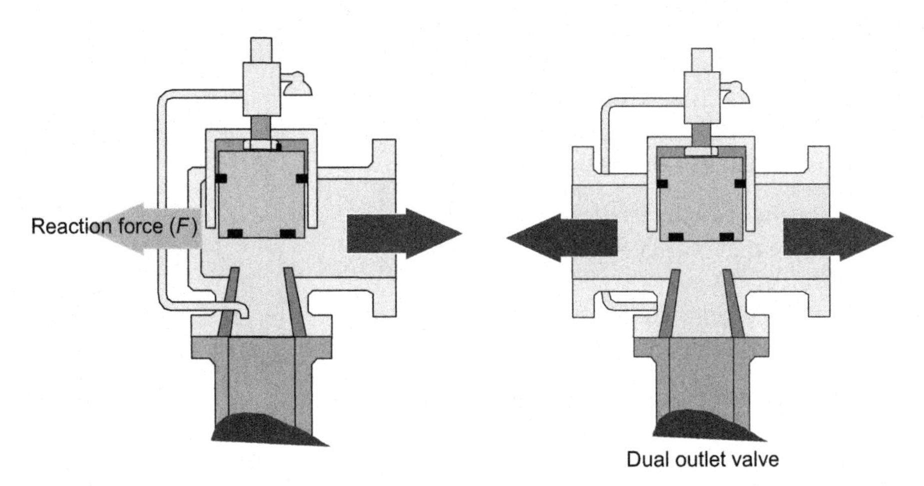

FIGURE 6.23
Dual-outlet valves lift the reaction forces

highly recommended on larger valves and for higher pressures on compressible fluids, typically above 50 barg.

The reactive force for SRVs on gases and vapours is calculated as follows, per API 520, Part II, section 2.4:

$$F = \frac{W\sqrt{\frac{kT}{(k+1)M}}}{366} + A_0 P_2$$

where

F = Reaction force (lbs)
W = Flow (lbs/h)
k = C_P/C_v
T = Flowing temperature (degrees Rankine)
M = Molecular weight
A_0 = Area of outlet at the point of discharge (in^2)
P_2 = Static pressure at point of discharge (psig)

The first of the two components of the reaction force equation (flow) accounts for the change in momentum of the process through a flowing and right-angled SRV.

The second component (area) accounts for the 'jet engine' effect of the exhaust jet to atmosphere.

It is important to notice that P_2 (pressure at point of discharge) is not the atmospheric pressure. It is the pressure in the exhaust pipe just prior to being vented to atmospheric pressure and is calculated as follows.

$$P_2 = 0.00245 \frac{W}{d_2} \sqrt{\frac{T}{kM}}$$

where

d_2 = Internal diameter of the discharge connection (in)

So P_2 is not atmospheric and, as can be deducted from both formulas, the discharge reaction forces decrease if the discharge pipe is made larger (d_2 increases, therefore P_2 decreases, and P_2 is proportional with the reaction force F). Based on the same principle, the reaction forces also decrease if, for instance, the exit to atmosphere through a tail pipe is 'scarf cut' at an angle, because this results in an enlarged, oval exit area.

Even though the reaction force due to the exhaust jet to atmosphere is significantly greater than the change in momentum component, the API makes no distinction for SRVs discharging into a closed header system compared with discharging through a tail pipe or direct to atmosphere.

The reaction force for a liquid SRV is much less than for gas and is normally of very little concern. Liquids are not compressible and do not expand when lowered in pressure as do gases and vapours. Nevertheless, care needs to be taken when the liquid is flashing during relief.

In Europe in the late 1980s, manufacturers who were not necessarily ASME VIII approved or complying with API issued a simplified (empirical) method for calculating reaction forces.

After extensive testing of the formula, it was found that this empirical formula was about 10% safer than the traditional API method:

$$F = K_f A P_1$$

where

F = Reaction force (daN)
K_f = Correction coefficient depending mainly on the pressure drop due to the type of nozzle inlet, something that is ignored in the API
A = SV orifice area (cm^2)
P_1 = Valve set pressure + allowed overpressure + 1 bar (bar)

The main K_f factors are as follows:

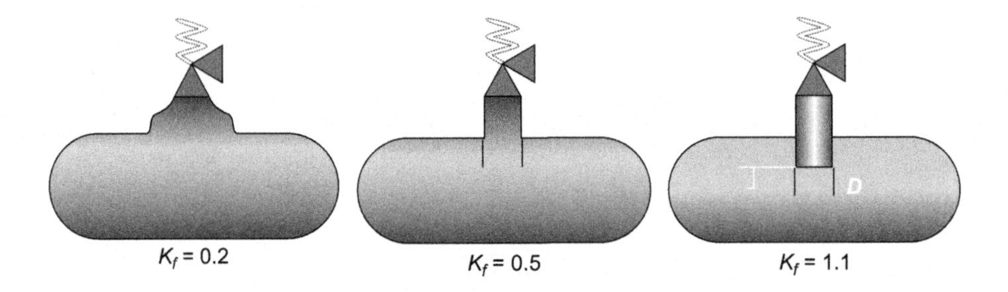

Some installation tips when high reaction forces are to be expected:

FIGURE 6.24
Correctly braced tail pipe

Depending on the calculated reaction forces, it is recommended to brace the discharge piping or tail pipe (Figure 6.24).

Figure 6.25 shows the preferred tail pipe configuration to atmosphere in order to minimize the bending moment at the base of the SRV inlet riser.

A tail pipe to atmosphere must be configured to eliminate or minimize the bending moment on the nozzle connection. It is important to note that the preferred type of stress on the inlet piping is compression, not tension as caused by bending.

For high-flow SRVs, dual outlets help to balance the discharge reaction force as the flow splits equally to each outlet, with no significant bending moment on the inlet piping. Figure 6.26 shows a well-braced, dual-outlet pilot-operated SRV on a main gas supply line. Note in particular the tie bar between the two outlet tail pipes. Owing to the change in the direction of the flow through the elbows, there is a slight tendency for each tail pipe to bend outward; the tie bar prevents this phenomenon and thus prevents added stresses on the nozzle.

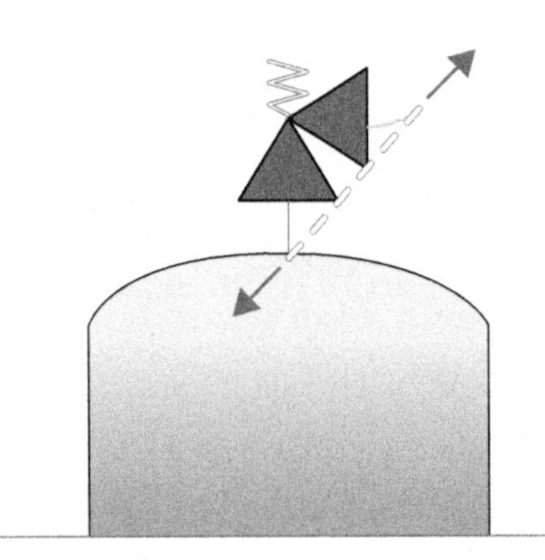

FIGURE 6.25
Orientation of reaction force

Another phenomenon caused by reaction forces, but one that is usually neglected, is the installation of a threaded SRV with tail pipe (Figure 6.27). If the outlet tail pipe is tilted in the wrong direction, it can cause the SRV to rapidly unscrew at the inlet when it relieves, making it come off and go airborne.

This has happened several times and is quite a rude surprise, to say the least.

FIGURE 6.26
Well-braced dual-outlet valve

6.4 TEMPERATURE TRANSMISSION

Depending on the installation and piping arrangements leading to the SRV, the temperature at the valve inlet can be very different from the process temperature. In particular, this needs to be taken into consideration when selecting the valve material. Here we need to make a distinction between continuous exposure to a temperature and short exposure, such as during a relief cycle. Where and how the valve is installed has an important influence on its continuous temperature exposure. The considerations in selecting the valve are fundamentally different, depending on its continuous temperature exposure; its temperature exposure during relief is not as critical a factor, as this should (hopefully) be as short as possible.

A rule of thumb for a non-insulated inlet riser is that there be a temperature difference of 55°C per linear 30 cm of riser. With the SRV closed, there is very little product circulating within the inlet piping, allowing it to cool down or heat up (in case of cold applications) rapidly.

The non-isolated riser itself becomes a quite effective air-to-air heat exchanger. In hot process, it is cooled down and in cold process is heated. Note that only the nozzle and the main valve seat material are continuously exposed to the process temperature.

FIGURE 6.27
Threaded valves with tail pipe

6.5 INSTALLATION GUIDELINES

While it is obviously impossible to address every installation mistake ever made, here is a short summary of the most frequent installation mistakes encountered in the field.

FIGURE 6.28
Inlet piping problem

■ Example A

While at first sight the installation shown in Figure 6.28 seems reasonable, we can see that the size of the inlet riser piping is smaller than the SRV inlet size, which will lead to excessive pressure drop once the valve opens. The valve will chatter and be damaged or even destroyed prematurely. The valve will also not flow its rated capacity and is therefore a hazard.

In addition, the length between the vessel to protect and the valve is too long and is connected with a pipe smaller than the SRV inlet connection, adding significantly to the pressure drop. It is clear that the pressure drop on this valve will be far above the 3% recommended by API. It would be wise to verify valve size as a first step, because the valve for this application could be undersized.

While here the tail pipe has the right construction and orientation, it is unfortunately not equipped with a drain. Therefore dirt can accumulate and create excessive backpressure, and can enter the valve and damage the seat surface, causing valve leakage after operation. The size of the tail pipe at the connection is smaller than the valve outlet. This will provide excessive built-up backpressure. Again, these considerations must be taken into account when sizing the valve, as the valve may otherwise be undersized. ■

FIGURE 6.29
Inlet piping problem

■ Example B

The system shown in Figure 6.29 is equipped with a rupture disc protection upstream of the SRV. Here caution is required, for the many connections in this configuration are subject to possible leakage, which influences the correct opening of the valve and creates pressure drops.

The isolation valve used is a non–full bore type which creates additional pressure drop. All valves isolating an SRV should be full bore types.

The isolation valve has no locking device. It is always recommended to have a locking device on each isolation valve so it can be blocked in open position. It is also highly recommended that the valve be equipped with a clear position indicator, so that when the valve is in open position can be verified easily, and the system is protected. Too often closed valves in front of SRVs are found.

On a positive note, here the outlet tail pipe is the same size as the valve outlet size and is equipped with a drain system. ■

■ Example C

The outlet of the tail pipe shown in Figure 6.30 is directed downwards, which will not only create excessive built-up backpressure but is also not safe for personnel and equipment. The flow of an opening SRV can achieve high velocities and can be toxic, hot and flammable. It is highly recommended that the outlet be directed to the sky. ■

FIGURE 6.30
Outlet piping problem

■ Example D

The outlet of the tail pipe shown in Figure 6.31 is far too near the walking deck (almost under the walking surface). It would be inadvisable to walk on this deck when this SRV goes off. When discharging, it will blow against the deck, which can possibly create built-up backpressure. ■

FIGURE 6.31
Outlet piping problem

■ Example E

The tail pipe of the construction shown in Figure 6.32 is far too long to be without bracing. The moment arm is too long and will create excessive reaction forces on the inlet nozzle once the SRV starts discharging. Also, with excessively long tail pipes, vibrations will occur during discharge, causing mechanical stress. ■

■ Example F

Most major manufacturers will ship SRVs from their facilities covering all possible orifices of the valve in order to avoid ingress of dirt before the valve gets installed. Plastic or wooden flange protectors cover the flanges, and the vents of the bellow valves are protected by a plastic plug or, better yet, by a vent plug with filter. In conventional valves, the vent is plugged with a solid steel plug or the bonnet is integral. In Figure 6.33, the plastic vent protection was not removed. All plastic protection on the valve needs to be removed; there are no plastic accessories on SRVs. This may not be clear because these plastic protectors are also used to protect the valve when it gets painted and might seem to be an integral part of the valve.

FIGURE 6.32
Tail pipe problem

Leaving protective test gags on valves falls into the same category of mistakes. Some manufacturers supply test gags as standard and some end users require test gags to protect the valve not only during hydraulic testing but also during transportation. It is of course important to remove any test gags before the commissioning of the plant or process. An SV with a test gag installed is the same as a blind flange and does not protect the system in any shape or form. ■

■ Example G

It is obvious that good piping practices need to be followed at all times. The bolt for the SRV shown in Figure 6.34 is insufficient and the valve is held on by only a couple of threads. The forces on an operating SRV are extremely high, and inadequate bolting may cause the valve to come loose and go airborne.

FIGURE 6.33
Valve configuration problem

FIGURE 6.34
Connection problem

The corrosion on this SRV flange is rather extensive and it would be recommended to overhaul this valve completely. ■

■ Example H

Despite the clear notice that the valves shown in Figure 6.35 are balanced bellow valves, the vent hole in the bonnet is plugged. This valve will not work properly and is a great installation hazard. ■

■ Example I

Figure 6.36 shows an installation where a rupture disc is supposed to protect the SRV (for example, from corrosive or polymerizing fluids). The gauge, however, shows pressure between the rupture disc and valve, which means that the rupture disc has ruptured and is not protecting the SRV. Modern technology provides electronic signals alarming the control room when the connection piece between th rupture disc and SRV is pressurized. Too long an exposure of the SRV to, for instance, polymerizing fluids can block the valve and prevent it from opening when an upset occurs. ■

FIGURE 6.35
Plugged bonnet on a bellows valve

FIGURE 6.36
Rupture, bursting disc

FIGURE 6.37
Isolated open bonnet

■ Example J

Open bonnet valves, as shown in Figure 6.37, are used on steam or high-temperature applications. The open bonnet cools down the spring so that it retains its characteristics. Isolating or covering the open bonnet will cause the spring to heat up and can change the opening characteristics, resulting in erratic opening. ■

■ Example K

The installation in Figure 6.38 is equipped with two full-bore isolation valves that serve as changeover valves. This is done to ensure that one redundant valve is always on standby so the other valve can be maintained without having to shut down the process; one valve can remain open while the other is isolated. In such cases, both valves need to be interlocked so that when one is opened, the other automatically closes. If not interlocked, both valves could accidentally close, jeopardizing the safety of the system. In this example, no interlock system is provided. Interlock systems can be manual or automatic; specially designed changeover valves are also available on the market. ■

FIGURE 6.38
No interlock system

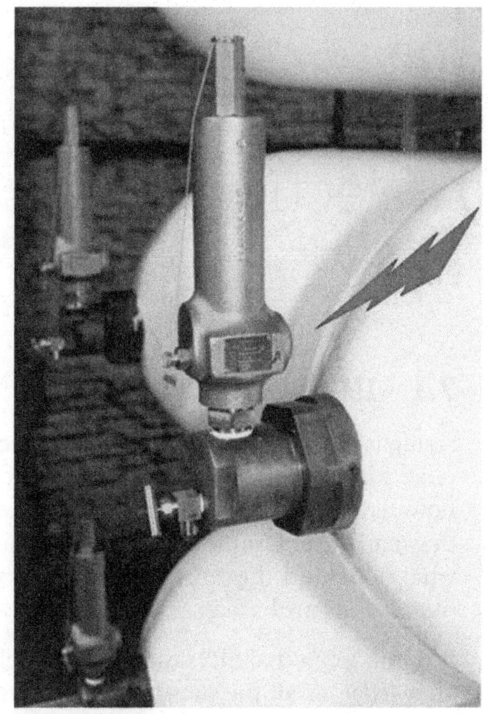

FIGURE 6.39
Wrong outlet direction

■ Example L

In Figure 6.39, the SRV outlet is situated only a couple of centimetres away from the protected vessel. When the valve opens, the high velocity of the fluid can damage the vessel and create valve instability due to turbulence close to the outlet. Also, here the plastic protection cap at the outlet has not been removed. ■

As can be seen from the examples, an onsite inspection of a process plant and its SRV installations can unearth a wide variety of installation errors that could jeopardize the process. Merely having a mandatory SRV installation does not make a system safe. True safety is a combination of valve sizing and considering all environmental factors, as well as transportation, installation and, last but not least, maintenance.

CHAPTER 7

Sizing and Selection

7.1 INTRODUCTION

Sizing is probably the most important component in selecting the right safety relief valve (SRV) for the job, ensuring optimal safety of the process. Because nowadays sizing is done almost exclusively with software, some available on the market and some supplied by manufacturers, it is important to know what is behind the software and to look into the formulas on which these calculations are based.

It is important that SRVs ultimately be selected by people who have a complete knowledge of all the pressure-relieving requirements of the process to be protected, as described in Section 2.3, taking into account the connecting piping and installation conditions of the valve (Section 7.6). They must also understand local code requirements and be aware of what is available on the market in order to select the correct valve for the correct application, ensuring a safe system.

Most reputable manufacturers can assist the end user in sizing the valve based on his input of the relevant technical data (an API example of necessary data is shown in Appendix K), but it is ultimately the responsibility of the user to select the valve based not only on the process data but also on all other factors. Many manufacturers offer reliable software programs for valve sizing, but most are unfortunately based exclusively on their own specific products, making it difficult to compare valve brands.

Frequently, SRV sizes are determined by merely matching the size of an existing available vessel nozzle or the size of an existing pipeline connection. This operating method is extremely dangerous and does not comply with the codes.

Correct and comprehensive SRV sizing and selection is a complex, multistep process that should preferably follow a step-by-step approach:

1. Evaluate each piece of equipment in a process for potential overpressure scenario (see also Sections 2.3 and 13.4).

2. Establish an appropriate design basis for each vessel that needs protection based on the different overpressure cases described in Section 2.3. Choosing the correct design basis requires assessing alternative scenarios to find the credible worst-case scenario.
3. Calculate the size of the required SRV based on the design basis.
4. Having established the required size for the SRV, look at the peripheral conditions and the application in order to select the correct SRV type.

If possible, the sizing calculations should be based on current methods and incorporate such considerations as two-phase flow and reaction heat sources.

This section addresses the SRVs as individual components. Detailed system design aspects pertaining to ancillary piping systems are covered separately, in Section 6.1, although this is not exhaustive of all possible configurations available. Design issues can be further addressed by analysis using standard accepted piping engineering principles. Covering all the possibilities is not within the scope of this book, although inlet and outlet piping aspects are covered in Section 6.1, including where relief device inlet and outlet piping are subject to important guidance by the ASME Code.

Sizing SRVs involves determining the correct orifice of a particular valve type for a required relieving capacity, as discussed in Chapter 2.

The methodology for sizing SRVs is as follows:

- Establish a set pressure at which the SRV is to operate based upon operational limits of the process and the code (see Appendix H and Chapter 4).
- Determine the relieving capacity (Chapter 2).
- Select the size and type of valve that will flow that capacity within the limits of the code and suitable for the particular application (Chapter 9).

SRVs are usually sized by calculation but sometimes also by selecting from a capacity chart in a manufacturer's literature. This last method is only to be used if the reliability of that chart can be demonstrated by the manufacturer. Here we will only discuss sizing by calculation, the more scientific method.

This section will try to assist the reader in sizing SRVs. In the first part, sizing data are given in English units (United States Customary System, USCS) consistent with the requirements of ASME Section VIII, API Recommended Practice 520, for pressures above 1.03 barg (15 psig). Metric units are given only for the most common sizes.

The basic formulas and capacity correction factors contained in this book reflect current state-of-the-art safety relief valve (SRV) sizing technology.

The following is a suggested minimum list of service conditions which must be provided in order to properly size and select an SRV:

1. Fluid properties
   a. Fluid and state
   b. Molecular weight
   c. Viscosity
   d. Specific gravity
      i. Liquid (referred to water)
      ii. Gas (referred to air)
   e. Ratio of specific heats (C or $k = C_p/C_v$). Use 315 if unknown.
   f. Compressibility factor (Z). Use 1 if unknown
2. Opening conditions
   a. Operating pressure (psig or kPag max)
   b. Operating temperature (°F or °C max)
   c. Maximum allowable working pressure – MAWP (psig or kPag)
3. Relieving conditions
   a. Required relieving capacity
      i. Gas or vapour (lbs/h/SCFM or kg/h/Nm^3/h)
      ii. Liquid (GPM or l/min)
   b. Set pressure (psig or kPag)
   c. Allowable overpressure (%)
   d. Superimposed backpressure (psig or kPag) and specify constant or variable
   e. Built-up backpressure (psig or kPag)
   f. Relieving temperature (°F or °C)

After determining the required orifice area necessary to flow the required capacity, the appropriate valve size and style may be selected. It should have a nominal effective area equal to or greater than the calculated required effective area. API effective areas and coefficient of discharge for different manufacturers can be found in their respective catalogues or in their comprehensive sizing software.*

The rated coefficient of discharge for an SRV, determined per the applicable certification standards, is generally less than the effective coefficient of discharge used in API RP 520 (particularly for vapour service valves where the effective coefficient of discharge is typically around 0.975). This is particularly true for valves certified per the rules of the ASME Boiler and Pressure Vessel Code, where the average coefficient from a series of valve test results is multiplied by 0.9 to establish a rated coefficient of discharge (as seen earlier in Section 3.6). For this

* Updated approved manufacturers' data can be found in the so called NB-IV 'Red Book' issued by the National Board (www.nationalboard.org)

reason, the actual discharge or orifice area for most valve designs is greater than the effective discharge area specified for that valve size per API 526.

When a specific valve design is selected for the application, the rated capacity of that valve can be determined using the actual orifice area, the rated coefficient of discharge and the equations presented in this book. This rated relieving capacity is then used to verify that the selected valve has sufficient capacity to satisfy the application.

The effective orifice size and effective coefficient of discharge specified in the API Standards are assumed values used for initial selection of an SRV size from configurations specified in API 526, independent of an individual valve manufacturer's design. In most cases, the actual area and the rated coefficient of discharge for an API-lettered orifice valve are designed so that the actual certified capacity meets or exceeds the capacity calculated using the methods presented in API 520.

There are, however, a number of valve designs where this is not so. When the SRV is selected, therefore, the actual area and rated coefficient of discharge for that valve must be used to verify the rated capacity of the selected valve and to verify that the valve has sufficient capacity to satisfy the application.

This noncoherent publishing by manufacturers of K (nozzle flow coefficient) and A (nozzle flow area) is leading to a lot of confusion. Just remember that the capacity is directly proportional to $K \times A$, and manufacturers can show any K and any A as long as $K \times A$ is equal to or less than the ones certified by the National Board.

In order to clarify all this, let's take a practical example:

Spring valve:
Published values: $K = 0.975$, $A = 8.30\,\text{cm}^2$ (J orifice, API 526)
Actual values: $K = 0.865$ and $A = 9.37\,\text{cm}^2$

This is acceptable because $0.975 \times 8.30 \approx 0.865 \times 9.37$.

Pilot-operated safety relief valve (POSRV):
API published data: $K = 0.975$, $A = 8.30\,\text{cm}^2$ (J API 526)
ASME actual data: $K = 0.878$, $A = 9.65\,\text{cm}^2$

$0.975 \times 8.30 = 8.09 < 0.878 \times 9.65 = 8.47$
In this case, this means that sizing with the correct ASME data may allow selecting a smaller valve.

It is important to verify that the SRVs used in a code-driven environment are manufactured and tested in accordance with the requirements of the ASME Boiler and Pressure Vessel Code and that their relieving capacities have been tested and certified, as required by The National Board of Boiler and Pressure

Vessel Inspectors or approved per Pressure Equipment Directive (PED) requirements by a certified notified body.

Again, it should be emphasized that SRVs must be selected by those who have complete knowledge of the pressure-relieving requirements of the system to be protected and the environmental conditions particular to the specific installation. Selection should not be based on arbitrarily assumed conditions or incomplete information. Valve selection and sizing (or sizing verification) is the responsibility of the system or process engineer and the user of the equipment to be protected. If unsure, seek professional guidance.

To size the SRV, calculate the minimum area necessary in order to flow the required flow. When selecting the SRV, choose the next API orifice letter up; in practice, this is always a safety factor with normal API valves. Nevertheless, some manufacturers can customize valves to the exact required flow area (A). Oversizing valves is not good practice. Undersizing valves is simply dangerous.

7.2 GAS AND VAPOUR SIZING

The basic formulas used for sizing the SRVs are all based on the perfect gas laws in which it is assumed that a gas neither gains nor loses heat (*adiabatic*) and that the energy of expansion is converted into kinetic energy. Few gases, however, behave this way completely, and deviation from the perfect gas law becomes greater as the gas approaches its saturation point. To correct for these deviations, API introduced various correction factors such as the gas constant and compressibility factors (see Appendices C and D). Correction factors also account for the effects of backpressure, subsonic flow, and so forth. Worldwide, API is by far the most complete in making use of different correction factors and is therefore judged a safe sizing practice. Sizing for gases or vapours can be done either by capacity weight or volume.

We can divide the sizing formulas into two general categories based on the flowing pressure with respect to the discharge pressure.

In the first category, the ratio of P_1 (inlet) to P_2 (outlet) is approximately 2 or greater. At that ratio, the flow through the valve orifice becomes sonic; that is, the flow reaches the speed of sound for that particular fluid. Once the flow becomes sonic, the velocity of the fluid remains constant (cannot go supersonic). No decrease of P_2 will increase the flow in any shape or form. This is also sometimes referred to as 'choked flow'.

The second category covers subsonic flow, which occurs when the downstream pressure of the valve nozzle exceeds the critical flowing pressure. Under these conditions, the flow will decrease with an increasing backpressure, even though the upstream pressure will remain constant. The backpressure at which subsonic flow occurs varies with the flowing media and is calculated as follows:

$$P_{2(\text{critical})} = P_1 \left(\frac{2}{k+1} \right)^{\frac{k}{k-1}}$$

where $k = C_p/C_v$ is fluid dependable.

A flow in which the Mach number is less than 1 is known as subsonic flow. Typically, if the free stream Mach number is less than about 0.8, then the flow is subsonic. Subsonic flows are characterized by the absence of shocks, which appear only in supersonic flows.

Now for sonic flow, also called choked flow: If subsonic flow is accelerated in a converging duct to Mach 1, and then the duct is further converged, which is potentially the case in a PRV, sonic or choked flow will occur.

Volumetric flow rates of different gases are often compared to equivalent volumes of air at standard atmospheric temperature and pressure. The ideal gas law works well when used to size fans or compressors. Unfortunately, the gas law relationship, PV/T = constant, is frequently applied to choked gas streams flowing at sonic velocity. A typical misapplication could then be the conversion to standard cubic feet per minute in sizing SRVs. Whether the flow is sonic or subsonic depends mainly on the backpressure on the SRV outlet. In the API calculations, this is taken into account by the backpressure correction factor.

Backpressure correction factors highly depend on the design of the valve; it is impossible to provide generalized figures. In principle, manufacturers should provide their own data, and per EN 4126, they should be able to demonstrate them by tests.

Unlike European norms, API 520 has always published 'typical' backpressure correction factors in its code. These curves serve only as a guide and represent a sort of average for a number of manufacturers. API states that they can be used when the make of the valve is unknown (which is rather unlikely) or for gases and vapours when the critical flow pressure point is unknown.

As with EN 4126, however, API 520 also recommends contacting the manufacturer for this data.

When contacting the manufacturer, it is also recommended you ask for the test data supporting the numbers being published or issued. Interestingly, when you start comparing backpressure correction factors in several manufacturers' catalogues, you will notice that many just use the numbers given in the

API 520. Of course, these may be correct, but just to make sure, it is wise to double-check the test data, as suggested by EN 4126 (see Appendix B).

As a simplification, let's assume a theoretical nozzle which is inherently a convergence in the system. The pressure in that convergence is

$$P_C = P_1\left(\frac{2}{k+1}\right)^{\frac{k}{k-1}}$$

When BP (backpressure) < P_C, flow is sonic and the capacity depends only on P_1.

When BP > P_C, flow becomes subsonic and then capacity depends on ΔP.

Using English units, there are two formulas, depending on the units used:

$$A = \frac{W\sqrt{TZ}}{CKP_1K_b\sqrt{M}} \qquad A = \frac{V\sqrt{TGZ}}{1175CKP_1K_b}$$

where

W = Required relieving capacity (lbs/h)
V = Required relieving capacity (SCFM)
M = Molecular weight of the gas or vapour obtained from standard tables
G = Specific gravity of the gas or vapour obtained from standard tables
A = Minimum required effective discharge area (in.^{2})
C = Coefficient determined from an expression of the ratio of specific heats of the gas or vapour at standard conditions obtained from standard tables, or if the ratio of specific heats value is known, see in Appendix D 'Ratio of specific heats k and coefficient C'. Use C = 315 if value is unknown.
K = Effective coefficient of discharge. For most manufacturers $0.90 < K < 0.98$.
K_b = Capacity correction factor due to backpressure. For standard valves with superimposed (constant) backpressure exceeding critical values, see Appendix B for reference values, but consult the manufacturer's data also. EN 4126 requires physical testing of these values.
P_1 = Relieving pressure (psia) = Set pressure (psig) + Overpressure (psi) – Inlet pressure loss + Local atmospheric pressure (psia)
T = Absolute temperature of the fluid at the valve inlet, degrees Rankine (°F + 460)
Z = Compressibility factor (see Appendix C). Use Z = 1.0 if value is unknown.

Using Metric units, there are two formulas depending on which units are used:

$$A = \frac{13{,}160W\sqrt{TZ}}{CKP_1K_b\sqrt{M}} \qquad A = \frac{179{,}400Q\sqrt{TGZ}}{CKP_1K_b}$$

where

A = Minimum required effective discharge area (mm^2)
W = Required relieving capacity (kg/h)
Q = Required relieving capacity (Nm^3/min)
M = Molecular weight of the gas or vapour obtained from standard tables
G = Specific gravity of the gas or vapour obtained from standard tables
C = Coefficient determined from an expression of the ratio of specific heats of the gas or vapour at standard conditions obtained from standard tables, or if the ratio of specific heats value is known, see in Appendix D 'Ratio of specific heats k and coefficient C'. Use $C = 315$ if value is unknown.
K = Effective coefficient of discharge. For most manufacturers $0.90 < K < 0.98$.
K_b = Capacity correction factor due to backpressure. For standard valves with superimposed (constant) backpressure exceeding critical values, see Appendix B for reference values, but consult the manufacturer's data also. EN 4126 requires physical testing of these values.
P_1 = Relieving pressure (kPaA) = Set pressure (kPag) + Overpressure (kPa) – Inlet pressure loss + Local atmospheric pressure (kPaA)
T = Absolute temperature of the fluid at the valve inlet, degrees Kelvin (°C × 273)
Z = Compressibility factor (see Appendix C). Use $Z = 1.0$ if value is unknown.

Analysing these formulas, we can deduct that we get a bigger valve (A increases) with, as can be seen, the required flow $W(Q)$ and also T.

- Higher required flow rates
- Higher inlet temperatures
- Low set pressures
- Low overpressure
- Low C ($C_{air} > C_{nat\ gas} > C_{propane}$)
- A bad (low) nozzle coefficient which depends on the design of the valve
- High backpressure ($Kb < 1$)

7.3 STEAM SIZING (SONIC FLOW)

For steam service at 10% overpressure, we use the following formula based on the empirical Napier formula for steam flow. Correction factors are included to account for the effects of superheat, backpressure and subsonic flow. An additional correction factor, K_n, is required by ASME when the relieving pressure (P_1) is higher than 1500 psia (10.340 kPaA).

Using English units:

$$A = \frac{W}{51.5KP_1K_{sh}K_nK_b}$$

where

A = Minimum required effective discharge area (in.^{2})
W = Required relieving capacity (lbs/h)
K = Effective coefficient of discharge. For most manufacturers $0.90 < K < 0.98$
P_1 = pressure (psia) = Set pressure + Overpressure – Inlet pressure loss + Local atmospheric pressure (psia)
K_{sh} = Capacity correction factor, due to the degree of superheat in the steam. For saturated steam, use $K_{sh} = 1$. For other values, see Appendix E
K_n = Capacity correction factor for dry saturated steam at set pressures above 1500 psia and up to 3200 psia. See Appendix F
K_b = Capacity correction factor due to backpressure. For standard valves with superimposed (constant) backpressure exceeding critical values, see Appendix B (Table B.1) for reference values, but consult the manufacturer's data also. EN 4126 requires physical testing of these values

Using Metric units:

$$A = \frac{190.4W}{KP_1K_{sh}K_nK_b}$$

where

A = Minimum required effective discharge area (mm^2)
W = Required relieving capacity (kg/h)
K = Effective coefficient of discharge. For most manufacturers $0.90 < K < 0.98$
P_1 = Relieving pressure (kPaA) = Set pressure + Overpressure – Inlet pressure loss + Local atmospheric pressure (kPaA)
K_{sh} = Capacity correction factor, due to the degree of superheat in the steam. For saturated steam, use $K_{sh} = 1$. For other values, see Appendix E
K_n = Capacity correction factor for dry saturated steam at set pressures above 10,346 kPaA and up to 22,060 kPaA. See Appendix F
K_b = Capacity correction factor due to backpressure. For standard valves with superimposed (constant) backpressure exceeding critical values, see Appendix B (Table B.1) for reference values but consult the manufacturer's data. EN 4126 requires physical testing of these values

There is also a simplified formula that is sometimes used giving A in cm^2:

$$A = \frac{W}{52.5KP_1K_{sh}}$$

when $P_1 \geq 109$ Bara, the Napier coefficient must be taken into account.

7.4 STEAM SIZING – PER ASME SECTION I

ASME Section I SRVs are devices designed to protect power boilers during an overpressure event. Only the U.S. code addresses this sizing separately. PED, on the other hand, makes no distinction between fired and unfired pressure vessels and the method as per Section 8.3 can be used. Here we give the calculation only in metric units.

The proper design, sizing, selection, manufacturing, assembly, testing and maintenance for such AMSE I SRVs are all critical to obtain optimum protection.

Hereafter, relevant extracts from the ASME I Boiler Code which relate specifically to SRVs:

1. *Boilers-safety valve requirements (PG-67)*
   Boilers having more than 46.5 m² of bare tube and boilers having combined bare tube and extended water heating surfaces exceeding 46.5 m² as well as a design steam generating capacity exceeding 1814 kg/h must have two or more safety valves. If only two safety valves are used, the relieving capacity of the smaller must not be less than 50% that of the larger, so if only two valves are used, select valves so that each will relieve approximately half of the total boiler capacity.

2. *Superheater safety valve requirements (PG-68)*
   Boilers having attached superheaters must have at least one valve on the superheater. The valves on the drum must be large enough to relieve at least 75% of the total boiler capacity. It is good practice to size the superheater valve to relieve approximately 20% of the total boiler capacity to protect the tubes against overheating.

3. *Reheater safety valve requirements (PG-68)*
   Boilers having reheaters must have at least one safety valve on the reheater outlet capable of relieving a minimum of 15% of the flow through the reheater. The remainder of the flow through the reheater may be discharged by safety valves on the reheater inlet.

4. *Economizer SRVs requirements (PG-67) (Closed bonnet type valve)*
   Any economizer which may be shut off from the boiler, thereby permitting the economizer to become a fired pressure vessel, shall have one or more

SRVs with a total discharge capacity, in lbs/h, calculated from the maximum expected heat absorption in BTU/h, as determined by the boiler manufacturer, divided by 1000. SRVs in hot water service are more susceptible to damage and subsequent leakage than safety valves relieving steam. It is recommended that the maximum allowable working pressure of the boiler and the SRV setting be selected substantially higher than the desired operating pressure so as to minimize the times the safety valve must lift.

5. *Organic fluid vaporizer safety valve requirements (Dowtherm Service, PVG-12)*
   Safety valves shall be totally enclosed and shall not discharge to the atmosphere, except through an escape pipe that will carry such vapours to a safe point of discharge outside of the building. The safety valve shall not have a lifting lever and valve body drains are not mandatory. A rupture disc may be installed between the safety valve and the vaporizer. The required minimum safety valve-relieving capacity shall be determined from the formula:

$$W = \frac{0.75CH}{h}$$

where

W = Weight of organic fluid vapour generated per hour (kg)
C = Maximum total weight or volume of fuel burned per hour (kg or m^3)
H = Latent heat of heat transfer fluid at relieving pressure (J/kg)
H = Heat of combustion of fuel (J/kg or J/m^3)

In particular, on larger power boilers, the volume required to protect the system is important. Often, in direct fire applications more than one SRV is necessary to protect the system; in that case the sum of the SRV capacities marked on the valves shall be equal to or greater than W.

Per ASME I, steam sizing should be done at 3% overpressure. As PED does not make any distinction between fired and unfired vessels, here the sizing is at 10% overpressure.

Therefore, the following formula is used specifically for sizing valves for steam service at 3% overpressure. This formula is based on the empirical Napier formula for steam flow. API recommends some correction factors to account for the effects of superheat, backpressure and subcritical flow. In this particular calculation, an additional correction factor K_n is required by ASME when relieving pressure (P_1) is above 1500 psia.

$$A = \frac{W}{5.25KP_1K_{sh}K_n}$$

where

A = Minimum required discharge area (mm^2)
W = Required relieving capacity (kg/h)
K = Effective coefficient of discharge (typically between 0.85 and 0.90 depending on the manufacturer)
P_1 = Relieving pressure (mPaA). This is the set pressure (mPaG) + Overpressure (mPa) + Atmospheric pressure (mPaA)
K_{sh} = Capacity correction factor due to the degree of superheat in the steam. For saturated steam, use K_{sh} = 1.00. See Appendix E for other values
K_n = Capacity correction factor for dry saturated steam for set pressures above 10.34 mPaA and up to 22.06 mPaA. See Appendix F

Unlike other codes, ASME Section I, part PVG, specifically addresses the requirements for SRVs on very particular applications, more specifically, those used in organic fluid vaporizer applications. Here it is best to contact the manufacturer for assistance in selecting and sizing valves for this type of service.

For applications involving steam pressures that exceed 3200 psig (supercritical steam applications), it is again best to contact the SRV manufacturer for assistance in sizing and selection. Both sizing and, in particular, material selection at these extreme high ratings become critical, and not all manufacturers carry valves for this kind of application.

7.5 LIQUID SIZING

Here it should be noted that some manufacturers have their own formulas for their specific valves. In general, the following formula can be used conservatively in accordance with the rules set by the ASME Boiler and Pressure Vessel Code Section VIII.

English:

$$A = \frac{V_L\sqrt{G}}{38KK_WK_V\sqrt{P_1 - P_2}}$$

Metric:

$$A = \frac{0.196V_L\sqrt{G}}{KK_WK_V\sqrt{P_1 - P_2}}$$

where

A = Orifice area (in^2 or cm^2)
V_L = Required capacity (USGPM or m^3/h)

K_W = Backpressure correction factor (depends on manufacturer but for reference, see Appendix B)
K_V = Viscosity correction factor (assume 1.0 if unknown or Appendix G)
K = Effective coefficient of discharge (typically between 0.60 and 0.85 depending on the manufacturer)
G = Specific gravity of the liquid at flowing conditions (can be found in standard tables)
P_1 = Set pressure + Overpressure − Inlet pressure loss (psig or barg)
P_2 = Backpressure (psig or barg)

After the value of R is determined (in Appendix G), the factor K_v is obtained from the graph in Appendix G. Factor K_V is applied to correct the 'preliminary required discharge area'. If the corrected area exceeds the 'chosen effective orifice area', the above calculations should be repeated using the next larger effective orifice size, as the required effective orifice area of the valve selected cannot be less than the calculated required effective area.

7.5.1 Combination Devices

When combining two relief devices, they can be used after each other or in parallel. When used in parallel, for whatever reason, there is no need for a de-rating. However, the rated relieving capacity of an SRV in combination with a rupture disc is equal to the capacity of the PRV multiplied by a combination capacity factor to account for any flow losses attributed to the rupture disc which might depend on the type of installation.

Combination capacity factors that have been determined by test and are acceptable to use are compiled by The National Board of Boiler and Pressure Vessel Inspectors in the Pressure Relief Device Certifications publication, NB-18 or the so-called Red Book. This publication lists the combination capacity factors to be used with a specific rupture device and relief valve by manufacturer rupture device/valve models.

When a combination capacity factor that has been determined by test for the specific rupture disc and relief valve combination is not available, a combination capacity de-rating factor of 0.9 may be used.

7.6 TWO-PHASE AND FLASHING FLOW

While two-phase flow applications always existed, this topic became the subject of serious discussion only relatively recently, when the traditional method described in API, the so-called added areas, was cast into doubt. This led to various studies which we explain in this section. Although API makes some recommendations, which we will see later, it needs to be noted that as of today, no definitive method has proved to be the perfect solution in this complex matter.

Let's first define exactly what we are talking about: Two-phase flow describes a condition whereby a flow stream contains fluid in the liquid phase and at the same time in the gas or vapour phase. Flashing flow occurs as a result of a decrease in pressure, and all or a portion of a liquid flow changes into vapour. It is possible for both flowing conditions, two-phase and flashing, to occur simultaneously within the same application. The complexity of the issue results from the fact that this condition is never stable and constantly changes during a relief cycle.

Even today, how to calculate SRVs for mixed-phase flow conditions is the subject of serious debate. The difficulty exists in the fact that the gas and liquid ratios during relief are not necessarily constant.

Until around 2000, API recommended the exclusive method of calculating mixed flow, which consisted of calculating separately for gas and liquid and adding the results to arrive at a definitive effective area of discharge. This is also known as the 'added areas' method (Figure 7.1).

Initial discussions on the correctness and/or conservatism of the added areas method led to many organizations undertaking further research on the subject.

In this section, we will describe the techniques that may be used for calculating the required effective orifice area for an SRV application on two-phase flow based on the work done by the Design Institute for Emergency Relief

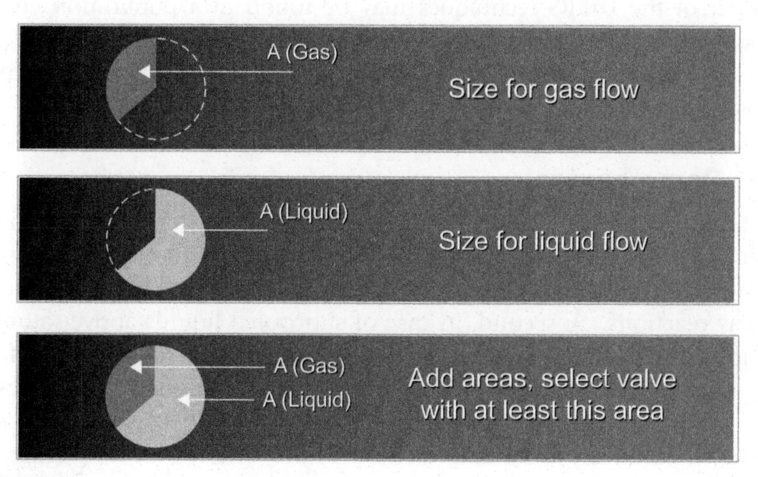

FIGURE 7.1
Traditional method for calculating two-phase flow

Systems (DIERS) and some others. Their research has revealed the complexity of this subject. Several techniques were developed in the early part of this century, but we will go into more detail on the one currently most accepted – the omega (Lueng) method.

An important conclusion from this research is how it became apparent that no single universally accepted calculation method would handle all applications, adding to the complexity of making recommendations. Some methods give accurate results over certain ranges of fluid quality, temperature and pressure and not on other process combinations. Also, complex mixtures require special consideration; furthermore, inlet and outlet conditions must be considered in much more detail than for single-component, non-flashing applications.

What has been demonstrated with this research is that, depending on the application, either the old or the new method is more or less conservative.

Figures 7.2 and 7.3 show the deviations for the four calculation methods versus the actual measured capacities for different fluid mixtures.

There are currently four ways to calculate mixed flows but the standards are not very clear as to which method exactly to apply. ASME mentions nothing on this subject except with regard to hot water: Appendix M, Figure 11-2 and others (such as PED) do not address this issue at all; only the single-phase calculation is addressed.

Therefore, it is necessary that those responsible for selecting SRVs for two-phase and flashing applications be knowledgeable and up to date on current two-phase flow technology, as well as familiar with the total system on which the valve will be used.

A number of the DIERS techniques may be found in a publication entitled, 'International Symposium on Runaway Reactions and Pressure Relief Design, August 2–4, 1995' available from the American Institute of Chemical Engineers, 345 East 47th Street, NY, NY 10017.

7.6.1 Some basics

Two situations call for two-phase/flashing flow: first, in case of upstream mixtures of gas and/or vapour and liquid, whether or not there is flashing, and mainly in HC processes (separator, flash column...) or in chemical processes (runaway reactions...); second, in case of subcooled liquids above saturation conditions of the outlet. This results in flashing and two-phase flow. A familiar example is water above 100 °C flowing into atmosphere. Liquid propane above −42 °C flowing to atmosphere will also flow in two phases. For more examples, we can refer to saturation curves of the different media (Figure 7.4). The following guidelines should be considered when sizing or selecting SRVs for two-phase and flashing flow.

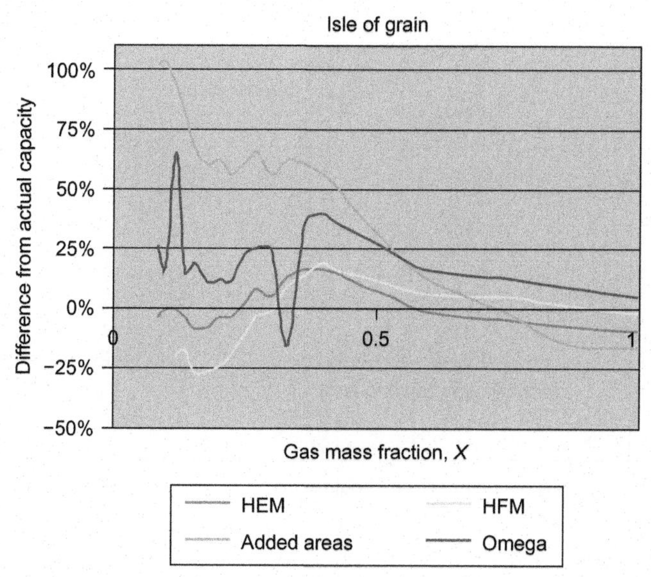

FIGURE 7.2

Deviations for a mixture 95% propane + 5% butane

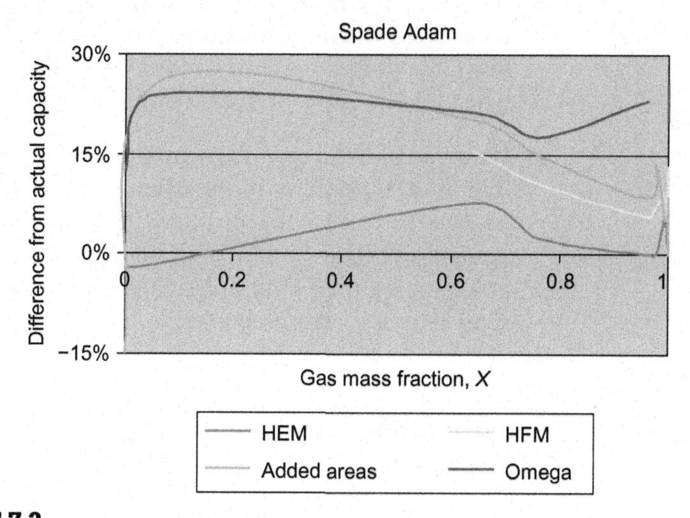

FIGURE 7.3

Deviations for mixture 64% to 84% methane + 11% to 30% propane + ethane butane

- The expected built-up backpressure must be considered at all times. If the backpressure is higher than the vapour pressure at relieving temperature, there is no flashing. If the backpressure is lower than the vapour pressure at relieving temperature, then we have a flashing condition (Figure 7.5).

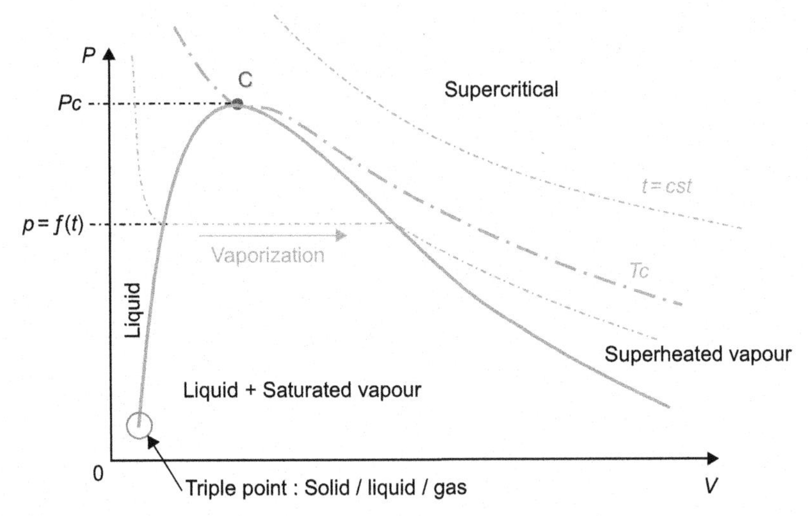

FIGURE 7.4
Typical saturation curve

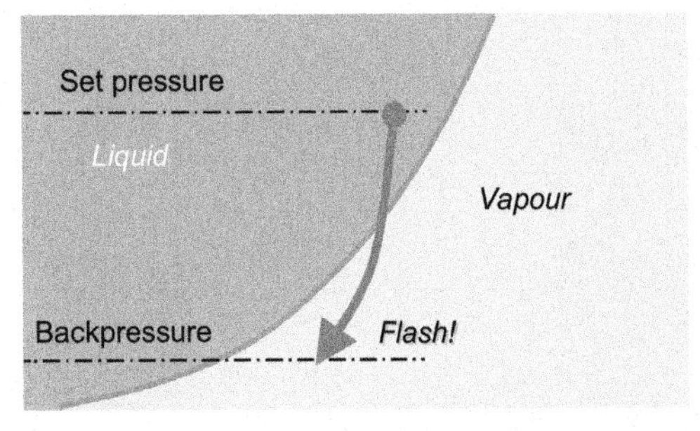

FIGURE 7.5
Flashing also depends on the backpressure

- A balanced bellows SRV or a POSRV may need to be considered when the pressure increases in the body bowl due to flashing-flow conditions – especially when it is excessive or cannot be accurately predicted.
- If the mass of the two-phase mixture at the valve inlet is 50% liquid or more, a liquid service valve construction must be considered. If the vapour content of the two-phase mixture is greater than 50% (mass), then a valve designed for compressible fluid service is recommended.

7.6.2 Two-phase liquid/vapour flow

(*Reference: API RP 520, Appendix D, January 2000*)

The method described hereafter is known as the 'Omega method' and has been developed by Dr. J. Leung. As already mentioned, it is not the only method available for sizing valves in difficult mixed-phase flows, but its relative simplicity gives it some advantage. Although the method is presented in API RP 520, Appendix D, appendices are not considered to be part of the 'body' of the standard. What is commonly referred to as the 'DIERS method' is usually the full Homogenous Equilibrium Model (HEM) method, of which the Omega method is a simplification. Detailed process simulations are required to perform the

HEM method, so it is beyond the capabilities of all SRV manufacturers, and must be done by process or chemical engineers who have access to special software process simulations.

As mentioned, the previous method (included in the body of the text of the API RP 520 sixth edition and deleted from the text of the seventh edition) consisted of sizing the valve separately for each phase and then adding the two required areas to determine the required nozzle area. However, many actual flow tests prove that this method led too often to undersizing the SRV, creating hazardous situations.

When sizing for dual-flow conditions, remember:

1. Both phases are at thermal and mechanical equilibrium (homogeneous equilibrium flow).
2. Consider ideal gas behaviour.
3. Heat of vaporization and heat capacity of the fluid are constant throughout the nozzle.
4. Vapour pressure and temperature follow the Clapeyron equation.
5. The flow is isentropic (reversible adiabatic).

7.6.2.1 *Omega*

This is not a physical parameter of the fluid mixture but is a convenient parameter that represents the compressibility or expansion of the mixture. It is defined by the formula:

$$\frac{\nu_n}{\nu_1} = \omega\left(\frac{P_1}{P_n} - 1\right) + 1$$

7.6.2.2 *Omega 9*

For each case, one formula to calculate the parameter ω (omega) uses the specific volume of the mixture at 90% of the considered pressure. This requires the process engineer to run a process simulation to obtain this specific volume after a flash calculation. For simplicity, this formula is called here the 'Omega 9'. It can be used in many circumstances.

Definitions

Boiling range: When there is a mixture of several components, the boiling range is defined as being the difference between the boiling point of the lightest and the heaviest component of the mixture at atmospheric pressure.

Mass fraction (χ): The mass proportion of gas and/or vapour in the mixture:

$$\chi = \frac{W_g}{W_1 + W_g} \quad \text{or} \quad \chi = \frac{\alpha \nu_1}{\nu_{g1}}$$

Void fraction (α): The volumetric proportion of gas and vapour in the mixture:

$$\alpha = \frac{Q_g}{Q_l + Q_g} \qquad \text{or} \qquad \alpha = \frac{\chi \nu_{g1}}{\nu_1}$$

Now we can consider different scenarios of flashing which can be determined by process simulations.

1. A simple two-phase system at the valve inlet either flashing liquid and its vapour or non-flashing liquid with non-condensable gas. A non-condensable gas is a gas that cannot be condensed under the process conditions, typically air, nitrogen, hydrogen, carbon dioxide, carbon monoxide or hydrogen sulphide.
   Example:
   i. Saturated liquid/vapour propane system flashing
   ii. Highly subcooled water (does not flash) and carbon dioxide. Highly subcooled means that the liquid is far from its boiling point and does not flash

API RP 520 divides this into two cases.

2. *Subcooled liquid flashing/no vapour at the inlet and no non-condensable gas:* This scenario can be applied to hot water and covers, therefore, the ASME VIII Appendix M, Figure 11-2.
   Example:
   i. Subcooled liquid natural gas flashing
   ii. Hot water

3. *Flashing two-phase system with non-condensable gas:* Flashing liquid with or without its vapour, and with non-condensable gas.
   Example:
   i. Saturated water/steam with air and the water flashes
   ii. Production crude oil with light components flashing and H_2S, CO_2...

Some formulas are of course sensitive to the units used. These include a conversion constant, a value which is given in Table 7.1 for some unit combinations. The formulas without such a conversion constant are not sensitive to the units used as long as the same units are used for the same data (all pressures, all specific volumes, etc.).

When shown, the numbering of the formulas is identical to the one used in the API RP 520 Appendix D.

7.6.3 Two-phase system with flashing or non-condensable gas

The two-phase system with flashing or non-condensable gas is applicable for flashing liquid and its resulting vapour or non-flashing liquid and non-condensable gas.

Table 7.1 Some constants and conversions used for the formulas

Units	Symbol	SI	Pseudometric
Pressure	P	Pa	Bar
Temperature	T	K	K
Gas mass flow capacity	W	kg/s	kg/h
Liquid volume flow capacity	V	m^3/s	m^3/h
Area	A	m^2	cm^2
Flux	G	$kg/s.m^2$	$kg/h.cm^2$
Specific volume	ν	m^3/kg	m^3/kg
Density	ρ	kg/m^3	kg/m^3
Specific heat	C_p	J/kg.K	kJ/kg.K
Latent heat	h	J/kg	kJ/kg
N1		2	200
N2		1	100
N3		1	113.84
N4		1.414	161
N5		1	1
N6		1	1

Fluids both above and below their thermodynamic critical point in condensing two-phase flow can be handled under this case as well.

7.6.3.1 Determine Omega

Flashing fluids:

IF

Boiling range < 83°C (or if there is only one component in the fluid). This relates to a differential temperature, therefore 83°C = 150°F = 83°K = 150°R.

And

The single-component fluid is far from its thermodynamic critical point ($T_1 \leq 0.9T_{cr}$ or $P1 \leq 0.5P_{cr}$). Note the 'or': if the relieving temperature, for example, is equal to the critical temperature but the relieving pressure is below 50% of the critical pressure, then the condition is verified.

Then

$$\omega = \frac{\chi_1 \nu_{\nu 1}}{\nu_1}\left(1 - \frac{N_1 P_1 \nu_{\nu l1}}{h_{\nu l1}}\right) + \frac{N_2 C_p T_1 P_1}{\nu_1}\left(\frac{\nu_{\nu l1}}{h_{\nu l1}}\right)^2 \quad \text{(F1)}$$

or

$$\omega = \frac{\chi_1 \nu_{\nu 1}}{k\nu_1} + \frac{N_2 C_p T_1 P_1}{\nu_1}\left(\frac{\nu_{\nu l1}}{h_{\nu l1}}\right)^2 \quad \text{(F2)}$$

For any other condition, we must use the Omega 9 formula:

$$\omega = 9\left(\frac{\nu_9}{\nu_1} - 1\right) \quad \text{(F3)}$$

Non-flashing fluids:

$$\omega = \frac{\chi_1 \nu_{\nu g1}}{\nu_1 k} \quad \text{(F4)}$$

7.6.3.2 Determining the critical conditions

Determine η_c from formula F1 or from the following formula:

$$\eta_c = [1 + (1.0446 - 0.0093431 \cdot \omega^{0.5}) \cdot \omega^{-0.56261}]^{(-0.70356+0.014685 \cdot \ln \omega)}$$

7.6.3.3 Determining the mass flux

If the flow is critical: $P_2 < (\eta_c P_1)$ then

$$G = N_3 \eta_c \sqrt{\frac{P_1}{\nu_1 \omega}} \quad \text{(F5)}$$

If the flow is subcritical: $P_2 \geq (\eta_c P_1)$ then

$$G = N_3 \frac{\sqrt{-2\left[\omega_s \ln \eta_2 + (\omega_s - 1)(1 - \eta_2)\right]}}{\omega_s\left(\frac{1}{\eta_2} - 1\right) + 1}\sqrt{\frac{P_1}{\nu_1}} \quad \text{(F6)}$$

$$\eta_2 = \frac{P_2}{P_1} \quad \text{(F7)}$$

Then go to the relevant formulas later to calculate the required area.

7.6.4 Subcooled liquid flashing

Applicable to subcooled flashing liquid, including saturated, without vapour or gas. Depending on the subcooling region of the liquid at the inlet of the valve, it will flash upstream (low subcooling) or downstream (high subcooling) of the nozzle throat.

7.6.4.1 Determine Omega

If all following conditions are met,

- Boiling range is <83 °C (or if there is only one component in the fluid).
- The single-component fluid is far from its thermodynamic critical point ($T_1 \leq 0.9T_{cr}$ or $P_1 \leq 0.5P_{cr}$).

Then the following formula can be used:

$$\omega_s = N_2 \rho_{l1} C_p T_1 P_s \left(\frac{\nu_{\nu ls}}{h_{\nu ls}} \right)^2 = \frac{N_2 C_p T_1 P_s}{\nu_{l1}} \left(\frac{\nu_{\nu ls}}{h_{\nu ls}} \right)^2 \quad \text{(F8)}$$

For all other conditions, we have to use the Omega 9 formula:

$$\omega_s = 9\left(\frac{\nu_9}{\nu_{l1}} - 1 \right) = 9\left(\frac{\rho_{l1}}{\rho_9} - 1 \right) \quad \text{(F9)}$$

7.6.4.2 Determining the subcooling region

Determine the transition saturation pressure ratio for low or high subcooling region:

$$\eta_{st} = \frac{2\omega_s}{1 + 2\omega_s}$$

Then

$P_s > \eta_{st} P_1 \Rightarrow$ Low subcooling region (flashing starts upstream of the nozzle)
$P_s \leq \eta_{st} P_1 \Rightarrow$ High subcooling region (flashing starts at the throat of nozzle)

where

P_s = (inlet vapour pressure at T_1 for single fluid) or (inlet bubble point pressure at T_1 for mixed fluids)

Determining the critical conditions in the low subcooling region:

Determine η_c from formula F3 or from the following formulas, with $\eta_s = \dfrac{P_s}{P_1}$:

If $\omega = 0.5$ and $\eta_s = 1$, then $\eta_c = 0.5$
If $\omega \neq 0.5$ and $\eta_s \neq 1$, then

$$\eta_c = 2\left(\frac{1 + \eta_s - \sqrt{\eta_s^2(1 - 0.5\ln\eta_s) + \eta_s + \dfrac{1}{\eta_s} + 0.5\ln\eta_s + 1}}{\eta_s - \dfrac{1}{\eta_s}}\right)$$

If $\omega \neq 0.5$, then

$$\eta_c = \eta_s \frac{2\omega_s}{2\omega_s - 1}\left(1 - \sqrt{1 - \frac{2\omega_s - 1}{2\eta_s\omega_s}}\right)$$

Then

$P_2 \geq \eta_c P_1 \Rightarrow$ Subcritical flow
$P_2 < \eta_c P_1 \Rightarrow$ Critical flow

In the high subcooling region:

$P_2 \geq P_s \Rightarrow$ Subcritical flow (i.e. all liquid flow in this case)
$P_2 < P_s \Rightarrow$ Critical flow

Determining the mass flux in the low subcooling region:

If the flow is critical, use η_c for η.
If the flow is subcritical, use η_2 for η.

where $\eta_2 = \dfrac{P_2}{P_1}$

$$G = N_3\sqrt{\frac{2(1-\eta_s)\left[\omega_s\eta_s\ln\dfrac{\eta_s}{\eta} + (\omega_s - 1)(\eta - \eta_s)\right]}{\omega_s\left(\dfrac{\eta_s}{\eta} - 1\right) + 1}}\sqrt{\frac{P_1}{\nu_{l1}}} \qquad \text{(F10)}$$

or

$$G = N_3\sqrt{\frac{2(1-\eta_s)\left[\omega_s\eta_s\ln\dfrac{\eta_s}{\eta} + (\omega_s - 1)(\eta - \eta_s)\right]}{\omega_s\left(\dfrac{\eta_s}{\eta} - 1\right) + 1}}\sqrt{P_1\rho_1}$$

In the high subcooling region:

If the flow is critical, use P_s for P.

If the flow is subcritical, use P_2 for P (API all-liquid flow standard).

$$G = N_4\sqrt{\frac{P_1 - P}{\nu_{l1}}} = N_4\sqrt{\rho_{l1}\left(P_1 - P\right)} \tag{F11}$$

Go to relevant formulas later for required area calculations.

7.6.5 Two-phase system with flashing and non-condensable gas

This method is applicable for flashing liquid with non-condensable gas with or without a condensable vapour. If the solubility of the non-condensable gas in the liquid is appreciable, the method in 1 should be used.

7.6.5.1 Determine Omega

If all following conditions are met:

- Boiling range is <83 °C (or if there is only one component in the fluid).
- The mixture is far from its thermodynamic critical point ($T_1 \leq 0.9T_{cr}$ or $P_1 \leq 0.5P_{cr}$).
- The mixture contains less than 0.1% in weight of hydrogen.
- The partial pressure of the vapour phase is less than 90% of the total pressure ($P_{v1} < 0.9\ P_1$), or the partial pressure of the non-condensable gas is greater than 10% of the total pressure ($P_{g1} > 0.1P_1$).

Then omega can be calculated as follows:

$$\omega = \frac{\alpha_1}{k} + N_2\left(1-\alpha_1\right)\rho_{l1}C_pT_1P_{v1}\left(\frac{\nu_{vl1}}{h_{vl1}}\right) = \frac{\alpha_1}{k} + \left(1-\alpha_1\right)\frac{N_2C_pT_1P_{v1}}{\nu_{l1}}\left(\frac{\nu_{vl1}}{h_{vl1}}\right) \tag{F14}$$

where

$$\alpha_1 = \chi_1\frac{\nu_{vg1}}{\nu 1}$$

If any of the above conditions is not met, we need to use the following Omega 9 formula.

$$\omega = 9\left(\frac{\nu_9}{\nu_1} - 1\right) \tag{F15}$$

7.6.5.2 Determining the critical conditions

If omega was determined using formula F14, then

$$\eta_{vc} = [1 + (1.0446 - 0.0093431 \cdot \omega^{0.5}) \cdot \omega^{-0.56261}]^{(-0.70356+0.014685 \cdot \ln\omega)}$$

$$\omega = \frac{\alpha_1}{k} f_o \eta_{gc}$$

$$\eta_{gc} = [1 + (1.0446 - 0.0093431 \cdot \omega^{0.5}) \cdot \omega^{-0.56261}]^{(-0.70356+0.014685 \cdot \ln\omega)}$$

Partial pressure of the gas: $y_{g1} = \frac{P_{g1}}{P_1}$

Determine the critical pressure: $P_c = [y_{g1}\eta_{gc} + (1 - y_{g1})\eta_{vc}]P_1$

If omega was determined using the Omega 9, then

$\eta_c = [1 + (1.0446 - 0.0093431 \cdot \omega^{0.5}) \cdot \omega^{-0.56261}]^{(-0.70356+0.014685 \cdot \ln\omega)}$

$P_c = \eta_c P_1$

Then

$$P_2 \geq P_c \Rightarrow \text{Subcritical flow}$$
$$P_2 < P_c \Rightarrow \text{Critical flow}$$

7.6.5.3 Determining the mass flux

If omega was determined using formula [F14], then

If the flow is critical,

$$G = N_3 \sqrt{\frac{P_1}{v_1}\left[\frac{y_{g1}\eta_{gc}{}^2 k}{\alpha_1} + \frac{(1 - y_{g1})\eta_{vc}{}^2}{\omega}\right]} \qquad \text{(F16)}$$

If the flow is subcritical,

Use the following equations to determine η_g, the non-flashing partial pressure ratio, and η_v, the flashing partial pressure ratio

$$\eta_v = \frac{1}{2(1 - y_{g1})}\left\{\eta_2 - y_{g1} + \frac{\omega - \sqrt{\left[(\eta_2 - y_{g1})\left(\omega - \frac{\alpha_1}{k}\right) + \omega\right]^2 - 4\omega\eta_2(1 - y_{g1})\left(\omega - \frac{\alpha_1}{k}\right)}}{\omega - \frac{\alpha_1}{k}}\right\} \qquad \text{(F17)}$$

$$\eta_g = \frac{1}{2\left(\gamma_{g1}\right)}\left\{\eta_2 + \gamma_{g1} - \frac{\omega - \sqrt{\left[\left(\eta_2 - \gamma_{g1}\right)\left(\omega - \frac{\alpha_1}{k}\right) + \omega\right]^2 - 4\omega\eta_2\left(1 - \gamma_{g1}\right)\left(\omega - \frac{\alpha_1}{k}\right)}}{\omega - \frac{\alpha_1}{k}}\right\} \tag{F18}$$

Determine the non-flashing mass flux:

$$G_g = N_3 \frac{\sqrt{2\left[\frac{\alpha_1}{k}\ln\frac{1}{\eta_g} + \left(\frac{\alpha_1}{k}\right) - 1\left(\eta_g - 1\right)\right]}}{\frac{\alpha_1}{k}\left(\frac{1}{\eta_g} - 1\right) + 1}\sqrt{\frac{P_1}{\nu_1}}$$

Determine the flashing mass flux:

$$G_v = N_3 \frac{\sqrt{2\left[\omega\ln\frac{1}{\eta_v} + (\omega - 1)(\eta_v - 1)\right]}}{\omega\left(\frac{1}{\eta_v - 1}\right) + 1}\sqrt{\frac{P_1}{\nu_1}}$$

Then

$$G = \sqrt{\gamma_{g1}G_g^{\,2} + \left(1 - \gamma_{g1}\right)G_v^{\,2}} \tag{F19}$$

If omega was determined using Omega 9, then

If the flow is critical,

$$G = N_3\eta_c\sqrt{\frac{P_1}{\nu_1\omega}} \tag{F20}$$

If the flow is subcritical,

$$G = N_3 \frac{\sqrt{2\left[\omega\ln\frac{1}{\eta_a} + (\omega - 1)(\eta_a - 1)\right]}}{\omega\left(\frac{1}{\eta_a - 1}\right) + 1}\sqrt{\frac{P_1}{\nu_1}} \tag{F21}$$

where $\eta_a = \frac{P_a}{P_1}$

Go to Section 7.6.5.4 for required area calculations.

7.6.5.4 Determination of the required areas of the SRV

Depending on whether the total required flow is given in mass units or volumetric (for liquid phase only at inlet) units:

$$A = N_5 \frac{W}{K_{2\varphi} K_{bw} K_c G} \qquad \text{(F7,F12)}$$

or

$$A = N_6 \frac{\nu_{\rho l1}}{K_{2\varphi} K_{bw} K_c G} \qquad \text{(F12)}$$

K_c is the combination correction factor if a rupture disc is installed upstream of the SRV (if there is no rupture disc, $K_c = 1$).

For $K_{2\varphi}$ reputable manufacturers present $K_{2\varphi}$ values for each valve model available for two-phase flow service. $K_{2\varphi}$ is based on P_2/P_1.

As it is usually accepted that the Omega method tends to be overconservative (nozzle area larger than it should be), it is recommended to use the actual ASME certified data for K and A.

Symbol units	Description	'Pseudo'metric units
A	Required discharge area	cm^2
C_p	Liquid specific heat at constant pressure at inlet conditions	kJ/kg.°K
G	Total mass flux of mixture	$kg/h.cm^2$
G_g	Non-flashing mass flux	$kg/h.cm^2$
G_v	Flashing mass flux	$kg/h.cm^2$
h_{vl1}	Latent heat of vaporization of the liquid phase at inlet conditions. It is the difference between the vapour- and the liquid-specific enthalpies	kJ/kg
h_{vls}	Latent heat of vaporization of the liquid phase at the saturation point at T_1. It is the difference between the vapour- and the liquid-specific enthalpies, at P_s and T_1	kJ/kg

(Continued)

Symbol units	Description	'Pseudo'metric units
K	Ratio of the specific heats of the vapour phase. If unknown, a value of 1.0 can be used	–
$K_{2\varphi}$	Nozzle coefficient for two-phase flow. There is no ASME certified nozzle coefficient for two-phase flow. $K_{2\varphi}$ curves are published for each valve model suitable to be used on two-phase flow and depend on manufacturer's design	–
K_{bw}	Backpressure correction factor K_{bw} is accounted for in the published $K_{2\varphi}$ curves for each model	–
K_c	Combination correction factor for a valve installed in combination with a bursting disc. If no disc is installed, take $K_c = 1$. If a disc is installed but the combination correction factor is not known, take $K_c = 0.9$	–
P_1	Absolute flowing pressure at inlet = set pressure + overpressure + atmospheric pressure	Bara
P_2	Absolute backpressure at outlet = total backpressure + atmospheric pressure	Bara
P_c	Absolute critical (flow) pressure	Bara
P_s	Saturation or vapour pressure at inlet temperature, T_1. For a multicomponent mixture, use the bubble point pressure at T_1	Bara
T_1	Absolute temperature at inlet	°K
V	Required liquid volumetric flow	m^3/h
ν_1	Specific volume of the mixture at inlet. The specific volume is the inverse of the density: $\nu = 1/\rho$	m^3/kg
ν_9	Specific volume of the mixture evaluated at 90% of the flowing pressure at inlet temperature, T_1, after a flash calculation. The flash calculation should preferably be carried out isentropically, but isenthalpic flash is sufficient	m^3/kg
ν_{v1}	Specific volume of the vapour at inlet	m^3/kg
ν_{vg1}	Specific volume of the vapour, gas or combined gas and vapour at inlet	m^3/kg
ν_{vl1}	Difference between the vapour- and the liquid-specific volumes at inlet = $\nu_{v1} - \nu_{l1}$	m^3/kg
ν_{vls}	Difference between the vapour- and the liquid-specific volumes at the saturation point at T_1	m^3/kg
W	Total required mass flow of mixture	kg/h
χ_1	Vapour, gas or combined vapour and gas mass fraction (quality) at inlet = $W_{vapour+gas}/W_{mixture}$. Where $\chi_1 = \alpha_1 \times \nu_1/\nu_{vg1}$	–
y_{g1}	Gas mole fraction in the vapour phase at inlet	–
α_1	Void (vapour–gas volume) fraction. Where $\alpha_1 = \chi_1 \times \nu_{vg1}/\nu_1$	–
η_2	Backpressure ratio, P_2/P_1	–
η_c	Critical pressure ratio. The method to determine the critical pressure ratio varies with the two-phase system scenario. Refer to the detailed two-phase flow sizing explanations	–

(Continued)

Symbol units	Description	'Pseudo'metric units
η_s	Saturation pressure ratio = P_s/P_1	–
η_{st}	Transition saturation pressure ratio = $2_{\omega s}/(1 + 2_{\omega s})$	–
ρ_9	Density evaluated at 90% of the saturation (vapour) pressure at inlet temperature, T_1, after a flash calculation. For a multicomponent mixture, use the bubble point pressure at T_1. The flash calculation should preferably be carried out isentropically, but an isenthalpic flash is sufficient	kg/m^3
ρ_1	Liquid density at inlet	kg/m^3
ω	Omega factor, expansion factor of the mixture	–
ω_s	Omega factor for flashing subcooled liquids	–

7.6.5.5 Determination of Omega 9

If the mixture is not flashing (a subcooled liquid with a non-condensable gas), there is no need to use the Omega 9 formula.

If there is flashing, Omega 9 will have to be used if the fluid/mixture meets the following conditions:

- It is a complex hydrocarbon mixture.
- It is close to its critical point ($T_1 \geq 90\%$ of T_c and $P_1 \geq 50\%$ of P_c. Note that if only one of these conditions is satisfied, the 'natural' omega can be used).
- It has a boiling range greater than 83 °C or 83 °K.
- It is a supercritical fluid in condensing phase.
- It also contains a non-condensable gas and contains more than 0.1% of H_2, or if the partial pressure of the gas is lower than 10% the total pressure (or the partial pressure of the vapour phase is higher than 90% the total pressure).

Hereafter, a non-exhaustive list of examples of some applications:

Omega 9:

- A gas/oil separator (complex HC mix)
- Most HC mixes on oil/gas production (complex HC mix, non-condensable gases usually with high H_2 content)
- Supercritical ethylene

Natural Omega:

- Liquid gases flashing or boiling (LNG, LIN, liquid butane, etc.)
- Hot water
- Liquid propane and nitrogen and/or methane and/or CO_2

Table 7.2 shows the data requirements for sizing SRVs on two-phase flow.

Figures 7.6 and 7.7 are useful graphs showing the critical ratios for two-phase flow and liquid flashing.

A frequently heard concern is how much liquid is allowed in a mixed-phase service in order to still be able to use a normal pop action SRV, either spring-operated or pilot. There are so many variables involved here that we cannot give a true 'cast-in-stone' answer for this problem.

However, a conservative approach, which has been successful for many years in my personal experience, is to limit the liquid phase to about one-third by volume. Anything above that will need at least a liquid trim for a spring-operated valve or a modulating pilot as a solution to this problem. Oil/gas separators are a typical application where this rule of thumb can be applied.

Table 7.2

Data Requirements for Sizing Safety Valves on Two-Phase Flow
per "Omega Method" (Ref API RP 520 7th ed, app D)
(Fill in the cells in the column corresponding to the flow case)

Data	Conditions	Symbol	Liquid + its vapour flashing, OR Gas + liquid no flash		Liquid only flashing		Liquid flashing + Gas with or without vapour		Units (1)	Comments
Omega Formula in API 520 app D:			D1/D2 & D4	D3	D8	D9	D14	D15		
Set pressure		p								Gauge pressure
Overpressure		OP							% of set	%
Backpressure		p2								Gauge pressure
Atmospheric pressure		Pa								Absolute pressure
Flowing pressure	= p + OP + Pa	P1								Absolute pressure
Flowing temperature		T1								
Critical pressure of fluid		Pc								Absolute pressure
Critical temperature of fluid		Tc								Temperature
P1° 50% *Pc* AND *T1* > 90% *Tc*	Test (2)		NO (3)		NO		NO			
Nominal boiling range > 83°C	Test (2)		NO (4)		NO		NO			
> 0.1% weight H_2	Test (2)						NO			
*Pg*1 < 10% *P1*	Test (2)						NO			
Gas + vapour flow	at P1, T1									
Liquid flow	at P1, T1									
Specific volume liquid	at P1, T1	*vl1*								or density
Specific volume vapour only	at P1, T1	*vv1*								or density
Specific volume liquid	at T1, Psat	*vls*								or density
Specific volume vapour only	at T1, Psat	*vvs*								or density
Specific volume gas + vapour	at P1, T1	*vvg1*								or density
Specific volume of fluid at 90%	of P1	*v9*								or density
Specific volume of fluid at 90%	of Psat	*vs9*								or density
Saturation pressure	at T1	*Psat*								absolute or gauge
Latent heat of vapour of liquid	at T1, or at Psat	*hvl1*								= Vapour Enthalpy – Liq. Enthalpy
Liquid specific heat at P_{st}	at P1, T1	*Cp*								
Gas only partial pressure	at P1, T1	*Pg*1								Absolute pressure
Ratio of specific heat of vapour + gas		*k = Cp/Cv*								req'd for *Kb* factor
Molecular weight of gas + vapour		*M*								for react.force and gas flow unit

(1) Please indicate the units clearly, particularly if absolute (e.g. psi A) or gauge (e.g. psi G).
(2) If answer to any one of the applicable test is "Yes", then fill in the right column.
(3) Only in the case of flashing single-component systems.
(4) Only for flashing systems.

Data not required (dark shading)
Data not really necessary, or calculated (light shading)

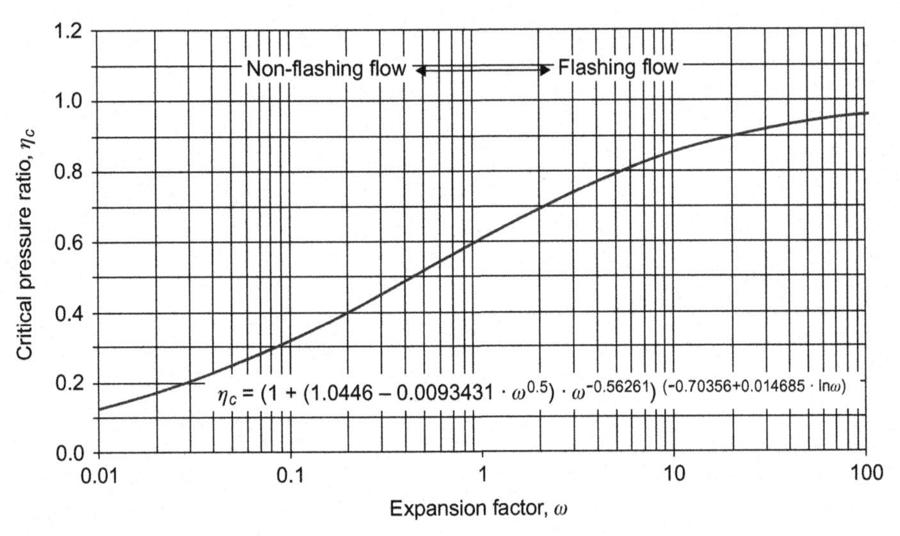

FIGURE 7.6
Critical ratio for two-phase flow

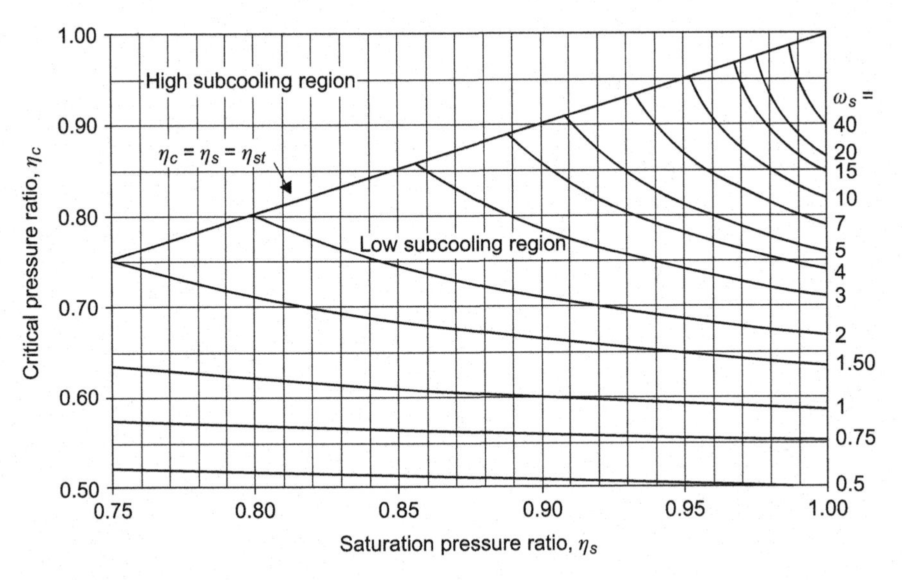

FIGURE 7.7
Critical ratio for flashing liquids

Especially when using a pop action pilot valve, which is typically designed for gas, it is vital that this POSRV also opens on mixed flow, which depends on the design of the pilot. It is recommended to always carefully check this as the valve may otherwise become extremely unstable. A primary concern is the valve closing on a very high percentage of liquid in the flow stream, then being called upon to open again.

Bottom line: For flow streams with more than one-third liquid by volume, use good-quality modulating pilots which will be more consistent in operation on gas, liquid or mixed phase service.

CHAPTER 8

Noise

In recent years questions with regards to noise from Safety Relief Valves (SRVs) have been increasing. Until not so long ago, end users only required noise calculation on control valves. SRVs were considered silent sentinels that opened only very occasionally – or preferably never. This, however, changed several years ago, primarily for four reasons:

1. Environmental concerns became more stringent.
2. Several grassroot plants were getting closer to urban regions.
3. Pressures in process systems increased to increase the output, so that systems were working much closer to the set pressure, which led to more frequent opening of SRVs.
4. Since SRVs opened more frequently, there was a realistic danger of acoustic fatigue on the system.

Historically, end users have asked SRV suppliers to control the noise of the valve during opening, in particular on boilers and large steam-generation units. Silencers have historically been the solution, mainly on steam applications. In normal process applications, the use of silencers is not yet very common (Figure 8.1).

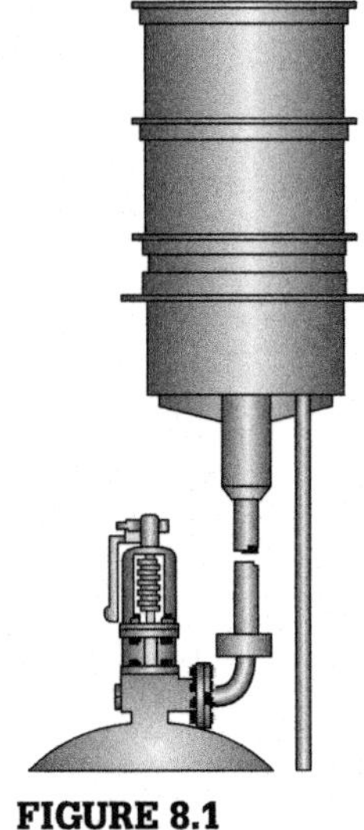

FIGURE 8.1
Vent silencer

The correct design of a silencer is critical not only for noise reduction but also for correct functioning of the SRV. First, we will describe the silencer in a little more detail and then handle the theoretical background.

When SRVs are called upon to operate, the valve discharge noise can be intense, reaching levels which may be considered harmful to operating personnel (see details later) or surrounding urban environments – that is, levels above those prescribed by most federal, state, environmental and local regulations. Unlike control valves, however, the noise of SRVs is not constant and should be of very limited duration, a consideration to be taken into account in the whole discussion around noise generated by SRVs.

Most SRVs are by definition designed for a rapid full lift operation. The resulting generated noise has distinct characteristics depending on the specific process conditions and the type of valve used and for which the silencer must be specifically designed. Silencers are designed to break up the shock wave occurring when the SRV first opens and to attenuate the steady state noise which follows.

Figure 8.2 shows a general cross-section of a silencer. Although different suppliers might have slightly different designs, the general principle is the same. The inlet of the silencer is designed to provide a high acoustical impedance to the valve-opening shock wave and low impedance to steady flow. The following stages employ expansion, reaction and absorption principles to provide the desired levels of attenuation.

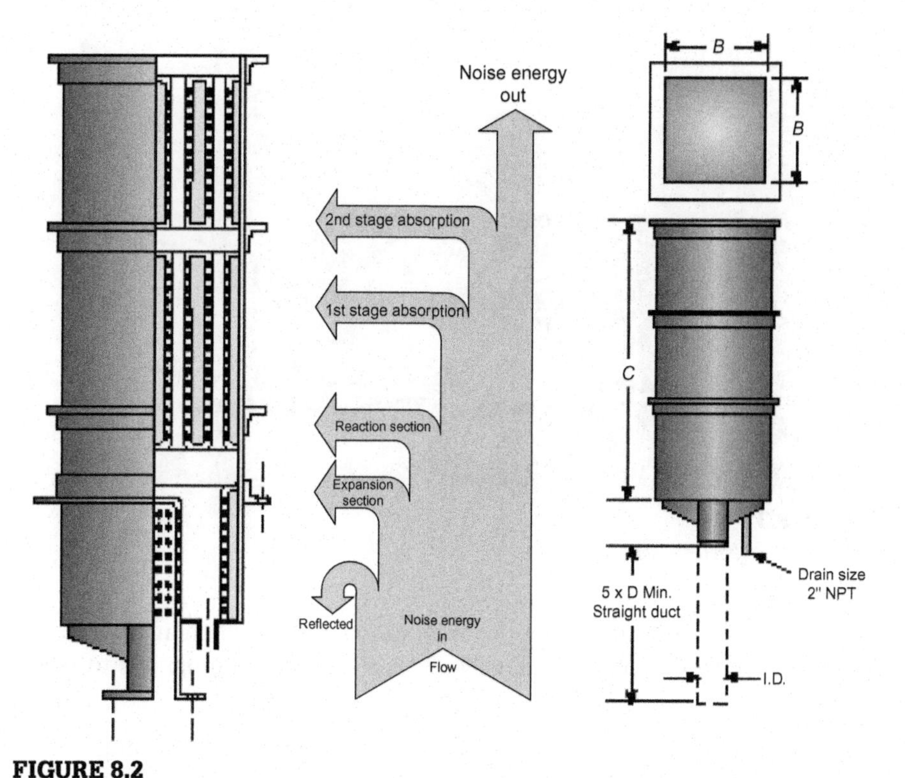

FIGURE 8.2
Silencers breaking the noise in stages

Noise attenuation is achieved by combining several steps. Part of noise energy is:

1. Reflected back
2. Attenuated in the expansion section of the silencer

3. Attenuated in the reaction chamber
4. Attenuated in the absorptive sections

To prevent regeneration of noise, the exit velocity should be limited to 9.1 km/h or less at full flow via a cage principle.

This particular design is most effective in reducing SRV discharge noise to safe levels and equally effective in reducing noise from superheater vent stacks, process waste gas stacks and other sources discharging steam, air or gases into the atmosphere (Figure 8.3).

FIGURE 8.3
Silencers on a steam generation unit

In fact, there are quite a number of variables to consider when an SRV opens: valve size, pressure, type of fluid, valve type, supporting piping, just to name a few.

As an engineer, when it comes to evaluating the noise in an opening SRV, it is preferable to have a model or other method of solution in place. Surely, to have 'no available model' shows absence of prior thought on this specific subject. Some models that do exist show simply lack of thought, but in any case, it is better to have a simple model than none at all.

As of today, the only internationally recognized method known to approximate the noise level caused by SRVs is described in the API RP 521.

This method is based on characteristics of simple supersonic nozzles discharging gas directly into the atmosphere, and therefore it can rightly be considered as approximate, and (although probably not always) quite conservative. Some consideration, however, should be given to the fact that SRVs are not just simple nozzles and that the design of, for instance, the body bowl could significantly influence the noise generated by the valve.

In using the control valves standard (IEC 534-8-3, ISA-S75.17 or VDMA 24 422) to achieve more accuracy, we always encounter enormous difficulties. It impossible to ensure that using this method to calculate noise level is more accurate, or even as accurate, as simply using the API RP 521 method.

The major issue is that by trying to apply the control valves' standards to SRVs, we need to find a way of 'modelling' the SRV characteristics with control valves' parameters. These parameters (particularly, d_j, F_p, x_T...) have not been established for SRVs and, in any case, would probably be irrelevant. Second, control valves are used and built such that the speed of the fluid at the outlet of the valve is always kept well below sonic speed (the referred standard put an upper limit at 0.3 Mach). On an SRV, however, outlet speeds far exceed the speed of sound (supersonic or even hypersonic speed) because the purpose of an SRV is to relieve the fluid as quickly as possible, preferably with no pressure recuperation.

Hereafter, we will find examples of a simple model, old information and issues that are not well defined, but with which we will have to work with at the present time. Quite frankly, the issue of noise from SRVs is not well covered in the general literature and lacks research.

Some important questions are: What is an SRV? Which one is it (there are many types)? What is its design? What body thicknesses are used? What fluid is it used on? What are the relieving pressures? What is the shape of its body bowl? How often, long and loud is its noise?

An SRV may release the process fluid directly to the atmosphere or release it via a pipe to a flare, scrubber, header or some other equipment. An SRV is always actuated by the upstream pressure and is usually characterized by what is described as a 'pop' action upon opening in the case of compressible fluids. It is important to recognize that one should not expect a gentle release of gas proportional to valve lift, regardless of the design of the valve; one possible exception is when a modulating pilot-operated SRV is used.

The noise from most spring-operated SRVs on compressible fluids can be expected to be in the range between 150 and 190 dB(A). However, the noise is very dependent on size and set pressure, to mention only two of the variables we touched on earlier.

Because the operation frequency of an SRV cannot be predicted, let's estimate a figure of 'once in a hundred years' for the operational frequency of a single SRV. Thus on a plant with hundred SRVs, an average noise of 170 dB(A) might be heard once a year, hopefully for only a few minutes. The noise is never constant and usually changes constantly and decays with time as the pressure decreases. The noise is greatest while the pressure drop across the valve induces sonic velocities in the valve. The higher the differential pressure between inlet and outlet, the higher the noise.

However, what is probably worse is that an SRV may also make very loud noise due to 'chatter'. The rapid movement from the disc on the seat is due to flow instability in the valve because of insufficient blowdown, usually caused by incorrect installation; oversizing of the valve; faulty inlet piping under the SRV; a working pressure constantly close to set pressure or another factor described in detail in this book. We could define these emergency releases of gas as transient noise sources but, contrary to the correct popping of a valve, these may last much longer. Chatter is destructive to any valve, leading to leakage, which then itself produces constant noise in the valve. Noise decreases as the erosion on disc and seat becomes greater. Obviously, metal-to-metal valves are much noisier than soft-seated or modulating valves.

Generally three main noise-related criteria are suggested in the evaluation of SRV noise:

1. Noise received at the local community
2. Acoustic fatigue of the components and associated pipework
3. The possibility of a worker who is 'close' to an opening SRV sustaining hearing damage due to noise

We focus on the third point as we believe this is the most significant noise issue for end users. Noise limits set by process plants in order to avoid the risk of hearing damage from transient sources such as SRVs or control valves can be expected to be in the region of 100 to 125 dB(A) sound level (SL). The 115 dB(A) limit in API EA 7301 can be considered typical. This SL must be measured or estimated at the worker location (or expected location), that is at ground level, on platforms, on ladders and on stairs. Sometimes this is also very close to an SRV or its open vent, especially when people are testing valves *in situ* or operating lifting levers.

SRV noise is primarily radiated from the associated piping, supports and equipment as well as from the wall of the SRV itself. In case of open-vent systems, the majority of the noise exists at the open vent and at the end of a tail pipe. Noise may also be generated due to the mixing of high-velocity gas with surrounding air.

8.1 RISK OF DAMAGE TO HEARING

The following data are based upon a presentation and paper presented by Eur Ing MDG Randall from Foster Wheeler Energy, 'Energy on PSV noise – criteria, limits and prediction'.

Permanent hearing damage can result from one very loud event; a series of loud events; or days, weeks and years of relatively loud noise in a work environment. Figure 8.4 shows how selected percentages of a male population are gradually affected by continuous noise for 8 hours a day over a number of years of work. Figures show the effect of noise on hearing level for a population of males during part of a working lifetime after starting at age 20.

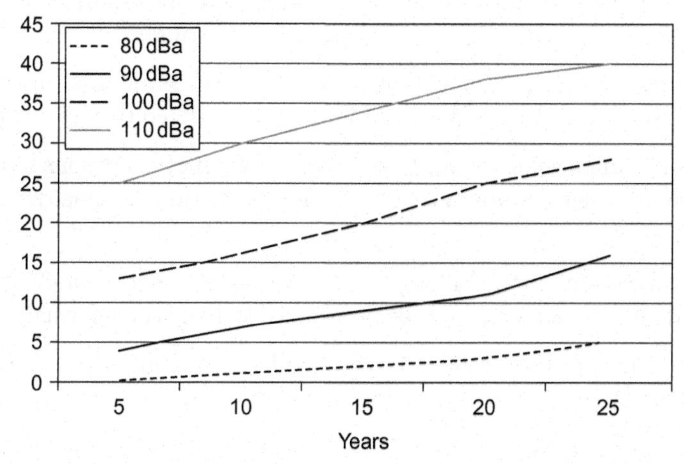

FIGURE 8.4
Permanent hearing damage due to noise as a function of number of years of exposure

Some social impact may be said to start at 30 dB(A) hearing level, with hearing loss perhaps occurring at 50 dB(A) and above.

It has been investigated whether an exposure of, on average, 80 dB(A) for 8 hours of work each day results in about 5 dB(A) hearing level loss in a male population average (mean) after 25 years (age 45). At an average level of 110 dB(A), the male population mean reaches a hearing level loss of 30 dB(A) in 10 years.

The basis of these charts is the discovery that hearing loss is a function of both the noise level and the cumulative time of exposure. In the late 1900s, the normal limit of exposure to continuous noise was 90 dB(A) for 8 hours or its equivalent. Now we see companies and governments seeking a limit of 85 dB(A) for 8 hours.

Just as a reference, a vacuum cleaner at 1 m distance produces about 70 dB(A) and a diesel truck at 10 m distance produces about 90 dB(A). The threshold for severe discomfort is about 120 dB(A). Pain starts at 130 dB(A).

Damage to hearing caused by one very loud event, as in the case of an SRV, may however be different in onset and character to the observed noise-induced hearing loss from lower but more continuous levels. In this situation, damage to the inner ear is caused to the whole organ rather than to certain cells.

Neither all countries nor all standards organizations require that SRVs be treated as a source of sound which has to be limited, mainly because their operation should, in principle, be very occasional. Instead of limiting the noise level, some countries and standards organizations require that SRVs be installed at safe distances from workers or other persons.

As an example, let's look at NORSOK, a Norwegian standards organization. Its view is that noise from SRVs should not be considered during design:

> The noise limits shall not apply to design emergency conditions e.g. near safety valves, firepumps or outdoor areas during full emergency flaring, etc.

Since there are no international or general codes related to noise on SRVs, it is for the individual designer, wherever the pressurized system is being installed, to:

1. Find the appropriate eventual local codes and standards.
2. Decide how they are to be complied with.

In my opinion, the following are the only two representative documents which include noise limits that seem appropriate for review of SRV noise:

1. *The API Medical Research Report EA 7301. (Ref 2)*: This document dates from 1973 and sets a limit of 115 dB(A) for steady sound and 140 dB (peak) for impulsive noise. These limits were based on the data in the US OSHA 1970 Act.

2. *The 86/188/ECC. (Ref 4)*: This directive states that if a maximum value of the unweighted instantaneous sound pressure level (SPL) is greater than 200 Pa, 'suitable and adequate' ear protectors which can be reasonably expected to keep the risk to hearing to below the risk arising from exposure to 200 Pa must be used. It is on this directive that the U.K.'s Noise at Work regulations are based.

It is the plant designer's or contractor's responsibility to:

1. Make a prediction of the expected sound in order to provide the overall noise expectations of an installation.
2. Have a series of design options ready for the SRV that approaches or exceeds the limit.

Limiting noise levels is not restricted to the SRV alone (Figure 8.5). Design of piping and, in particular, the isolation downstream and pipe supports are generally greater contributors to noise than the SRV itself.

FIGURE 8.5
Noise generation of an SRV is not limited to the noise from the SRV itself

So how can SRV noise at known worker positions be calculated?

Luckily, most current SRV sizing programs include a noise calculation. However, it is perhaps interesting to know in detail some methods used in the industry. Standard methods are available for calculation of valve noise heard at a certain distance.

Generally, two methods for calculation of SRV noise are suggested:

1. The (1995) IEC standard for control valve noise prediction
2. Sections 4.3.5 and 5.4.4.3 of API 521, which appear to be based on the 1950s method described by Franken

SPL at a point, typically 30 m distant from an SRV vent of sound power level (PWL) is calculated by using the following equation:

$$SPL = PWL - 10\,Log(4\pi r^2) \qquad \text{(N1)}$$

where

$r = 30$ (distance)
$SPL_{30} = PWL - 41$

Sound power level is calculated by use of the equation:

$$PWL = 10\,Log\left(\frac{\frac{1}{2}MC^2\eta}{10^{-12}}\right)$$

or

$$PWL = 10\,Log\left(\frac{1}{2}MC^2\eta\right) + 120 \qquad \text{(N2)}$$

where

$½MC^2$ = Kinetic power of the choked flow through the valve

η = Acoustic efficiency associated with the transformation of some of the kinetic power to sound power

M = Mass flow rate

C = Speed of sound in the choked gas

Thus

$$C^2 = \frac{kRT}{MW}$$

Franken shows that η ranges between about 10^{-5} and 10^{-2}.

With the above two equations, the Franken discussion leads to an equation for the SPL at 30 m:

$$SPL_{30} = 10\,Log\left(\frac{1}{2}MC^2\eta\right) + 79 \qquad \text{(N3)}$$

Franken provides a graph of η versus pressure ratio.

The API formulation for SPL at 30 m is similar:

$$SPL_{30} = 10\,Log\left(\frac{1}{2}MC^2\right) + 10\,log\,\eta + 79 \qquad \text{(N4)}$$

where the value in decibels of the term $10\,\text{Log}[\eta] + 79$ is to be evaluated from the graph (dB versus pressure ratio) published in the Recommended Practice.

The ordinate scale on the API graph is labelled '$L = L_{30} - 10\,\text{Log}(½MC^2)$'; however the simple derivation given above displays its physical basis.

The curve that is the result of plotting jet acoustic efficiency η against pressure ratio is shown in Figure 8.6. Both the API ($10\,\text{Log}[\eta] + 79$) and Franken ($\eta$) ordinate scales are given for this characteristically shaped curve.

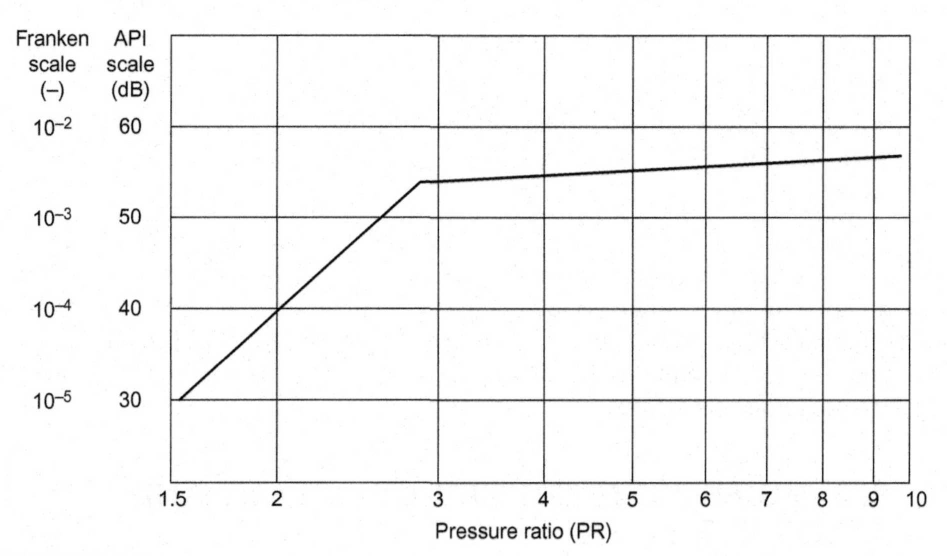

FIGURE 8.6
Acoustic efficiency of a choked jet

Most SRV vendors have simplified things further. By ignoring smaller SRVs with pressure ratios of less than 3, and with the assumption that one value of acoustic efficiency, say 4×10^{-3}, can be used for pressure ratios above 3, a formula results for the SPL at 30 m which is:

$$SPL_{30} = 10\,Log\left(\frac{1}{2}MC^2\right) + 55 \qquad \text{(N5)}$$

Other methods used for calculating the SPL of SRVs can be found in some technical literature. However, we could not find extensive research on this area. Note, however, that no one method appears to have been adopted by the 'valve industry' and that no international standard exists solely for the calculation of SRV noise. Therefore, it is important to know that when noise levels are questioned, the exact methods of calculations are comparable.

8.2 PLANNING AN ACCIDENT

Here is an issue for discussion raised by the nature and use of SRVs. How much planning and design time should be given to noise which is an accidental byproduct of accidents and emergencies? It may be argued that fire and like cases are emergency or accidental situations and are not/cannot be avoided by legislation or planning guidelines. Where noise occurs as a result of an accident (say an explosion), it is accepted as part of that accident. Where noise occurs as a result of an emergency, it is generally accepted as 'accidental' and thus not subject to planning measures.

We are not aware of any published planning legislation, regulation or guidance that specifically requires control of noise in accident or emergency situations.

Chatter, however, is due to incorrect installation or valve selection and cannot be seen as an accidental occurrence. The noise can sometimes be worse than an opening valve, but this is impossible to calculate. Here, there are no official guidelines to avoid wrong valve selection or installation. Therefore, making a correct selection and avoiding wrong installation are also paramount in order to avoid noise from SRVs.

8.3 NOISE FROM THE SRV, OPEN VENT AND ASSOCIATED PIPE

Preferably, SRVs are installed outside, but not infrequently in, for instance, power, gas and petrochemical plants; designers are forced to install them both inside and outside buildings.

Most of the noise always comes from an open vent; the next most important source is the downstream piping. Less obvious areas of noise radiation are the upstream pipe if it is of any length, and finally the body of the valve itself, although the body usually emits less noise than downstream piping.

Measurements are required to find the relationship between upstream and downstream pipe SPLs, but a rule of thumb is a 10-dB(A) decrease across a valve downstream to upstream.

These different 'sources' play different parts depending upon whether the SRV is mounted outside or inside, whether the open vent is inside or outside and how much upstream and downstream pipe is inside or outside. The major installation variations are shown in Figure 8.7.

When silencers or pipe insulation are suggested, attention must be paid to the sources that remain after the treatment is applied. The noise of a dominant source only masks lesser sources while it is dominant.

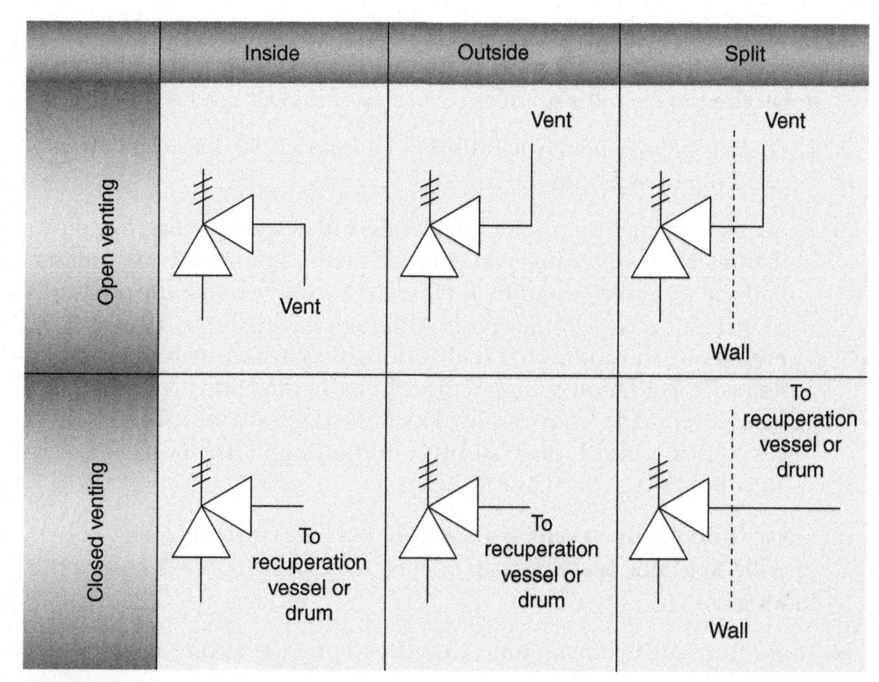

FIGURE 8.7
Possible venting systems

8.4 NOISE CALCULATIONS

According to an article from Mr. Randall, Foster Wheeler Energy, he believes the responsibility for equipment noise to rest with the equipment vendor. No concession is made where equipment noise radiates from connecting pipework.

According to some end users, however, SRV vendors are only expected to provide data on the noise produced by their equipment and not on the noise radiated through associated pipework (not in the vendors supply) and the suggested means of reducing or controlling noise to an agreed level at a specified distance.

Most SRV vendors will refuse any liability when it comes to noise caused by complete installations, as they do not in any shape or form control the selection or installation of such material, nor do they control the work procedures and piping designs.

They can however provide some recommendations to help end users determine how much noise radiates from the various parts of the system; these recommendations can be considered as a rule of thumb and some are:

- Noise from the valve. This can be provided by the SRV vendor as SPL 1 m from the downstream pipe and 1 m downstream of the valve.

- Noise from the downstream pipe. This is basically the same number as in point 1 hereafter, but more accurately there will be some more loss down the long lengths of pipe. Two rules of thumb can be given here:
  1. There is approximately a 3-dB(A) loss for each 50 diameters of gas-filled pipe away from the source.
  2. Acoustic insulation on the valve body and at least the first five pipe diameters downstream is an effective method of noise treatment to half the loss given in point 1. Note here, however, that the quality of insulation is very important. Although having no particular preference, the author has had reasonably satisfactory results with Refrasil® insulation, a high temperature thermal and acoustic insulation made from basalt, glass fibre and silica which reduced a 6 × 8 in. control valve (steam conditioning) noise level from 105 dB(A) to as much as 85 dB(A).
- Noise from the open vent. Few SRV vendors are currently able to provide this data, but where an open vent is present, it is definitely the loudest source.
- Noise from the upstream pipe. In order to provide recommendation on this, a complete isometric must be performed, including consideration of pipe supports and the insulation of their fixtures. SRV vendors are usually not in a position to perform such calculations.

Noise from an SRV – An appreciation

To gain some more insight into the noise generated from an SRV, we can explore the range of PWLs (power watt levels = noise source) to be expected when the process fluid (remember, it must be gaseous) in the valve exists in a range of:

10 to 60 molecular weight (MW)
200 to 1200 (°K)
0.01 to 300 (kg/s)
$\eta = 0.004$
$\gamma = 1.3$
$R = 8300$

The selected range is only for discussion/reference purposes and is by far not exhaustive.

First, speed of sound is determined from knowledge of temperature and molecular weight of the gas. For the selected range, this is illustrated in the following table, which provides approximate data. Note that we are only considering gaseous fluids.

T (°K) MW	200	400	600	800	1000	1200
10	484	657	805	929	1039	1138
20	328	465	569	857	735	805
30	268	379	464	530	600	657
40	232	328	402	465	519	469
50	208	294	359	416	465	509
60	190	268	329	370	424	465

The calculated speed of sound in the gas can be thought of as that at the valve (i.e. its choke point).

Knowing the speed of sound and mass flow rate, we can now determine the PWL in the next table. The position of a phase change line, gas to liquid, may have to be determined where this is critical.

Flow (kg/s)	Speed of sound, *C* (m/s)					
	315	400	500	630	800	1000
0.01	123	125	127	129	131	133
0.03	128	130	132	134	136	138
0.1	133	135	137	139	141	143
0.3	138	140	142	144	146	148
1	143	145	147	149	151	153
3.16	148	150	152	154	156	158
10	153	155	157	159	161	163
31.6	158	160	162	164	166	169
100	163	165	167	169	171	173
316	168	170	172	174	176	178

The calculated PWL of the valve can be thought of as that part which goes down the tail pipe. Where the exit pipe leads to an open vent, we may need to evaluate a 'safe distance' or a distance before a 'community or environmental limit' is reached.

The following two tables provide approximate answers but should not be regarded as tools for final design. The next table indicates how far one has to be away from an SRV tail pipe if one is to be at or below some set value of SPL.

This may be for hearing damage risk calculations at close range or to avoid environmental sound pollution. The numbers given in the table are the hemi-spherical distance from PWL source to an SPL point, expressed in metres.

Vent PWL [dB(A)]	Limit SPL [dB(A)]						
	55	70	85	100	115	130	145
180	709431	126157	22434	3989	709	126	22
160	70943	12616	2243	399	71	13	
140	7094	1262	224	40			
120	709	126	22				
100	71						

In the next table, one can see how far away a person has to be from an SRV vent if noise is to be below, say, 130 or 115 dB(A). The table is for spherical radiation but includes no other attenuation mechanisms. This table, expressed in metres, can be used to provide a first estimate of the vertical length of vent pipe that is required where stack height is to be used as the main method of noise reduction.

Vent PWL [dB(A)]	Limit SPL [dB(A)]						
	55	70	85	100	115	130	145
180						89	16
160					50	9	2
140				28	5	1	
120		89	16	3	1		
100		9	2				

8.5 CONCLUSIONS

Different papers have been written on this subject, but the reality is that SRV noise calculation is very complex because:

1. The valve is usually closed.
2. The different dynamics happening during the opening of an SRV are seldom constant and very complex, so we can only consider the conditions at full lift as 'worst condition'.

3. The opening of the SRV is very dependent on the type of SRV selected (proportional relief valve or pop action). Yet the given formulas do not take into account design features of the SRV.
4. The way the SRV is mounted and/or supported on the process system is relevant.
5. Installation, supporting and isolation of the surrounding piping must be considered.
6. Length of upstream and downstream piping must be taken into account.

The most important factor of the above-mentioned points is that an SRV is normally closed and in normal designed processes should never open. In my opinion, this means that an SRV cannot be considered as continuously contributing to the environmental noise levels of a plant.

Nevertheless, excessive noise at the point of discharge may be generated aerodynamically by a full lift SRV discharging to atmosphere at a maximum emergency flow rate during an occasional upset of the process. As just described, it should be noted that the noise level is usually of short duration.

In any case, the noise intensity should only be referenced in areas where operating personnel normally work, which is very unlikely to be near an open valve outlet as this would mean incorrect installation practices had been applied.

In special cases such as multiple valve installation, the aerodynamic noise level may warrant more detailed determination. However, this is a decision of the system designer familiar with all the local conditions and regulations and, more importantly, implantation of the valves within the site.

Beraneck's 'Noise Reduction' paper is one basic reference on this subject, although many other papers have been written.

We believe his theory is most applicable to particular SRVs as he estimates the SPL at 30 m from the outlet resulting from an expanding jet at sonic velocity. This should be taken into consideration for incidentally occurring sound sources.

Beraneck's theory states that if we absolutely must know the level at 1 m from the source, 30 dB(A) can be added to the SPL at 30 m.

The following formula may be used if an extreme condition needs detailed calculation from point of discharge at 1 m with a pressure ratio P_1/P_2 across the valve:

$$L_a = 85 + 10\,Log_{10}\left(\frac{kWT}{3.42M}\right)$$

where

L_a = dB(A) exit noise level = SPL at 1 m
W = Lbs/h gas or vapour
k = Specific heat ratio
T = °R = °F + 460
M = Molecular weight

Noise considerations are becoming increasingly important in the environmental decision making of exploitation permits of industrial sites. This chapter has hopefully given some insight, but again it should be stressed that all calculations discussed here are based on a valve in full lift. Note that noise levels can be reduced by using modulating pilots that only go into full lift when absolutely necessary but modulate the pressures gently during normal operation or smaller process upsets.

CHAPTER 9

Safety Relief Valve Selection

These guidelines will provide some guidance in selecting a safety relief valve (SRV) for use in a specific application or process condition where the use of traditional spring-operated valves could be questionable.

It is always important not just to use a supplier who provides SRVs, but to use one who provides pressure relief solutions exactly adapted to the needs of the specific application. The solution should entail the most economical and safe, reliable, long-term operation.

Some particular situations may not be covered here, and some recommendations may lead to further discussion, but this is usually better than just going for the lowest priced solution when safety is an issue.

We will consider some typical concerns when SRVs must be specified.

9.1 SEAT TIGHTNESS

Operating pressures between 90% and 95% of set: Here the use of pilot-operated safety relief valves (POSRVs) or soft-seated, spring-operated SRVs should be considered. Metal-seated spring valves will not stay tight for long and usually get damaged after a couple of operations. On the other hand, soft-seated valves are limited in temperature and sometimes also in pressure (see Section 5.2.6.5).

Operating pressures above 95% of set: Preferably use modulating POSRVs or soft-seated valves. The choice for the more expensive modulating pilots depends on system pressure fluctuations. The more fluctuation close to or over set pressure, the more a modulating valve should be considered.

Operating pressures below 90% of set: All ASME VIII– and/or PED-approved valves can be used.

9.2 BLOWDOWN

Short blowdown necessary outside ASME I or VIII requirements (<7%): Some POSRVs or high-performance, soft-seated, spring-operated SRVs have large adjustable blowdown ranges. Some range from 3% to 30% blowdown adjustment. This is, however, only available with a limited number of suppliers. A normal standard ASME VIII spring-operated SRV may not reach full lift at 10% overpressure when it is adjusted for a very short blowdown. Blowdown can seldom be set shorter than 5% on conventional spring valves.

Long blowdown required due to inlet pressure losses above 3%: On gas service, a pop or modulating action POSRV can be used. On some pilot valves, blowdowns between 3% and 25% can be achieved. In case of very high pressure losses, pilot-operated valves with a remote sensor should be considered.

9.3 SERVICE TEMPERATURE

−197°C and below: Here, it is highly recommended to use soft-seated, pop action POSRVs or soft-seated, high-performance, snap action spring valves which operate beyond the ASME and PED recommendations. Look for a valve type which has:

- Bubble tightness close to set pressure
- Snap opening faster than 10% overpressure
- Very short adjustable blowdown

These types of valve are only available with a limited number of manufacturers.

In case of backpressure, using bellows should be avoided at all times. Bellows will react differently at cryogenic temperatures, and moisture can be trapped between the bellows and freeze which will cause erroneous operation of the valve.

Always consider that, in cryogenic applications the risk exists that moisture in the air can cause the valve to freeze up around the seat area when the valve shows the smallest leak. This can have the same effect as welding the disc down on the seat, preventing the valve from opening so that it acts like a blind flange (see Chapter 11 for details).

Below −70°C and thermal expansion: Typical applications here are LNG, LIN and LOX. Here again, we must avoid the risk of freezing, but in this case we also need to consider repeatable tightness; therefore, high-performance, soft-seated spring valves should be used. A metal seat may leak prematurely, freeze up and start acting like a blind flange.

Below −70°C and backpressures needing a balanced valve: Pilot-operated, soft-seated valves are the only valid solution here. Balanced bellows valves are not acceptable at these temperatures due to the high risk of bellows failure ('cold' working) and the risk of condensates freezing inside the bellows. This phenomenon is discussed in detail in Chapter 11.

Below −70°C and no backpressures: Pilot-operated valves are still preferred, but spring valves could be used, preferably soft-seated ones. It should be noted that metal-seated valves always present a high risk of freezing at these temperatures.

Fully cryogenic liquid service: Pilot-operated, soft-seated valves with a type of vaporizer which (relatively) warms up the fluid entering the pilot; or high-performance, soft-seated, spring-operated SRVs. The vaporizer and other accessories of a typical cryogenic configuration on a pilot-operated valve keep the pilot warm, which then works on vapour. In any case, these are applications that should be discussed with your SRV supplier. Some suppliers have done extensive tests on cryogenic applications and have experience to share on this specific application.

Service temperatures above 300°C: Metal-seated spring valves with special springs and adapted materials should be considered. Soft-seated valves can only be used when providing enough inlet pipe length (gases lose more than 150°C per metre of bare steel pipe), but then the potential inlet pressure losses must be taken into account. Pilot valves should be avoided, as frequent thermal expansions could cause galling in the long term in the small-tolerance components of the pilot.

9.4 WEIGHT AND/OR HEIGHT

When weight and/or height are a concern (mainly on the larger sizes 3–4 in. and up) – mainly on offshore platform applications, difficult accessible places or on ships – POSRVs should be considered. On larger sizes, their weight and height is reduced by half compared with normal spring-operated SRVs, as the pilot does not increase in size with the main valve. In extreme cases, the pilot can be installed remotely or on an extra-low bracket to reduce further the height of the valve.

9.5 BACKPRESSURE

Tables 9.1 and 9.2 show maximum backpressure percentages on gas/vapour and liquid applications, respectively.

Absolute backpressure is higher than the maximum acceptable pressure of the bellows: Always check the maximum acceptable backpressure for standard bellows and,

Table 9.1 Maximum backpressure percentages on gas/vapour applications

Backpressure Type	Effects on valves				Selection
	Value (% of set)	**Conventional**	**Balanced Spring Valve**	**Pilot Operated**	
Constant	<30% [1]	Set point increased by backpressure [3]	No effect	No effect	Conventional, balanced or POSRV
	30%–50%		Lift/capacity reduced (coefficient) [6]		
	>50% [2]	Set point increased by backpressure; flow becomes subsonic [4]	Generally unstable *Do not use*	Flow becomes subsonic [4]	Conventional or POSRV
Variable superimposed	<10%	Set point varies with backpressure [5]	No effect	No effect	Balanced or POSRV
	10%–30% [1]	Unstable			
	30%–50%	*Do not use*	Lift/capacity reduced (coefficient) [6]		
	>50% [2]		Generally unstable *Do not use*	Flow becomes subsonic [4]	POSRV only
Variable built-up	<10%	No effect	No effect	No effect	Conventional, balanced or POSRV
	10%–30% [1]	Unstable			Balanced or POSRV
	30%–50%	*Do not use*	Lift/capacity reduced (manufacturer coefficient) [6]		
	>50% [2]		Generally unstable *Do not use*	Flow becomes subsonic [4]	POSRV only

Notes:

[1]*This limit varies among different valve types.*

[2]*In extreme case, some spring valve models can perform with higher backpressures if a pilot-operated valve is absolutely not acceptable.*

[3]*Then the 'Cold Differential Set Pressure' (set pressure on the test bench) must be reduced by the amount of the backpressure to obtain the correct set pressure on the installation: CDSP = Set – BP.*

[4]*Because of the ΔP, the flow is not choked, but subsonic or subcritical. This obviously has an effect on the sizing of the valve (coefficient). Subsonic can occur at 25% to 30% backpressure: Always check first!*

[5]*The superimposed backpressure varies, so the set pressure of the conventional valve will vary proportionally. This is acceptable if the valve set pressure increased by the maximum backpressure is equal to or below the maximum allowable pressure of the protected installation.*

[6]*There is a coefficient for gas applications and one for liquid applications, which usually varies among valve types.*

Table 9.2 Maximum back pressure percentage on liquid applications

Backpressure Type	**Effects on Valves**				**Selection**
	Value (% of set)	**Conventional**	**Balanced spring valve**	**Pilot-Operated**	
Constant	<20% [1]	Set point increased by backpressure [3]	No effect	No effect	Conventional, balanced or POSRV
	20%–50%		Lift/capacity reduced (coefficient) [6]		
	>50% [2]	Set point increased by backpressure[4]	Generally unstable *Do not use*		Conventional or POSRV
Variable superimposed	<10%	Set point varies with backpressure [5]	No effect	No effect	Balanced or POSRV
	10%–20% [1]	Unstable			
	20%–50%	*Do not use*	Lift/capacity reduced (coefficient) [6]		
	>50% [2]		Generally unstable *Do not use*		POSRV only
Variable built-up	<10%	No effect	No effect	No effect	Conventional, balanced or POSV
	10%–20% [1]	Unstable *Do not use*	Lift/capacity reduced (coefficient) [6]		Balanced or POSRV
	>50% [2]		Generally unstable *Do not use*		POSRV only

Notes:

[1]*This limit varies among valve types.*

[2]*In extreme cases, some spring valve models can perform with higher backpressures if a pilot-operated valve is absolutely not acceptable.*

[3]*Then the 'Cold Differential Test Pressure' (set pressure on the test bench) will have to be reduced by the amount of the backpressure to obtain the correct set pressure on the installation: CDTP = Set – BP.*

[4]*Because of the ΔP, the flow is not choked, but subsonic or subcritical. This has obviously an effect on the sizing of the valve (coefficient). Subsonic can already occur at 25% to 30% backpressure: Always check first!*

[5]*The superimposed backpressure varies, so the set pressure of the conventional valve will vary proportionally. This is acceptable if the valve set pressure increased by the maximum backpressure is equal to or below the maximum allowable pressure of the protected installation.*

[6]*There is a coefficient for gas applications and one for liquid applications, which usually varies among valve types.*

if exceeded, use a bellows valve with high-pressure bellows which can accept the absolute backpressure. This solution is usually still more economical than using a POSRV in the lower sizes or special (piston) balanced spring valves. In the higher sizes, a pilot valve might have to be considered.

Absolute backpressure higher than the maximum pressure acceptable by the outlet flange: It needs to be checked if the supplier uses full rated body configurations or cosmetic configurations. Some suppliers have a #300 cosmetic outlet flange drilling on a #150-rated body.

9.6 ORIFICE SIZE – SIZING

Too large an orifice required for the available connection size: Some manufacturers can offer full-bore POSRVs where a larger orifice can be obtained in a smaller size valve by increasing the curtain area. Also some spring valves can be equipped with customized nozzles.

Multiple valves needed: Here, one has the option of installing multiple spring valves or going to a pilot-operated valve with full bore orifice, in which case fewer valves might be necessary. Pressure/size limits are usually much higher for pilot-operated valves than for spring-loaded valves. In some cases, one valve can replace three or more spring-loaded valves, without any special configurations.

Oversizing >40%: Sometimes, a valve size is imposed because of the connections available on the pressure vessel. When the smallest orifice available is at least 40% larger than what is required, this valve will definitely chatter. In that case, a true modulating pilot-operated valve should be selected. The valve should modulate from 0 up to full lift and then will not chatter, even if oversized. This may not be the case with some types of modulating valves which first 'pop' to about 30% lift before starting to modulate. This should be checked, as oversized valves will chatter and be damaged; they can also be very destructive for the surrounding piping, even causing it to rupture.

9.7 TWO-PHASE FLOW

A two-phase flow can be defined when gas/vapour and liquid capacities must simultaneously be evacuated via the same valve, or if the liquid at an upstream temperature is higher than its saturation temperature under the outlet pressure (some of the liquid will then 'flash').

If not specified, the valve is usually sized using the classic API 'add-on' method. More often nowadays, however, it is specified to size the valves using the Omega-DIERS method (see Section 7.6).

The liquid capacity is <30% total (volumetric capacity): A conventional gas valve can be quoted.

The liquid capacity is >30% total (volumetric capacity): If a spring valve is required, it should be a valve with liquid trim. Preferably, however, a modulating pilot-operated valve can be used. The modulating pilot ensures stability of operation whatever the phase and cannot be 'oversized' as it adapts its flow to the need of the system.

9.8 TYPE OF FLUID

Dirty service with particles in the fluid: A spring-loaded valve, preferably with elastomer soft seat is preferred (preferably no PTFE). An elastomer soft seat greatly limits seat damage. PTFE is hard and easily scratched and not good for particle-laden fluids. If a pilot-operated valve is preferred or necessary for the application, the pilot should be protected from the particles. There are different options and configurations to protect pilot valves nowadays, but the option depends on the 'dirtiness' of the fluid. This should be discussed in detail with the SRV vendor (also see Section 5.3.3.4).

Polymerizing fluid: Spring-loaded valves are most suitable here. However, electrical or steam heat tracing may be recommended to avoid any polymerizing damage inside the nozzle. Spring-operated valves are available that are designed with a steam envelope around the valve. Note, however, that when the volume of this envelope reaches a certain size, it becomes a pressure vessel in itself and requires separate PED/ASME approval. If a pilot-operated valve is preferred or necessary for the application – but only if absolutely necessary – the pilot should be protected from the fluid. Nowadays, there several systems are available to do this, but these should be discussed with the vendor, as designs can differ significantly.

Highly viscous fluid, waxy fluid, hydrates formation: Spring-loaded valves with heat tracing may be necessary for highly viscous fluid. Hydrates can form and block the small size valves (D, E, etc.) so that they too may require heat tracing. If a pilot-operated valve is preferred or necessary for the application, the pilot should be protected from the fluid or its effects (see Section 5.3.3.4).

9.9 RECIPROCATING COMPRESSORS

Many SRVs are installed close to reciprocating compressors. Reciprocating compressors can generate high-frequency pressure peaks and can be very destructive for downstream equipment. Sometimes, the frequency is so high that it is difficult to measure with traditional instrumentation and the valve has no time to react. If the frequency is lower, it can make the valve chatter under pressure surges. An oscillographic recording is shown in Figure 9.1.

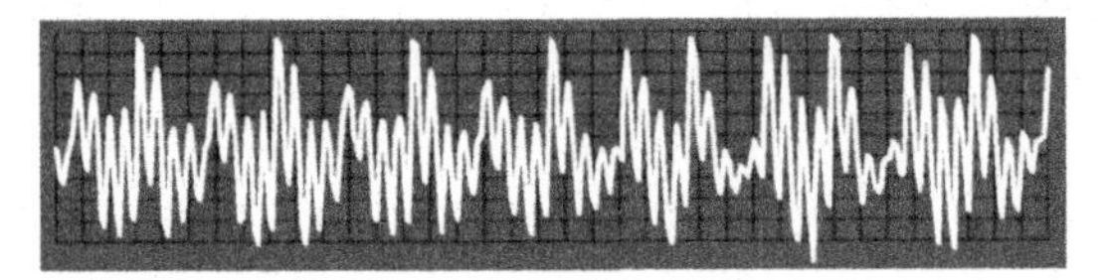

FIGURE 9.1
Pressure spikes generated by a reciprocating compressor

In these cases, it is possible to use a pilot-operated valve fitted with some sort of pressure spike compensator before the fluid enters the pilot. If the set pressure of the valve is lower or very close to the highest compressor pressure spike, the valve may leak or even pop. The pressure surge compensator dampens any spikes of pressure before they enter the pilot, so the valve reacts to the average pressure and remains stable and tight. This compensator can be built into the supply line to the pilot and has an effect on the pilot, as shown in Figure 9.2.

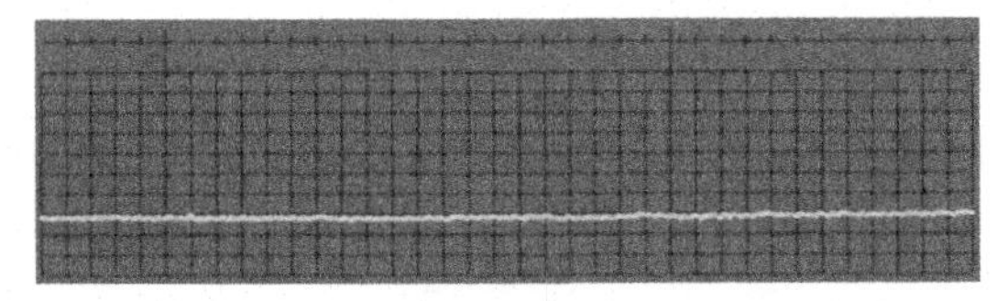

FIGURE 9.2
Pressure at the entry of the pilot valve after having passed a pressure spike compensator

9.10 LIQUID

Normal operation: Spring-operated valves can be used but need to be equipped with a trim suitable for liquid service so they can obtain nominal flow at 10% overpressure. When quick opening (and closing) is required, the operation will be unstable and will possibly cause water hammer. Alternatively, a modulating pilot valve can be used, preferably equipped with a filter. The volume of the filter slows the pilot, which may otherwise react too fast, creating instability and water hammer.

For operating pressures higher than 90% but no more than 92% of set pressure: Use a true modulating spring-operating valve (always ask for the valve's opening characteristics) or use a modulating pilot-operated valve for operating pressures up to approximately 95% of set pressure, preferably equipped with a liquid pulsation dampener. This dampens any spikes of pressure that usually exist in liquid flows and keeps the valve stable and tight.

9.11 MATERIALS

An important factor in selecting the correct SRV is choosing the correct materials for the application. It is obviously impossible to cover all possible applications, but we will provide some general guidelines here.

Sour gas applications: Here, NACE recommendations (see Section 4.4) should be followed. Most wrought grades are acceptable to NACE. Therefore, in general, duplex valves with bar stock nozzles and discs will meet NACE requirements. Some limited cast grades will generally comply also:

- ASTM A890 CE3MN UNS J93404
- ASME SA 351 CD3MWCuN UNS J93380
- ASME SA 351 CE8MN UNS J93380
- Z 6 CNDU 28.08 M, NF A 320-55 at 17 HRC MAX
- ASTM A351 CK3MCUN (6Mo)

High-temperature applications (>530 °C): As a general rule, a bellows should be used so that the spring gets protected from heat and does not shift its characteristics during eventual temperature cycles.

For body, bonnet and cap, up to 815 °C cast materials can still safely be used: ASME SA 351 CF8M with a minimum 0.04% carbon content and maximum 8% ferrite content. In any case, full-rated and integral castings are recommended. Alternatives are:

- ASME SA 351 CF8 SAME (with same chemical restrictions as CF8M)
- ASME SA 217 WC6 (up to maximum 593 °C)
- ASME SA 217 WC9 (up to maximum 593 °C)

Wrought materials can also be used up to 815 °C: ASME SA 479 Grade 316H with a carbon content between 0.04% and 10%. Alternatives are:

- ASME SA 479 Grade 304H (same chemical restrictions as 316H)
- ASME SA 479 Grade 347H (same chemical restrictions as 316H)

For the temperature range 530 °C to 815 °C, the following precautions should also be taken into consideration:

- *Recommended bolting material:* ASME SA 193 Grade B8 Class 2/ASME SA 194 Grade 8 – strain hardened.

- *Recommended bellows material:* INCONEL 625 Grade 2 (LCF) – annealed after forming or HAYNES ALLOY 230. ASTM B435 UNS N06230 is preferred above 650 °C but is rather expensive.

- *Recommended spring material:* INCONEL X750. Above 232 °C, 316SS springs should no longer be used. INCONEL X750 or chrome-plated springs should be considered. Note that cadmium plating used in the past is no longer allowed.
- *Recommended gasket material:* 316 STAINLESS STEEL.

CHAPTER 10

Maintenance and Testing

Some specialized literature regarding the maintenance of safety relief valves (SRVs) is available, but it is best to always consult the manufacturer's detailed installation and maintenance manuals on the subject, as some type of valves might require special attention.

It is highly recommended to always use genuine manufacturer's spare parts. In some parts of the world, this is even required by law when it comes specifically to SRVs. This enables everyone to keep track of the valve, which is installed under legal conditions, and will assure that the guarantee on the valve is not jeopardized. It also keeps its 'passport' up to date, and it can be verified if set pressure or backpressures eventually change.

10.1 MAINTENANCE FREQUENCY AND COST

10.1.1 Introduction

Many users are concerned about the maintenance frequency of their SRVs installed base. Many major companies have their own internal procedures or are following codes and recommendations. However, there are no strict legal requirements that touch on maintenance frequency of SRVs. Once in some boiler installations in the past, it was a rule, or better a habit, for the firemen to pop the valves daily or weekly by means of the lifting lever, an unenviable job as it was extremely dangerous for the personnel and also not very good for the valve itself. It also required a minimum operating pressure of 75% of set pressure, usually of very hot steam, something you do not want to be very close to. This procedure also led to a lot of accidents and so is not used very frequently anymore. Other companies only tested and maintained their valves after an accident or major pressure upset, and all the rest was/is between both these extremes.

While preventive maintenance on this safety component is extremely important, it should also be noted that a valve cannot be endlessly tested, popped, overhauled, and so forth, and that maintenance (and testing) also cause wear on the valve. Therefore, a good compromise must be found.

Regular inspections of SRVs are necessary to ensure safety. However, here we should define 'regular', taking into account that inspections are also costly and potentially destructive. A delicate balance between safety and cost must be obtained. This is a very complex problem involving multiple factors that include the individual valve's application, pressure, temperature, medium, age, size and type. The problem here is that it is difficult to generalize for the complete valve part.

Testing and maintenance is indeed a necessary evil but should not be done more than absolutely necessary.

Here we will present a simulation model for determining the inspection policy for SRVs in a typical petrochemical plant, based on experience. In my opinion, it minimizes the total inspection and repair costs without jeopardizing the safety. The model is simply a result of 20 years of observation in the chemical, petrochemical, power and oil and gas industries.

The maintenance frequency is, in my opinion, dependent on too many variables and combinations thereof (process conditions, environment, location, temperature variations, pollution, etc.) to completely generalize the recommendations. It is my experience that reliability of SRVs can only be determined on a historical basis for each individual installation, application, location and even type of SRV. The variety of applications and types of valve makes it impossible for us to correlate meaningful information with relation to such events as failure rates for all industries; however, we can track the individual valve and its 'health'. I would compare the frequency for inspecting the valves with regular doctor check-ups. For instance, the older you get, the more frequent will be your doctor check-ups; if you work in a dangerous environment, the more frequently you should see a doctor.

10.1.2 Maintenance cost

Although it is impossible to evaluate 'a general average' for maintenance costs per type of valve, what we do know out of experience is that the life cycle cost (LCC) of some valves is much higher than others. This not only relates to the frequency of inspection or maintenance but also to the total cost thereof (test, repair, spare parts and handling of the valve).

Although lacking mathematical proof, I have nonetheless experienced that a soft-seated valve needs about 2.5 times less maintenance than a metal-seated valve. This is logical considering the higher possible operating frequency of a spring-operated soft-seated valve or a soft-seated pilot-operated valve vs a metal seated valve. The maintenance frequency decreases even more when a modulating pilot valve is used because of the lesser forces that are created inside the valve during opening. Estimating the full maintenance cost and labour, however, is very difficult as it depends on size of valve, type, location, where it is installed, process, and so on.

Based only on the cost of spare parts, we can easily make a comparison:

Let's consider a 4P6 valve ('P' orifice, 4 in. inlet size and 6 in. outlet size):

A balanced bellows metal-to-metal valve spare part kit:

Gasket set: €35,-
Standard bellows: €1.203,-
Disc: € 239,-
Total: € 1.477,-

A soft-seated pilot-operated valve spare part kit:

Pilot soft good kit: € 45,-
Main valve soft good kit: € 110,-
Total: € 155,-
Price of a new standard balanced bellows valve 4P6: € 3.790,-
Price of a new pilot-operated 4P6 valve: € 3.950,-

Conclusion: The difference in purchasing price between a pilot valve and a balanced bellows metal-to-metal valve is already compensated after the first maintenance cycle of these valves. This is an ROI of less than 1 year if the attached recommended maintenance frequency is followed.

Another extremely important cost factor is the weight of the valve and especially where it is installed when it has to be removed for testing and maintenance. SRVs are usually mounted high up. Therefore, it is important to take weight and size into consideration when selecting a valve in order to keep down later maintenance costs. A pilot-operated safety valve is much lighter than a spring-operated valve from sizes 3 to 4 in. and up and can more easily be handled.

Example of a specific bulkiness comparison:

Rating and Size	Spring Valve (cm)	Pilot Valve (cm)	Height Saving (%)
2″ × 3″ – 600#	58	48	17
3″ × 4″ – 600#	86	51	41
4″ × 6″ – 300#	94	58	38
6″ × 8″ – 300#	105	66	37
8″ × 10″ – 150#	140	74	47

Example of a specific weight comparison:

Rating and size	Spring valve (kg)	Pilot valve (kg)	Weight saving (%)
2″ × 3″ – 600#	31	24	23
3″ × 4″ – 600#	70	42	40
4″ × 6″ – 300#	100	73	27
6″ × 8″ – 300#	210	120	43
8″ × 10″ – 150#	340	192	43

10.1.3 Maintenance frequency

The best way of effectively determining a maintenance schedule is by keeping a detailed log on the history of each valve. Inspection frequency should be based on criteria which is explained hereafter. We will demonstrate how to build a schedule of inspection activities based on historical data for each individually installed valve. This procedure starts with installation: An SRV is tagged and is given a 'passport' containing all of its data, process data and revision dates with comments and the spare parts used and when. Actually, the basis of this passport already exists when it leaves the manufacturer (on the tag plate) and the basic records of the valve are also kept with the manufacturer for later reference.

10.1.3.1 Rationale

SRVs are generally inspected at the same time as the elements to which they are fitted. Each item of surface safety equipment should be allocated an inspection grading 1, 2, 3 or 4 which indicates the maximum intervals that may elapse between two inspections.

Each valve should:

i. Initially be given an inspection grade '1' and be given its first inspection after a short service period, typically maximum 1 year. This first inspection tells a lot about the condition of the valve and the application it is used on.

ii. Subsequently, based on built-up knowledge of service conditions and surface safety system parameters and conditions following the first thorough inspection, the inspection grading should be reviewed, allocating either grade 1 or grade 2.

iii. Subsequently, based on extended knowledge gained of service conditions and surface safety system parameters and conditions following previous extensive inspections, be graded to inspection grade 1, grade 2 or grade 3.

iv. Subsequently, based on the further extended knowledge of service conditions and surface system parameters gained from the previous extensive inspections, be graded to inspection grade 1, grade 2, grade 3 or grade 4.

v. Have an inspection review of the kept records of the valve carried out during each inspection in order to determine the inspection grade to be allocated during that inspection period.

The purpose of this review is also to identify any changes in the system or service conditions that may affect the inspection grading of the item and to build up an inspection history of corresponding surface safety systems. It is recommended to return to grade 1 if the service and process conditions have changed significantly. The recommendation for process change before changing grade would be approximately ±5% to 7% from original process conditions.

10.1.3.2 Factors affecting selection of an inspection grading

The following factors can affect the selection of an inspection grading:

- Design constraints
- Operating constraints and conditions
- Legislative constraints
- Certifying authority requirements
- Modes of possible failure and consequences
- History of a particular item
- History of similar items in similar service
- Current inspection grade
- Period elapsed since previous inspection.

The above factors should be taken into account by the inspection engineer in the process of assessing which inspection grade each item should be awarded.

10.1.3.3 Inspection grade awards guidelines

Inspection grade 1: All surface safety systems should be awarded inspection grade 1 until a system history can be built up by a series of inspections or, at least until the first major inspection has been effected.

Inspection grade 2: This inspection grade may be awarded where the following conditions are met:

i. The valve was under grade 1 and opens within a tolerance band of ±5% of the cold differential set pressure (CDTP). The leakage rate is acceptable according to API 527 or to the company specifications. The internal conditions of the dismantled valve show no or minor defects.

ii. The valve was under grade 3 and fails to open or opens outside a tolerance band of ±5% of the CDTP. The leakage is outside the tolerance of API 527, the company specifications or the internal inspection of the valve shows defects which require further investigation, replacement or repair.

iii. The valve was under grade 4 and fails to open, the leakage rate is excessive and the internal inspection of the valve reveals serious defects such as galling and possible seizure which require further investigation, replacement or repair.

Inspection grade 3: This inspection grade may be awarded where the following conditions are met:

i. The valve was under grade 2 and opens within a tolerance band of ±5% of the CDTP. The leakage rate is acceptable according to API 527 or to the company specifications. The internal inspection shows no or minor defects.

ii. The valve was under grade 4 and fails to open or opens outside a tolerance band of ±5% of the CDTP. The leakage rate is outside the tolerance of API 527, the company specifications. The internal inspection reveals defects, which require further investigation, replacement or repair.

Inspection grade 4: This inspection grade should only be awarded where the following condition is met:

i. The valve was under grade 3 and opens within a tolerance band of ±5% of the CDTP. The leakage is acceptable according to API 827 or to the company specifications and the internal inspection of the valve shows no or minor defects.

Grading transfers should only be considered after extensive inspections. The first extensive inspection can form the basis on which a valve may be transferred to grade 2 only if the parameters for this grading are met.

Subsequent extensive inspections of the valve can form the basis on which a valve may progress through the inspection grading system, taking no more than one upwards step per inspection, only if the parameters for the grading step can be met.

In no case should the interval between inspections of safety devices exceed the interval between inspections of the pressure vessels involved.

The converse of this also applies, with the downgrading of valves if inspection results indicate the current grading parameters are not being met.

Sample inspections on the surface safety systems are not recommended. Each system must be inspected at the specified interval.

10.1.3.4 Inspection requirements and reporting

It is recommended that every complete replacement valve, either withdrawn from the stores or returned from a valve specialist, be inspected. New valves coming from an ASME or PED-approved SRV supplier should already be tagged and leaded and will most probably meet all code requirements without requiring further inspection. The only question is how transport could have affected the setting and operation of the valve. Careful verification of how the valve was packed and shipped is important. Therefore many users test all valves, regardless, before putting them on the system.

Actually it is good practice that any valve received from an off-site location should be bench tested to verify set pressure and leakage rate. However, please be aware that if the tag on a new valve is broken, the supplier forfeits on the warranty of the valve unless it is done by or in the presence of a notified body who is able to tag or lead the valve again.

The set pressure may then be adjusted accordingly and this inspection may be used for inspection grading purposes as described above.

The valves don't need to be stripped for this commissioning inspection unless problems are encountered requiring further overhaul.

Results of this inspection are also to be recorded and added to the 'Valve History Record' or passport.

From this point, a credible valve history must be established for each unit. This will create a reliable overall safety system for pinpointing trouble spots on surface safety systems and will facilitate future material selection or system modifications. Any system set up to record valve history and reporting should be simple to operate and clear to any user.

Each SRV in service will have its own unique history of problems, lifts and repairs. It is important that records of each stage of the valve's history are documented as these will form the valve history.

This valve history or passport will at a minimum contain:

- **i.** Original valve specification data
- **ii.** Valve commissioning report
- **iii.** All subsequent rectification/overhaul/recalibration reports
- **iv.** Any material changes/spring changes or changes of specification
- **v.** Eventual process changes exceeding 5% to 7% from the original

From this information, control sequences and planned maintenance routines may be designed and altered to suit the particular process area with which that specific valve is involved.

A database such as this is also important where valve interchangeability is required in order to determine a minimum stockholding for maintenance purposes. Valves of similar body and trim materials can possibly be utilized in many different locations and services, thus removing the need for one-to-one valve stocking. This system will also mean that valves can easily be sourced from non-essential systems in order to maintain essential system viability.

10.1.3.5 Inspection intervals and survey requirements

The following table gives a suggested interval, by grade, for SRVs with respect to the type of equipment on which they are fitted.

Recommended inspection intervals for Safety Valves by equipment type:

Equipment	Maximum interval (months)			
	Grade 1	Grade 2	Grade 3	Grade 4
Unfired pressure vessels	12	24	36	48
Fired pressure vessels	14	26	26	26
Steam receivers – offshore	12	24	36	48
Steam receivers – onshore	12	26	26	26
Heat exchangers	12	24	36	48
Air receivers	12	26	26	26
Storage tanks	12	24	36	48
Pig traps	12	20	20	20
Pressure piping systems	12	24	36	48

10.2 TRANSPORTATION AND DIRT

In new installations, many times the valve supplier is requested to grant long warranties because the time between order placement and installation can be very long. In such cases, although the valve may have never operated, it is not unusual that once the valve is put in service, it does not work correctly.

In most cases, after checking the test reports, it is ascertained that these valves did work correctly upon leaving the valve manufacturing plant, a fact sometimes witnessed by independent inspectors.

In many instances, the reason is very simple: transport and dirt! It is extremely important that both the transport and storage of these critical items are given due attention.

It has happened more than once that when called on site to check so-called 'defective' brand new valves, the valves to be installed were waiting under the sand or at least in very dirty circumstances, with flange protectors removed. It is very important that due precaution is taken to keep the valves clean inside at all times and that the inlet and outlet flanges are protected at all times. Dirt, whether in a valve or in associated piping, can be very damaging and can cause an SRV to become inoperative even before it has to perform its duty.

Also, make sure that the manufacturer is taking enough care when packing the valves for transportation. They should preferably be fixed in a wooden box and should always be transported in a vertical position. They should also be protected against violent shocks.

Before leaving the valve manufacturing plant, every valve supplied by an approved vendor is tested and suitably packed according to end user specifications. As long as all recommendations are followed on transport, (clean) storage and installation practices, the valves should work correctly. Unfortunately these important factors are often overlooked when selecting or buying new valves.

Just as a reference, we summarized the highlights of the excellent recommendations made in API RP 520 Part II on the installation of pressure-relieving systems in refineries in that respect:

10.2.1 Preinstallation handling and testing of pressure relief valves

Because cleanliness is essential to the satisfactory operation and tightness of a pressure relief valve, all necessary precautions should be taken to keep out all foreign materials.

- Valves which are not installed immediately after receipt from the manufacturer should be closed off properly at both inlet and outlet flanges; particular care should be taken to keep the valve inlet absolutely clean. Flange protectors, covering the whole inlet and outlet flange, should only be removed just before installation.
- Valves should be stored preferably indoor away from the ground and in a location where dirt or other forms of contamination are at a minimum.
- Do not permit valves, whether or not closed off, to be thrown on a pile or promiscuously placed on the bare ground awaiting installation.
- Valves should be handled carefully and not subjected to shocks.

If due consideration is not given to this point, considerable internal damage or misalignment can result and seat tightness might be adversely affected.

As far as valve installation is concerned:

- Thorough visual inspection before installation is imperative.
- Manufacturer's operation, maintenance and start-up manual needs to be adhered to for the specific valve type as all SRV types are not the same and may need different installation procedures.
- Remove all protective covers only just before installation.
- Clean carefully, especially the full inlet side of the valve.
- Make sure the flange surfaces are cleaned and do not show any damage or scratches.
- Use the correct flange gaskets as recommended by the manufacturer.
- Make sure gaskets do not, even partly, obstruct the inlet or outlet passage as when opening the valve, (gasket) pieces may be torn off and cause leakage. Obstructions in the outlet passage may cause excessive backpressure.

Inspection and cleaning of systems before installation:

- Because foreign materials passing into and through a safety relief valve are damaging, the systems on which the valve is tested and finally installed must be inspected and cleaned before installation of the valve. New systems especially are prone to contain welding beads or even rods, pipe scale and other foreign objects which will definitely destroy the seats during the first opening of the valve.
- Wherever possible, the system should be purged before installing the SRV.
- It is also recommended that the SRV is isolated during the pressure testing (hydraulic test) of the system, either by blanking or closing a full bore stop valve upstream. If gagging is used, extreme caution must be exercised to avoid damaging the valve by overtightening the gag and especially to ensure that the gag is removed after the test.

10.3 TROUBLESHOOTING SRVs

In the next chapter, we will try to enable you to do some first-aid troubleshooting of SRVs. The list is by far not exhaustive but lists the most common problems encountered over the last 30 years of experience.

Some of the following recommendations relate to both spring valves and pilot valves, while others only apply to pilot valves.

Information that you should always have at hand when you start working on an SRV or call the manufacturer for assistance:

- Obtain nameplate data (both on pilot and main valve in case of pilot-operated valves).
- Type no.
- Serial no.
- Set pressure.
- What is the service media?
- What are the service conditions (pressure and temperature)?
- Has the valve cycled?
- How is the valve installed?
- Detail on inlet and outlet piping (if any).

FIGURE 10.1
Take the pressure off the valve when doing repairs

First of all it is important to know that you should never work on a valve or make any adjustments when there is pressure under the valve disc. The forces, noise, and so forth, that could occur if a valve opened (especially on gas and steam) are life threatening. Remember to always purge the system before you start working on an SRV. Only if absolutely necessary use a test gag (Figure 10.1).

10.3.1 Seat leakage

Check	Cure
Operating pressure too close to set pressure	**1.** Increase set pressure if allowed by code or system design pressure. **2.** Use pilot-operated valves. **3.** Retrofit valve to soft-seated or high performance design.
Corrosion and erosion on trim (nozzle and disc) (Figure 10.2) **FIGURE 10.2** *Torsion on nozzle and disc*	**1.** Check material compatibility with the process. **2.** Use harder and different material seat/disc combination. **3.** Remove disc, lap and retrofit (pyrex glass recommended for quick re-lapping). **4.** Retrofit to compatible soft-seat design.

(Continued)

Check	Cure
Popping (set pressure) tolerance	**1.** Check nameplate and adjust setting in function of (new?) process temperature according to manufacturer's setting instructions on CDTP. **2.** Use a high performance relief valve if necessary; tolerances cannot be met with traditional valves.
Particles (or traces thereof) between seat and disc – especially frequent in pump applications	**1.** Use a harder and different seat/disc combination. **2.** Make sure a knife = edged seat arrangement is used. **3.** Use easy to retrofit and compatible soft-seat design – preferably O-ring design. **4.** Use bellow valves in case large particles are found or use a rupture disc at both valve inlet and outlet.
Galling of guide ID (internal diameter) or disc holder stem OD (outside diameter). They can be evidence of hang-up which results in 'sticking open' (and long blowdown)	**1.** Replace the internals and lap the disc. **2.** Use rupture disc under and after the valve (Figure 10.3). 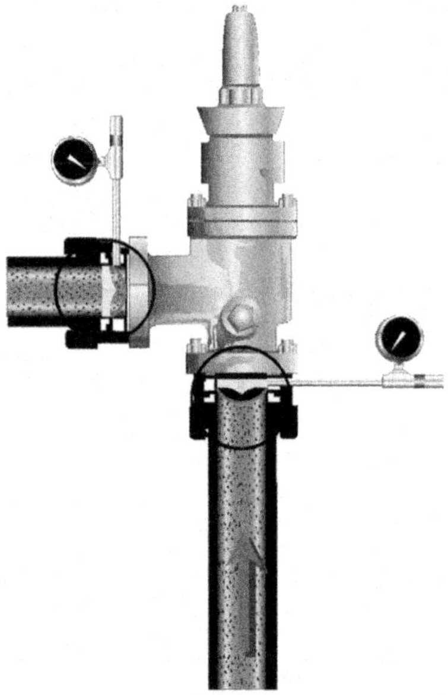 **FIGURE 10.3** *Rupture disc protecting valve inlet and outlet*
Look at the supports on the outlet piping (if any). Wrong alignment of outlet piping or thermal stresses can cause misalignment inside the valve, causing it to leak	**1.** Support outlet piping correctly and align. **2.** Use dual-outlet valves. **3.** Use expansion joints in outlet piping.
Vibrations on protected equipment	**1.** Eliminate system vibration. **2.** Increase differential between operating pressure and set pressure (set the valve higher if allowed by the design pressure). **3.** Use soft-seated valves.

(Continued)

Check	Cure
Check if valve is installed horizontally (Figure 10.4)  **FIGURE 10.4** *Valve mounted horizontally*	Clean valve and align vertically.
Alignment of the valve assembly	Jammed valves must be returned to the shop for reassembling and retesting.
Lift lever position or blockage	Put lift lever in correct position and make sure it is not in tension.
Nature of process media as gases with low molecular weight will leak faster or toxic media leaks will be detected earlier	Use soft-seated valves.
Disc on erosion/corrosion (Figure 10.5) 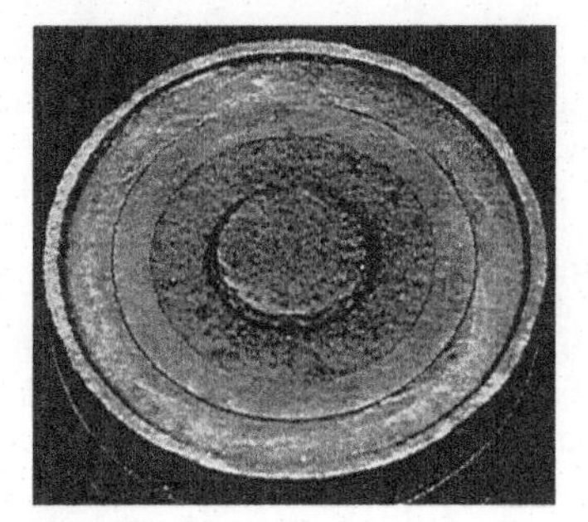 **FIGURE 10.5** *Disc attacked by corrosion (pitting)*	Especially on sulphur recovery units and heavy crude oil service, there could be excessive seat leakage of 316 SST trims caused by sulphide corrosion/pitting. In this case retrofit the valve with, for instance, an Inconel 718® nozzle and disc insert (with high nickel content). In any case, make sure that a proper material selection for the service is used.

10.3.2 Chatter

Chatter is also known as SRV instability and often results in numerous other SRV and/or system problems (Figure 10.6):

- Seat damage
- Galling/hang-up/SRV fails to reseat
- Bellows failures
- Damaged piping and even pipe rupture

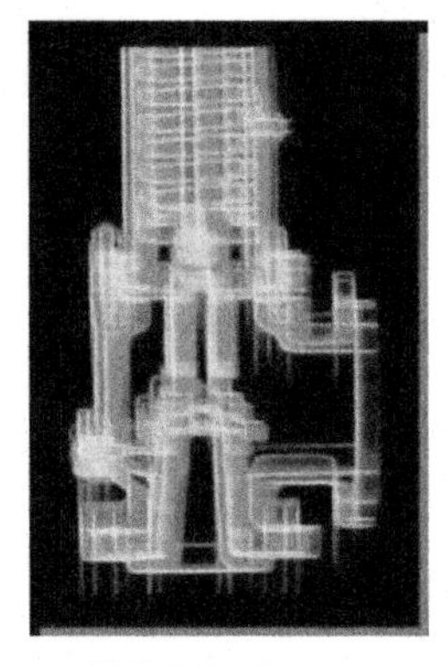

FIGURE 10.6
Chatter

Chatter always has very destructive effects on both the valve and the system and needs to be avoided at all times:

- *Damaged internal parts:* SRV violently travels/slams from fully closed to fully open many times a second.
- Valve has less than half of its rated capacity as it is closed half of the time.
- *Damaged companion piping:* SRV stress and inertia reversals caused by violent opening/closing cycles.
- *Pressure surges in liquid service:* Violent, multiple closing cycles cause liquid water hammer and place undue stress on companion piping, piping supports, internal components in pressure vessels, metering systems, and so on.

Check	Cure
Oversizing of the valve (as a rule of thumb chatter is starting at 25% above necessary/calculated capacity)	**1.** Use smaller size valves. **2.** Reduce the lift of the valve (only possible with pilot-operated safety relief valve, POSRV).
The condition of the disc	A damaged disc can give a chatter effect; in such cases, lap the disc or replace it.
Check if the spring used has the correct spring range for the set pressure. Set pressure needs to be preferably within 10% of the middle of the spring range	Change spring.
Outlet piping smaller than outlet size of the valve. This will cause violent turbulences when valve opens and the valve will start to chatter	Make sure outlet piping is at minimum the same size as the valve outlet size.
Pressure drop in inlet piping (3% is recommended with chatter starting around 10% depending on set pressure)	**1.** Rearrange inlet piping (Figure 10.7). **2.** Change valve location. **3.** Use remote sense (only possible with POSRV). **4.** Increase the blowdown above the pressure drop (only with special valves or POSRV). **5.** Install the valve as close as possible to the pressure vessel. **FIGURE 10.7** *Incorrect versus correct inlet piping*

(Continued)

Check	Cure
Constant pressure variations in inlet or/and outlet piping	**1.** Rearrange piping. **2.** Use balanced bellow valves if pressure variations occur in outlet piping. **3.** Use POSRV. **4.** Use multiple valves with different sizes in stagger set.
Process conditions for: **1.** Working pressure too close to set pressure **2.** Unavoidable pressure surges in the system **3.** Ventilators **4.** Compressors **5.** High output systems **6.** Etc.	Use POSRV.
Opening and closing tolerances too short	**1.** Change valve trim and retest. **2.** Use high performance valve.
Isolation valve in inlet	Isolation valve needs to be full bore and full open.

10.3.3 Premature opening

Check	Cure
Set pressure versus temperature in process	Use correct temperature correction factor as per manufacturer's recommendations while testing/setting the valve on the test bench.
Check if valve has been reassembled correctly after a maintenance cycle	**1.** Reassemble and retest. **2.** If the valve has already been reassembled several times, (typically 3–4 times), change the trim.
Vibration due to sharp edged inlet piping (Figure 10.8) Standing Vortices **FIGURE 10.8** *Sharp edged inlet piping*	Change inlet piping (Figure 10.9). Concentric reducer **FIGURE 10.9** *Concentric reducer eliminates extensive vibrations during opening*

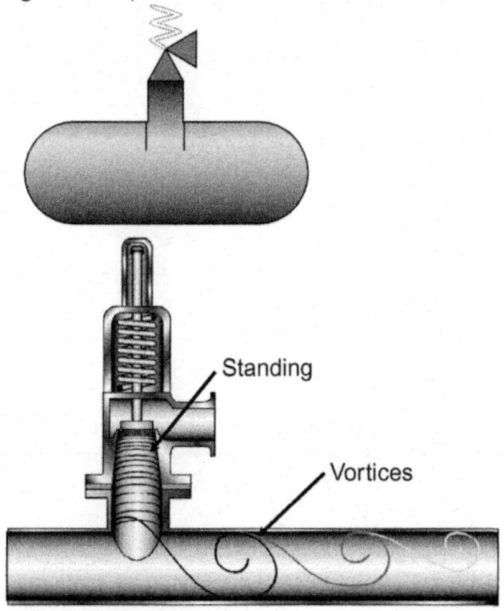

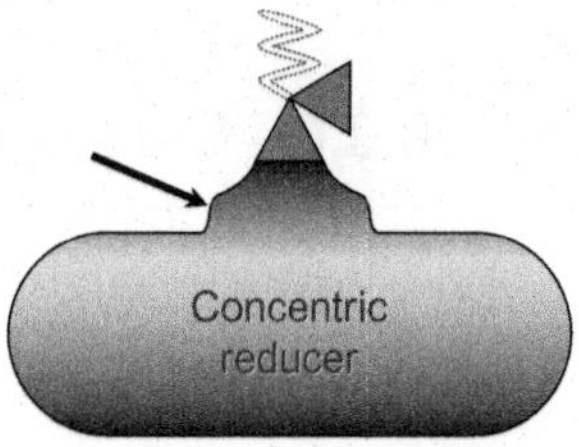

(Continued)

Check	Cure
Test report: valve could be tested on liquid and used in gas application	Test on correct medium.
No vacuums are created in the outlet piping (if any)	Use balanced bellows valves or POSRV.

10.3.4 Valve will not open

Check	Cure
Test gag	Remove test gag. Sometimes test gags are incorrectly used to protect valve during transport or during hydraulic testing and often their removal is forgotten.
Lift lever assembly	Remove lift lever assembly to test the opening of the valve.
Inlet and outlet piping alignment	Remove possible stresses on piping system.
Damaged internal parts (usually bent stem due to frequent full pop)	Change trim.

10.3.5 Valve opens above set pressure

Check	Cure
Test report	It could be that the valve was tested on gas and is used on liquid service so test the valve on liquid.
Backpressure	A conventional valve might see too much backpressure, in that case use a balanced bellows valve or POSRV.
In case of pilot valves, check for any obstructions in the supply piping to the pilot or clogged filters	Remove obstructions and clean filters.

10.3.6 Valve does not reclose

Check	Cure
Impurities between nozzle and disc	Try first hitting the valve with a hammer or disassemble and remove impurities.
In case of pilot valves check for any obstructions in the supply piping to the pilot or clogged filters	■ Remove obstructions and clean filters. ■ Change to a spring-operated valve.
In case of pilot valve, check the piston seal and other soft goods within both pilot and main valve	Change all soft goods.

10.3.7 Bellows failure

Bellows are the most fragile component in a spring-operated SRV and also the most expensive to replace. Usually, we can see one of the three different failure modes: mechanical, fatigue or corrosion.

Failure of bellows can be detected by medium leaking via the bonnet vent. As this is not always evident and detection systems not always very reliable, people have become very inventive in trying to detect bellows failure by putting whistles on the bonnet vent in order to detect leakages from the bonnet vent. The bottom line, however, is that bellows are a very vulnerable but, for its correct operation, very critical part of a spring-operated SRV. Bellow balanced valves need more frequent maintenance or at least checking in order to assure proper operation. The system might have an SRV installed but with the bellows invisibly ruptured, the SRV has no purpose whatsoever.

Bellows are also an expensive part of the valve maintenance cycle and also an expensive spare part as such. Alternatively a pilot valve can be considered. This might be more expensive in the initial purchase but has usually a lower LCC (life cycle cost).

When selecting a bellows valve, it is important to pay some special attention that the material selection is in accordance with the process conditions. Some SRV manufacturers use as standard bellow material INCONEL alloy 625LCF-UNS N06625 (ASME SB0443). This material is not perfect either but, compared to simple stainless steel, has an enhanced resistance to mechanical fatigue and sour gases; it is commonly used in refinery FCC systems for expansion joints.

10.3.7.1 Mechanical failure

Excessive pressures in the discharge system beyond the bellows limit will immediately deform the capsule. Bellows have different pressure limits depending on the manufacturer. They must be checked carefully when selecting a manufacturer.

Figure 10.10 shows the typical cause of mechanical failure due to excessive pressure in the outlet.

FIGURE 10.10
Mechanical bellow failure

If the pressures in the outlet are expected to be beyond bellows limits, some conditions may require specially designed 'heavy'-ply bellows or use pilot-operated SRVs.

In rare cases, mechanical failure to bellows can occur due to impurities which could be present in the outlet piping when the valve is closed.

10.3.7.2 Fatigue failure

This is usually caused by violent cycling of the SRV many times a second. In many instances, this causes the bellows flange to be sheared from the bellows capsule. This way the bellows can even become detached from the disc holder.

Even though the valve may be stable during a relief cycle, the bellows itself can still flutter during a relief cycle due to the high turbulence in the body bowl or vibration due to compressors or pumps in the system. This rapid flexing of the bellows convolutions will result in cold working of the bellows material and welds (Figure 10.11).

FIGURE 10.11
Bellow failure due to fatigue

FIGURE 10.12
Bellow failure due to a combination of H_2S attack and fatigue

The next bellows failure is also caused by fatigue due to chatter, but this is also in combination with H_2S/steam service and a 460°C operating temperature (Figure 10.12).

The next is also a bellows failure due to fatigue. Indications of galling on the disc holder OD and/or guide ID are evidence that the valve has cycled frequently or has been chattering (Figure 10.13).

10.3.7.3 Corrosion failure

This is due to chemical incompatibility of the bellows material with the process fluid. The resulting compromised mechanical strength of the bellows can produce small holes or shears or even the complete collapse of the bellows due to the external pressure acting on it.

While some corrosion of thick section parts of the valve may be acceptable, corrosion on the thin section bellows is absolutely not acceptable. Therefore, a bellows made of more costly corrosion-resistant material than the rest of the valve is often required (Figure 10.14).

FIGURE 10.14
Bellows failure due to corrosion

FIGURE 10.13
Bellows failure due to fatigue caused by chatter

Typical causes are high sulphur or chloride content (chemical attack). Independent lab analysis is usually required to confirm the exact causes in order to remedy the problem in the future, which can also be very costly.

The X-ray shown in Figure 10.15 indicates surface fractures of bellows caused by chemical attack: sulphide stress corrosion cracking (SSC).

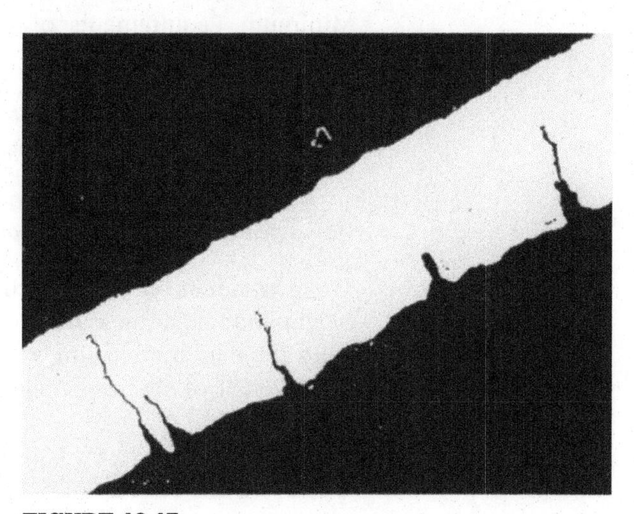

FIGURE 10.15
X-ray of a bellows subject to chemical attack

10.3.8 Springs

Springs are another critical component in SRVs as they determine the correct opening. Therefore a correct spring selection, usually done by the valve manufacturer when establishing his bill of material (BOM) in the valve is very important. Always check if the spring selection is correct. To start, their adjustment range should be at least ±10%.

When the setting is lower, we could see a smaller lift and hence a smaller flow. When the setting is higher, we will see a more 'snappy' opening and a slow blowdown.

Although the quality of the springs over the years has improved tremendously, they are subject to corrosion and it is important that their material selection is done properly for the process or that they are coated or treated correctly. Do not confuse, for example, aluminium paint with the higher specification (for NACE–level 2) six layers of cold-sprayed aluminium (called aluminized springs).

Cadmium coating was for a while a very good solution for many applications but due to environmental issues can no longer be used. These days other standard satisfactory spring coatings are replacing cadmium depending on the application:

- Nickel plating
- Deconyl coating
- Zinc phosphate
- Nylon
- Xylan

ASME Section VIII Division 1, 1992 Edition Pressure Relief Devices. Ug-136 Minimum Requirements of Pressure Relief Valves specifies the following regarding springs:

- *The design shall incorporate guiding arrangements necessary to ensure consistent operation and tightness.*
- *The spring shall be designed so that the full lift spring compression shall be no greater than 80% of the nominal solid deflection (NSD).*

The permanent set of the spring (defined as the difference between the free height and height measured 10 minutes after the spring has been compressed solid three additional times after presetting at room temperature) shall not exceed 0.5% of the free height.

10.4 TESTING

While frequent testing of the valves is necessary as seen before, everyone wants to reduce the downtime of the process. When a plant shutdown is planned, the valves can be dismantled and brought to the maintenance shop for testing on the test bench and eventual overhaul.

Testing procedures that are applied are usually according to the API 527 method as described in Section 4.2.

However, a plant does not always want to shut down when their SRVs need, for instance, intermediate testing because of a presumed failure. In that case, there are two options. The first solution is to install the valves on a change-over valve (Figure 10.16). In this system, one has a full redundancy of the SRV application, but this can become rather expensive. A calculation must be made weighing the initial investment against the eventual losses of a process shutdown. The investment might be justified if, otherwise, critical parts of the plant have to be shut down during each overhaul of the SRVs.

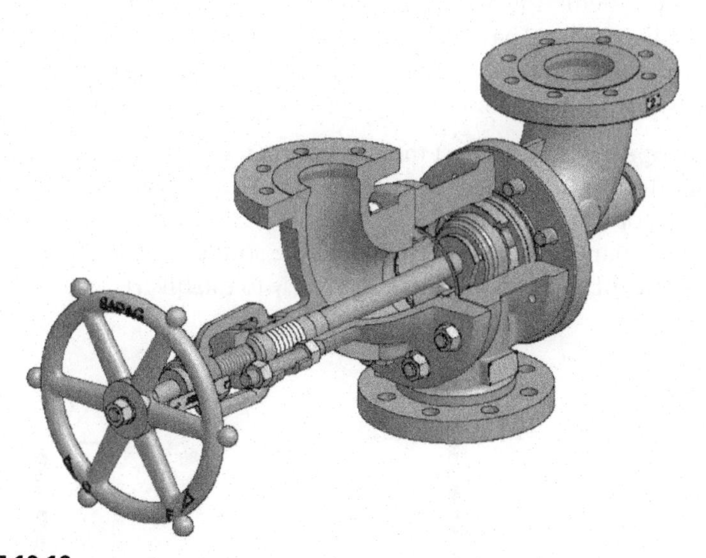

FIGURE 10.16
Traditional changeover valve

Besides their cost, the traditional changeover valves available on the market have a potential major disadvantage: pressure drop. This pressure drop, which is usually between 15% and 20%, needs to be taken into account when sizing and selecting the valve, which can make the whole system even more expensive.

Next to the traditional changeover valves, some suppliers offer selector valves which comply with the API requirements and only have a maximum pressure drop of 3%. This low-pressure drop is obtained by means of their special

FIGURE 10.17
Special shaped changeover valve with only 3% pressure drop

shape, as can be seen in Figure 10.17. These are, however, also more expensive in the initial investment.

When both of the above solutions are too much of an investment, we will have to go to *in situ* testing of the valves, which is also for spring-operated valves sometimes a cumbersome and expensive solution. Different service organizations offer the *in situ* testing on spring valves. Most are based on the same principle where set pressure is actually calculated based upon an external force applied on the spring (hydraulic or pneumatic).

10.4.1 *In situ* testing of spring-operated SRVs

While some systems for performing *in situ* testing of spring-operated SRVs differ slightly, most are based on the same principle and we will explain the principle of one of them (Figure 10.18). For all of them it is always required to know exactly:

- Inlet operating pressure at the time of the test.
- The effective (measured – not per API) seat or pressure area (not always easy to obtain).
- The auxiliary load applied to oppose the spring force at the time of effective lift of the valve (not always easy to establish).

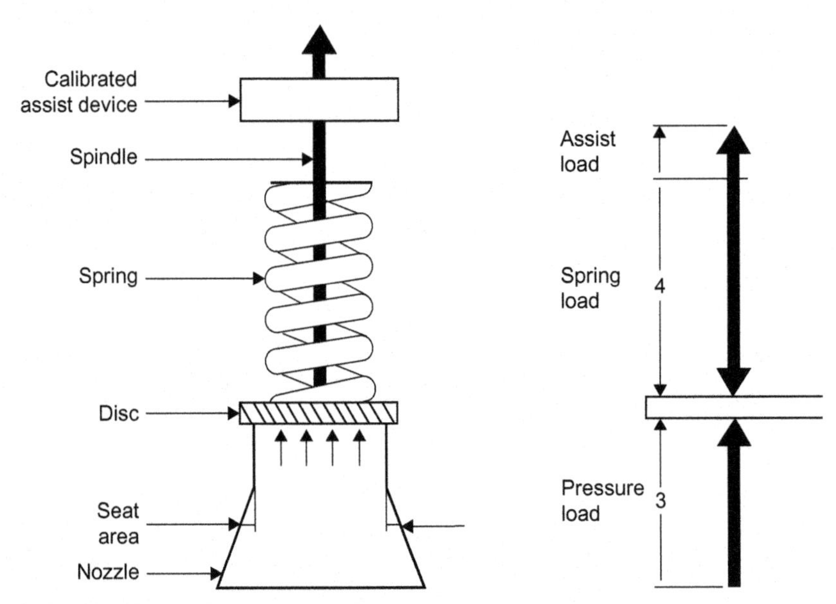

FIGURE 10.18
In situ *testing principle of spring-operated valves*

Only when all above are exactly known, can an estimated set pressure be calculated.

The general principle is as follows:

Forces at set pressure: Spring load = Pressure load + Auxiliary load

Test benches generally provide a relative accurate test pressure but of course are tested on CDTP and not on actual service conditions. On the other hand, *in situ* testing is based on calculations with several variables that need to be all measured correctly at the exact same time.

The functioning of a spring valve is based on a pre-stressed spring. When the pre-stress force is reached, the disc starts to move and the valve opens. With lifting equipment attached to the valve spindle, the valve is opened and the pre-stress force measured, taking into account the above factors (seat area and operating pressure).

The auxiliary force is usually controlled by software which is different from supplier to supplier. The spindle is then lifted typically 1 mm. A force sensor traces the force necessary to lift the valve spindle. Isochronically with the lift force measurement, the pressure of the system is traced.

Set pressure is then calculated via a simple formula:

$$P_{set} = P_o \frac{F_s}{A_s}$$

where

P_o = Operating pressure
F_s = Force on spindle
A_s = Effective seat area.

For each test parameter, usually a graph is created so that one can identify set pressure point and blowdown. Note, however, that not all *in situ* tests available on the market are fully computer-controlled and some are also based on the operator's experience and ability to know exactly when the upwards auxiliary force is equal to the spring force.

The best proven accuracy of such a system is between 3% and 5%. An investigation on the subject was carried out by BP Amoco Exploration by E. Smith and J. McAleese from the City University of London. These findings can be found on the 'Valve World' website: http://www.valve-world.net/srv/ShowPage.aspx?pageID=639.

While this *in situ* testing is an excellent alternative to shutting down the process, one needs to take some considerations into account:

- The accuracy and understanding of the effective seat area and also any manufacturing tolerances.
- The accuracy and tolerance of the lift assisting device and its calibration.
- There is always a possibility of fluid deposits on the SRV seat which can distort the measurements.
- If the operating pressure (which assists the lift) is less than 75%, the assist force to be applied can be too high and can deform the valve spindle if the right precautions are not taken. This happens frequently when SRVs are *in situ* tested before start-up when system pressure is still zero.
- If the valve has been overhauled and parts changed, this needs to be recorded and taken into consideration (change of seat, spring, rework of disc or nozzle, etc.).
- The test cannot check leakage.

The danger of the above precautions is that they do not necessarily give conservative results and that the operator may falsely conclude that the SRV set pressure is too low and adjust the valve beyond its acceptable tolerance; this is usually too high.

As a conclusion, it is obvious that this method is not as good as regular maintenance cycle testing, preferably according to the manufacturer's or service specialist's recommendations. It can quickly result in erroneous determination of the set pressure since it needs many factors to be correct at the same time which are not always fully controllable on site (exact operating pressure at the moment of test, test device limitations, manufacturer's limitations and the accuracy of the history of the valve). Unfortunately, there are few alternatives and therefore it is recommended that only experienced personnel carry out these tests.

10.4.2 *In situ* testing of pilot-operated SVs

In the case of pilot valves, the *in situ* testing is somewhat easier. It is in fact only the pilot that needs testing as it is the pilot which controls the operation of the valve. The main valve only serves to create the necessary capacity. This means that with little volume (average air volume necessary per test is around 0.4–0.6 l), both set pressure and blowdown can be tested and adjusted on site while the valve is in operation.

The only tools necessary are a check valve assembly mounted on the pilot, which are usually standard options with the manufacturer, a test gauge and a pressure bottle. Pressure from this bottle is applied to the pilot until set pressure is reached. This will cause the dome to vent and the valve to open. In order to lift the piston, a minimum of operating pressure is necessary.

Set up as per the below schematic:

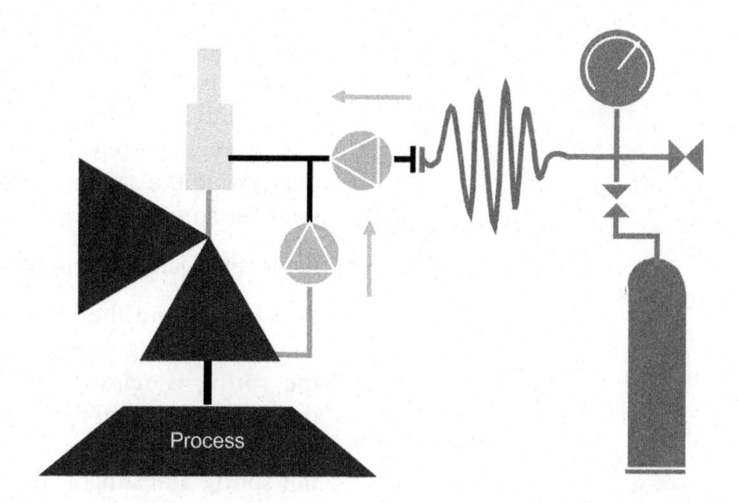

10.5 MAINTENANCE PROCEDURE

We will describe step by step the maintenance procedure of a spring-operated SRV when removed from the system. Where original equipment manufacturer's maintenance, installation and adjustment instructions are available, these are to be utilized in preference to this generic guide.

Before removing a valve from the system make sure that system is purged and that there is no longer pressure on the system.

Once removed, check if the lead seal which is attached to the cap is intact. This normally gives the guarantee that the information on the tag plate is correct and that the valve is not tampered with.

Make sure to check the records of the valve and that they have been kept up to date during the maintenance cycle.

Once the valve is removed from the process, check the inlet and outlet flanges for damage and make sure the valve is cleaned from eventual toxic or other substances.

The components which are most subject to damage and need to be verified carefully are the spring (and its washers), the bellows (if applicable), O-rings (if applicable), nozzle, disc and stem.

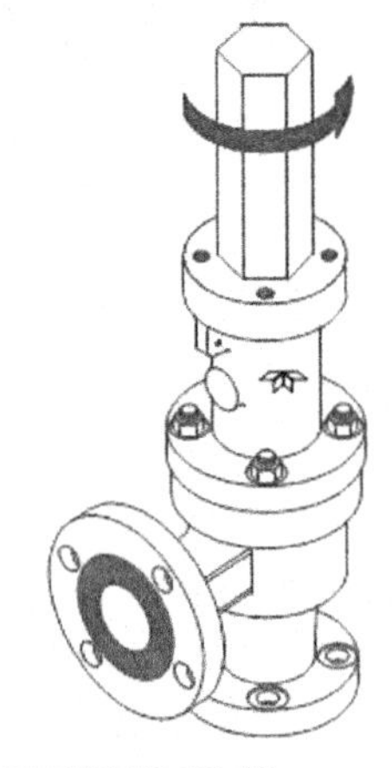

FIGURE 10.19

Once in the valve shop, the sequence for the valve maintenance operation is as follows.

Remove the cap and cap washers if any. If the valve has a lifting lever device, remove the lifting lever assembly (Figure 10.19).

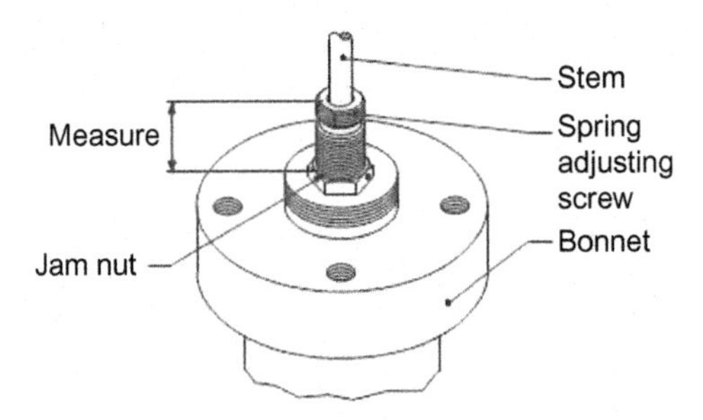

FIGURE 10.20

Since the valve will need to be retested and reset once reassembled, it is wise to measure the exact location of the spring adjusting screw for reference before unscrewing the adjusting screw to relax the spring (Figure 10.20).

Remove the jam nut (Figure 10.21).

First loosen and then remove the spring adjusting screw so that the spring is relaxed. Note that the stem and bonnet should be held firmly when loosening the spring adjusting screw as the spring is still under compression (Figure 10.22).

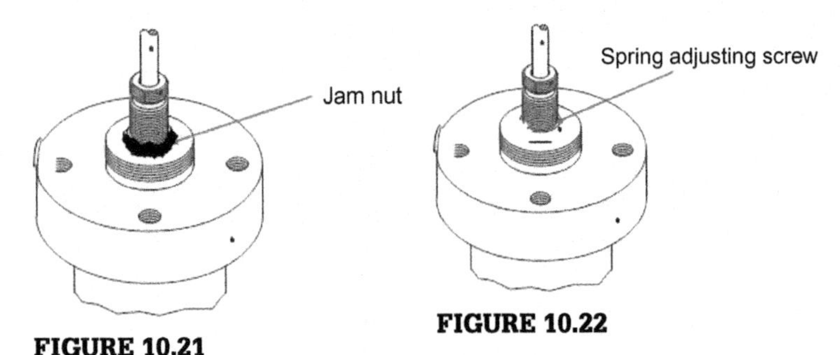

FIGURE 10.21

FIGURE 10.22

Loosen, preferably in a cross pattern, the body/bonnet nuts and remove the bonnet. Exercise care when lifting the bonnet as the spring and spindle will then be free to fall aside. Make sure to mark the body and bonnet for reference when reassembling the valve. It is best to reassemble the valve exactly the same way as it came from the manufacturer, as perfect alignment of the internals is paramount for good operation of the valve. Inspect the bonnet (Figure 10.23).

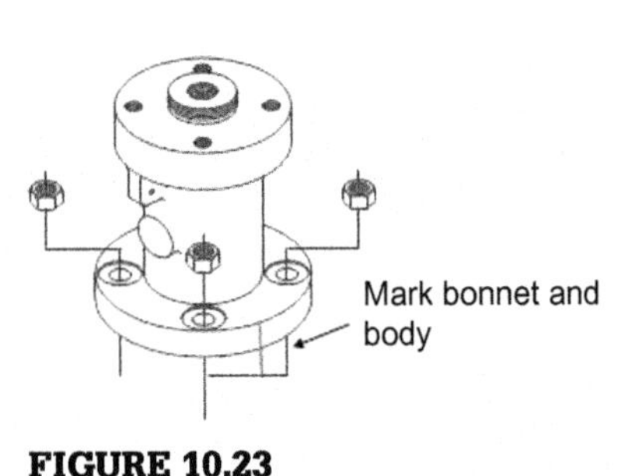

FIGURE 10.23

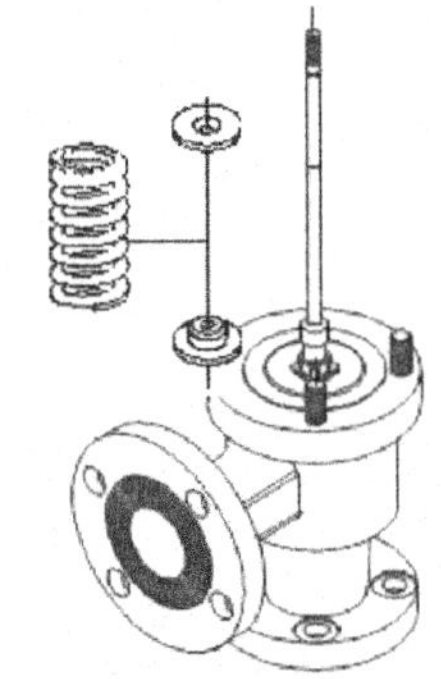

FIGURE 10.24

Remove the spring, spring washers and bonnet gasket. The spring and its washers are usually a unique set (Figure 10.24). Almost each spring has its unique set of washers. If you change the spring also replace the washers. Also spring washers are not always interchangeable between ends of the spring they belong to.

Inspect the spring and spring washers carefully – the spring for corrosion or eventual cracks. Make sure eventual spring coating is still intact. The spring washers should be inspected for abnormal damage.

Never reuse the bonnet gasket but always replace with new.

Remove the stem assembly and check to see that the stem is not damaged by galling or deformation (Figure 10.25).

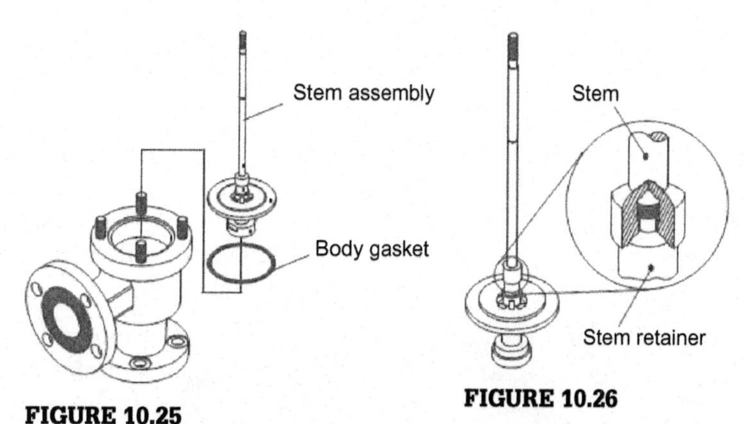

FIGURE 10.25

FIGURE 10.26

Only if there is a suspected problem, remove the stem from the stem retainer (Figure 10.26). How they are fixed might be different from design to design, but for quite a few designs it is enough to lift up the stem and rotate counterclockwise.

Remove the blowdown ring lockscrew from the body (Figure 10.27). Remove the blowdown ring from the nozzle and make sure you register the number of turns to unscrew the nozzle ring so it can be fitted in the same position when reassembling the valve. This is important because with small test benches without enough capacity, the blowdown cannot be set.

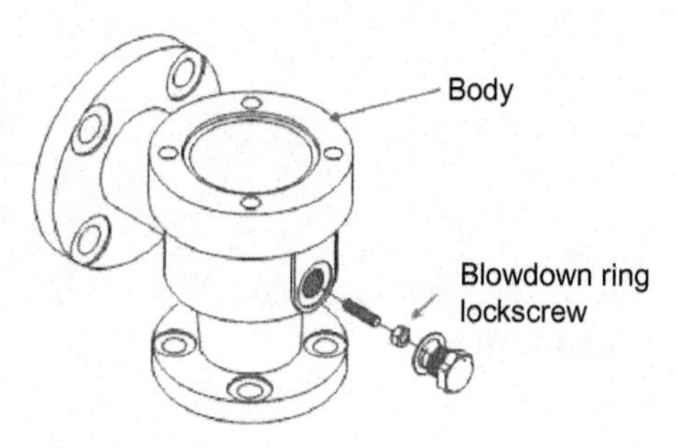

FIGURE 10.27

Lift the sleeve guide off the stem retainer (Figure 10.28).

In case of a bellows valve, remove the bellows at this point also.

If some parts are difficult to remove due to the presence of corrosive of foreign materials, soaking them for a while in a suitable solvent may be required.

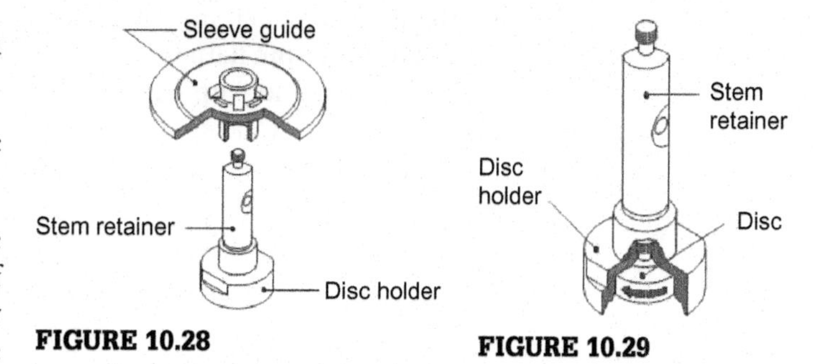

FIGURE 10.28

FIGURE 10.29

Remove the disc from the disc holder (Figure 10.29). Depending on the design, this can be cumbersome. Some designs have holes in the top of the disc holder where you can push the disc out; others you need to use a screwdriver to click the disc out. Others have a lockscrew fitted in the middle of the disc which can be loosened to remove the disc. In any case, make sure not to damage the disc or disc surface unless you plan to completely replace it. Make sure you know how to remove the disc or else refer to the manufacturer's instructions.

If the disc is not too much damaged, it can be lapped, preferably on a professional lapping machine. If not, it can be done by hand as will be described later. Note, however, that a disc cannot be repaired endlessly. The material taken away during lapping will start to influence the tolerances in the valve and possibly the set pressure.

FIGURE 10.30
Lapping block

Lapping blocks are made of a special grade of annealed cast iron (Figure 10.30). Normally, there should be a block for each orifice size. Each block has two perfectly flat working sides, and it is essential that they retain this high degree of flatness to produce a truly flat seating surface on either the disc insert or the nozzle. Before a lapping block is used, it should be checked for flatness and reconditioned after frequent use on a lapping plate.

Experience has proven that medium coarse, medium fine and polish lapping compounds will properly condition any damaged pressure relief valve seat except where the damage is too advanced and requires re-machining. Check with the valve manufacturer how much material can be taken away before it starts to affect the valve's tolerances and possibly also affect set pressure. Different grades of lapping compounds are available on the market, depending of the required flatness. Needless to say, in case of SRVs, the better the flatness, the better the tightness. Where micro finishing is desired, a diamond lapping compound is recommended. Extreme care should also be exercised to keep the lapping compound free from any foreign material and not to expose it too much to the atmosphere.

Different grades of lapping compounds should never be used on any one block or reconditioner because the coarser compound gets into the pores of the iron and scratches the surfaces, preventing a good lapped surface from being obtained. It is therefore recommended that a complete set of reconditioners and blocks be used for each grade of lapping compound used.

Different individuals have different methods of lapping valve seats, but certain essential steps must be taken to get satisfactory results. The following procedure is suggested for lapping nozzles by hand.

It is not recommended practice to lap the disc insert against the nozzle, although in extreme circumstances this might be the only option available.

Lap each part separately against a cast-iron lapping block. These blocks hold the lapping compound in their surface pores and must be recharged frequently. Lap the block against the seat. Never rotate the block continuously, but use an oscillating motion. Extreme care should be taken throughout to make certain that the seats are kept perfectly flat.

If considerable lapping is required, spread a thin coat of medium-coarse lapping compound on the block. After lapping with the medium-coarse compound, lap again with a medium-grade compound. Unless much lapping is called for, the first step can be omitted.

Next, lap again using a finer-grade compound. When all nicks and marks have disappeared, remove the compound completely from the block and the seat. Apply polish compound to another block and lap the seat. As the lapping nears completion, only the compound left in the pores of the block should be present. This should ultimately give a very smooth finish. If scratches still appear, the cause is probably dirty lapping compound. These scratches should be removed by using compound free from any foreign material.

Disc inserts should be lapped in the same way as the nozzles. The disc insert must always be removed from the disc holder before lapping.

SRV seats must be lapped to a micro finish using special compounds. Prior to super finishing, the valve seats should be lapped flat and to a fine surface finish in accordance with the standard practice as described above. A 3-μm-size diamond lapping compound should be used as described in the following procedure.

Clean the lapping block carefully using a suitable solvent prior to applying the diamond compound. Apply dots of 3-μm-size diamond lapping compound on the lapping block approximately ½ in to 1 in apart, circumferentially on the face of the lapping block, and if necessary, apply a drop of lapping thinner to each dot of compound. Lap the valve seat, keeping the lapping block against the seat and applying slight downwards pressure. During the operation, the lapping compound may begin to get stiff and movement of the lapping block more difficult. Remove lap from lapped surface and add a few drops of lapping thinner to the lapping block, replace on surface being lapped and continue to rotate, exerting no downwards pressure.

Be very careful as the lapping compound cuts very quickly and therefore the lapping block must be checked periodically to be sure the block remains flat and that a groove is not worn in the lapping block due to the lapping operation. While lapping, the lapping block should glide smoothly over the surface being lapped. Indications of roughness in lapping is indicative of contaminated compound; the lapping block and seating surface should regularly be thoroughly cleaned with a suitable solvent and the lapping operation always repeated.

Continue this for approximately 1 minute, then remove the lapping block and clean the lapped surface and the block with a suitable solvent and wipe with a clean, dry, soft, lint-free cloth. If the surface is still in an unsatisfactory condition, change lapping block and repeat the above process until a satisfactory surface is obtained.

Ensure that the equipment is always kept in a clean environment.

Finally, use an optical light source to measure flatness of disc/nozzle. Special instruments are available on the market.

Once the disc insert is lapped or replaced, carefully clean the nozzle surface. If defects are found, it is recommended that the nozzle be removed from the body. This can be cumbersome in some valve designs and is not recommended if not absolutely necessary. To remove the nozzle, turn the valve body over, taking care not to damage the bonnet studs. Turn the nozzle counterclockwise by using the wrench flats (if any) on the nozzle flange or a round nozzle wrench designed to clamp onto the nozzle flange.

To lap nozzle seats, the same precautions as with the disc are to be taken into account and the following procedure can be used:

1. Ensure that the work area is clean.
2. Select the correct compound for the first lapping sequence.
3. Squeeze a small amount of the compound on various spots of the lap.
4. With the side of the lap containing the compound facing you, hold the lap such that all five of your fingers point towards you and extend approximately 2 to 3 cm beyond the surface edge of the lap. Then invert the lap and place it flat onto the nozzle seat (avoiding any downwards pressure) and proceed with a circular oscillating and turning action as with the disc.
5. Proceed with a finer compound if necessary, same as with the disc lapping.

Keep the following in mind when lapping valve discs or nozzle seats:

- Discs typically require lapping with 320 grit, 500 or 900 grit and finally 1200 grit. Nozzle seats, because of their smaller surface area, require lapping only with 320 and 1200 grit. The exception is if either part has been severely damaged and has the metal surface eaten away. In such cases, a 220 grit is preferred over the 320 grit compound.

- Using a 7× measuring magnifier and flashlight for inspections (rather than just the naked eye) may save steps in the overall lapping procedure. If after the first lapping the magnifier reveals that most surface imperfections have gone, it is possible to polish immediately with the 1200 grit.

External parts such as the valve body, bonnet and cap should be cleaned by immersion in a bath such as hot Oakite® solution or equivalent. These external parts may be cleaned by wire brushing, provided the brushes do not damage or contaminate the base metals. Only clean stainless steel brushes should be used on stainless steel components. For the ultimate best results, these parts could also be pickled and passivated.

The internal components such as the guide, disc holder, disc insert, nozzle, guide ring and spindle should be cleaned by immersion in a commercial high-alkaline detergent.

Guiding surfaces may be polished using a fine emery cloth. The bellows and other metal parts may be cleaned using acetone or alcohol, then rinsed with clean tap water and dried.

If the inspection shows that the valve seats are very badly damaged beyond lapping, re-machining will be necessary or it may be advisable to just replace these parts. If re-machining is to take place, then original equipment manufacturer's dimensions must be consulted for critical machining dimensions and tolerances.

The valve spring should be carefully inspected for evidence of cracking, pitting or deformation. The bellows in a bellows-style valve should be inspected for evidence of cracking, pitting or deformation that might develop into a leak.

The bearing surfaces on the guide and disc holder should be checked for residual product build-up and any evidence of scoring.

Once everything has been carefully inspected, the valve can be reassembled in reverse order as the disassembly took place, keeping in mind that a correct alignment of the trim is paramount. Lubricate the spindle point thrust bearing and disc insert bearing with, for instance, pure nickel Never Seez®. Special attention should be given to the guiding surfaces, bearing surfaces and gasket surfaces to ensure that they are clean, undamaged and ready for assembly:

a. Screw the nozzle into the valve body and tighten with a nozzle wrench.

b. Screw the nozzle ring onto the nozzle, making certain that it is below the top surface of the nozzle seat. Try to locate the reference point so that the original blowdown setting is not affected.

c. Assemble the disc insert into the disc holder.

d. Either assemble the disc holder and guide by sliding the guide over the disc holder, while holding the disc holder or install the guide into the body. Wipe the nozzle seat with a clean cloth and then place the guide and guide ring in the valve body. The guide should fit snugly in the body without binding. On guides with vent holes, the holes should face the outlet. Wipe all dust compound, and so forth, from the disc seat and place the disc and spindle assembly in the guide.

e. Place the spring and washers onto the spindle and if applicable assemble the spindle to the disc holder. Some low-pressure valves are provided with a small piece of pipe, which fits over the spindle and between the two spring washers. This is a lift stop and protects the spring from excessive deflection. It is carefully fitted for the particular pressure range of the given spring in the valve and should not be used interchangeably with springs of other pressure ranges. Also make sure the washers go in exactly the same position as before disassembly.

f. Lower the bonnet into place, using care to prevent any damage to the seats or spindle. The bonnet is automatically centralized on the guide flange but must be tightened down evenly to prevent unnecessary strain and possible misalignment.

g. Screw the adjusting bolt and locknut into the top of the bonnet to apply force on the spring. Screwing the adjusting bolt down to the predetermined measurement can approximate the original set pressure.

h. If applicable, set the nozzle ring to minus 2 notches and the guide ring level. This is a test stand setting only and will assist when calibrating the valve giving a good indication of lift on the test stand.

i. Tighten the set screws on the control rings. The set screw pin should fit into a notch on the ring so as not to cause binding.

j. The valve is now ready for testing.

Before fitting the cap the valve can be retested. Using API 527 criteria is a good practice.

CHAPTER 11

Cryogenic Applications

While the use of safety relief valves (SRVs) for high temperatures on power boilers is well regulated in the ASME Section 1, a field where a massive number of SRVs are used is very poorly covered: Cryogenics! In particular, this applies to the booming industry of LNG in an era when people change from oil to the much more environmentally friendly natural gas.

Many other gases are also liquefied under cryogenic conditions, as their volume in liquid state is, depending on the gas, 500 to 900 times smaller than in the gas state.

The operation pressures depend on the temperature (saturation point) $P \sim T$. During the gas handling process (e.g. liquefaction), pressures can be relatively high; at very low temperatures, the storage tanks are almost atmospheric (50 to 350 mbarg).

Typical industries where we find cryogenic applications are the natural gas industry, (export, import, peak-shaving facilities), refineries (LPG), air separation (industrial gases – nitrogen, argon, oxygen, helium, etc.), marine (transport of LNG, LPG).

The saturation temperatures of some gases requiring care because of their potential cryogenics state are given below:

Fluid	Formula	Temperature (°C)
I-Butane	$CH(CH_3)_3$	−12
Vinyl chloride (VCM)	CH_2:CHCl	−13
Ammonia	NH_3	−33
Propane	C_3H_4	−42
Ethylene	C_2H_4	−103

(Continued)

Fluid	Formula	Temperature (°C)
Methane(≈LNG)	CH_4	−162
Oxygen	O_2	−183
Argon	Ar	−186
Nitrogen	N_2	−196
Hydrogen	H_2	−253
Helium	He	−269

Their pure cold state requires selection of correct materials, and some gases like liquid oxygen are also dangerous and cannot stand any dirt or oil. Therefore, when valves are used on liquid oxygen, they need to be oxygen cleaned – a special process which removes all traces of dirt under ultra-clean circumstances. It is important that suppliers are equipped to do this, as this requires special 'clean rooms'. The consequences if this does not happen correctly can be devastating for the valve and its surrounding environment, as can be seen in Figure 11.1.

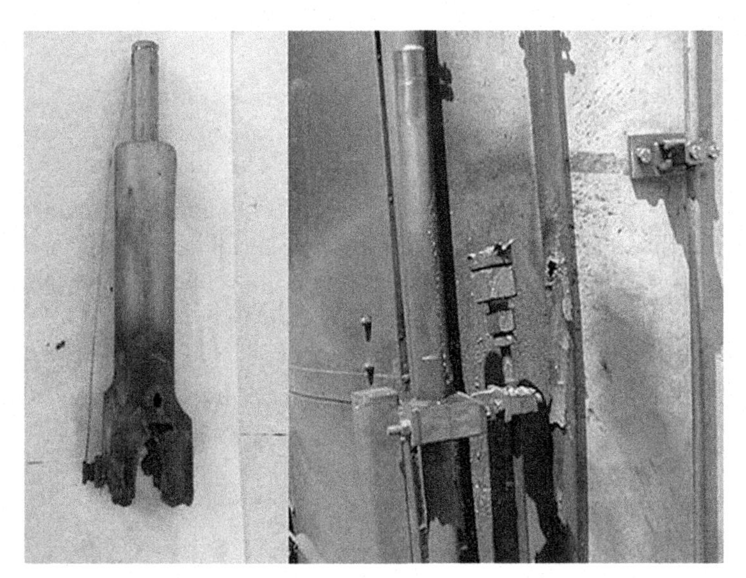

FIGURE 11.1
Insufficiently cleaned valve on cryogenic oxygen service

In the cryogenic processes, there are three main applications:

- Thermal relief
- Process
- Storage

Since storage is, as described earlier, only at low pressure and hence not code related (under 0.5 barg), we will only focus on the first two applications.

11.1 THERMAL RELIEF

The main requirement for thermal relief valves in cryogenic conditions is to reduce their freezing risks at any cost (Figure 11.2). Therefore, we must select valves with low simmer, a rapid pop/snap opening and high seat tightness. We need to reduce unnecessary product loss, so again low simmer and preferably a short blowdown is required.

FIGURE 11.2
Thermal relief valve on cryogenic service

The highest risk is when the cryogenic medium (which is usually ultra dry) comes in contact with the moisture present in the atmosphere, for instance, due to a simmering or leaking valve. The cold medium will freeze up the moisture around the seat and disc and in a short time it will become a solid block of ice around the seat area, which will prevent the valve from operating properly, or at all, creating a very dangerous situation.

Most metal-seated thermal relief valves are designed to operate proportionally and most have a fixed blowdown. Therefore, traditional metal-seated thermal relief valves are not recommended for use on cryogenic applications. There are very special resilient-seated cryogenic SRVs on the market (some initially developed for NASA), bubble tight up to 98% of set pressure, with a snap opening at 101% of set pressure and adjustable blowdowns between 3% and 25%. If you do not want to spend the money for these very special valves, then at least consider a soft-seated relief valve with a snap action.

11.2 PROCESS

Even more important is the use of SRVs for the cryogenic process. Because of my experience in the particular field of LNG, we will take this particular industry as an example, but most of what is presented here is also applicable to other cryogenic applications.

The safety of an installation handling LNG is paramount for obvious reasons. While it has become common practice since the mid-1960s to protect low-pressure LNG storage tanks with pilot-operated safety relief valves (POSRVs), many cryogenic installations under higher pressures are still relying on spring-loaded valves for the last level of pressure protection. One of the main advantages of spring-loaded SRVs is that their design is simple and proven in the general industry. However, their use to protect installations handling cryogenic LNG may be the cause of some serious concern in particular cases.

11.2.1 Conventional spring-operated SRVs on cryogenic service

On LNG liquefaction or re-gasification plants, many SRVs will be installed to protect equipment and personnel against the dangers of the same overpressures discussed earlier in this book. The valves considered here, however, must operate on cold cryogenic gas or liquefied gas. Just for reference, 'cold' is arbitrarily defined as any service below $-30\,°C$, and 'cryogenic' as any service below $-100\,°C$.

In both services, one essential concern is seat leakage. Not only is leakage unacceptable (seat erosion increasing the leak, potential risk of explosion or fire, economic loss, etc.) but on cold or cryogenic services, a leak will create icing around the opening area of the valve. Because of the constant flow of gas (or liquid), the seating parts of the valve will start to cool down, and the temperature of the whole valve will drop. Any moisture in the surrounding atmosphere will condense and freeze on the cold parts to form a solid block of ice, preventing the SRV from opening when necessary.

This can be a long process that will hopefully never happen. However, the catastrophic consequences of such a phenomenon mean that it must never be ignored.

So when a conventional spring valve is used, extra care must be taken to ensure seat tightness. A soft seat is generally the easiest solution to increase the seat tightness. The resilience of a soft seat, even greatly diminished by the cryogenic temperature, will still provide more 'softness' than a metal seat and therefore improve the tightness to reduce the risk of icing.

Furthermore, a correctly designed soft seat usually will guarantee a good repeatability of this tightness, even after many cycles of operation when metal-seated valves have already lost their tightness because of the repeated hammering of the two metal parts against each other.

11.2.2 Balanced bellows spring-loaded SRVs on cryogenic service

Onshore, the icing risk is often mitigated by connecting the exhaust of all these SRVs to a dry flare header or a recovery system. Only dehydrated hydrocarbon services are connected to a dry flare. In doing so correctly, the SRV system exhaust is free of moisture, which eliminates the risk of icing. However, even if nothing goes wrong in keeping the system moisture free, this still transfers the problem to the bonnet of the balanced bellows type SRV.

Many valves and other equipment are connected to the same header system, so the pressure in this system can vary greatly, causing a permanent variable superimposed backpressure on the SRV, as described in Section 3.4. This variable backpressure will act directly on the top of the disc of a conventional SRV and add itself to the original set point (opening pressure) of the SRV.

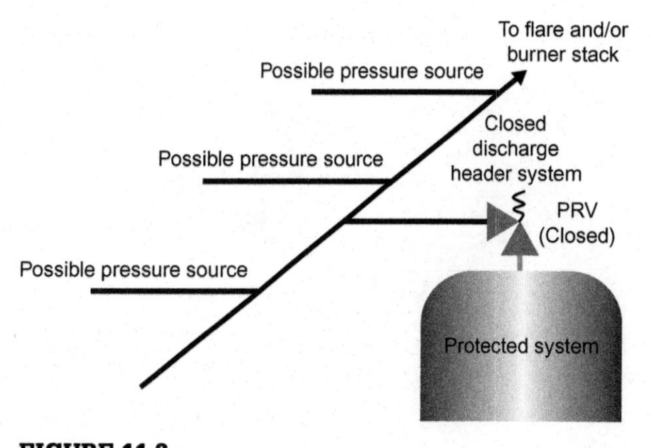

FIGURE 11.3
SRV connected to a header or flare system

If the pressure in the dry flare header exceeds the 3% allowed by most codes, the SRVs connected to it may then open at a pressure higher than what is allowed by this code or law, or even at a pressure that would exceed the acceptable tolerance on the design pressure of the protected equipment (Figure 11.3).

Therefore, as seen earlier, to avoid this highly hazardous situation, a balanced bellows spring-loaded valve must be used. This type includes a bellows assembly as described in Sections 5.2 and they subtract the disc of the valve from the influence of the backpressure so that the valve opens at the correct predetermined pressure (Figure 11.4).

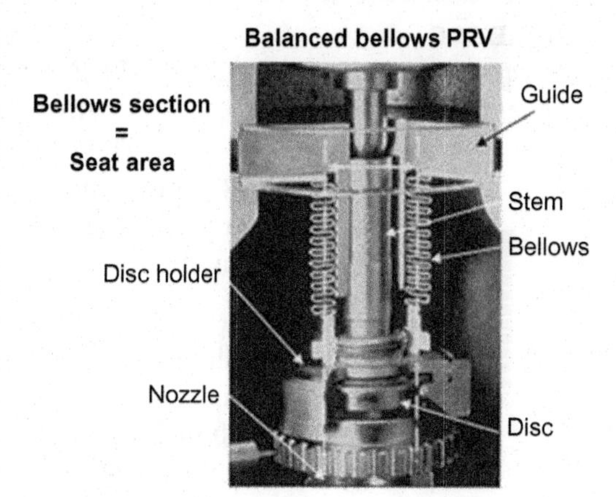

FIGURE 11.4
Bellows design

However, to work properly, the volume inside the bellow is vented to atmosphere into a bonnet with an open vent. This can cause a problem on cold or cryogenic service as now icing may occur inside the bellows due to atmospheric moisture in contact with the low temperature of the valve, and may block the opening of the SRV. This could be avoided by heat tracing the bonnet of the valve. However, in addition to the problems and costs directly linked to the heat-tracing system itself, this does not eliminate another potential risk of using bellows valves on cold and cryogenic services. The bellow will still always get cold (first, by conduction with the disc and then, during a relief cycle) and lose its resilience. It may crack on the first opening of the valve, which will then start to leak heavily by the bonnet vent. The valve will not be protected against the backpressure anymore and ice will form all over the valve (Figure 11.5).

FIGURE 11.5
Ruptured bellows due to loss of resilience due to cryogenic exposure

11.2.3 POSRV on cryogenic service

Because of the uncertain reliability of balanced bellows spring valves on cold and cryogenic services, POSRVs have offered a good alternative for these applications for more than 35 years now.

The POSRV uses the system pressure as closing force so that its seat tightness is at its maximum, close to the opening pressure – completely the opposite from a spring-loaded SRV. With the addition of a properly designed and selected soft seat, the tightness on cold or cryogenic service will be reinforced to avoid any leakage and icing risk.

As tightness is a particularly important feature for any SRV to be installed on an LNG application, a POSRV is even more of an ideal choice for cryogenic applications since it is balanced against (high) backpressures without the use of vulnerable bellows.

11.2.3.1 Considerations for POSRVs on LNG and cryogenic applications

There are, however, several factors to consider when installing a POSRV on a cold or cryogenic application. The main issue is that to operate properly, the POSRV (and particularly its pilot) relies on correctly chosen seals and O-rings which can accept extreme low temperatures.

In the main valve itself, the use of various plastic compounds and special techniques of sealing to improve the resilience of the compound at low temperatures have been proven by decades of good and reliable service by only a handful of suppliers. It is evident that especially the main valve should withstand the cryogenic conditions without any problem.

However, with many manufacturers, all the attempts to use special techniques in the pilot without the use of soft seats have so far proved unsuccessful after testing (erratic set and reseat pressures at least). So, as of today, the use of O-rings in the pilot is unavoidable, and therefore the pilot must be kept away from low temperatures as much as possible.

There are many proven ways of keeping the pilot 'warm'. One, which must not be overlooked on site, is that if the main valve is insulated like its associated piping, the insulation must be such that the pilot is kept out of the insulation, in the open air.

In any case, it is imperative that the pilot be of a non-flowing design. A pilot is said to be 'non-flowing' when it does not pass any fluid when the main valve state (opened or closed) does not change. The main valve may be open and flowing, but the non-flowing pilot will not flow until the valve has to reclose, in which case the pilot will flow the necessary quantity of fluid to re-pressurize the dome to close the valve. So at each cycle, no matter how long, the non-flowing pilot will only flow the volume of the dome in two ways: 'out' to open the valve and 'in' to reclose it. This can vary greatly from manufacturer to manufacturer, but dome volumes can be around 1 to 1.5 l for a large 8 in × 10 in POSRV, and down to less than 1 cm^3 for a small 1 in × 2 in valve. So by selecting a non-flowing pilot, the quantity of cryogenic fluid that could pass through the pilot and potentially cool it down is greatly reduced. It should, in any case, be verified that its volume is small enough to avoid icing of the pilot.

Various techniques are employed by POSRV manufacturers to isolate the pilot further from the cold medium and to ensure that no trace of liquefied cryogenic or cold gas reaches it. One easy way is by thermally isolating the pilot bracket, as shown in Figure 11.6.

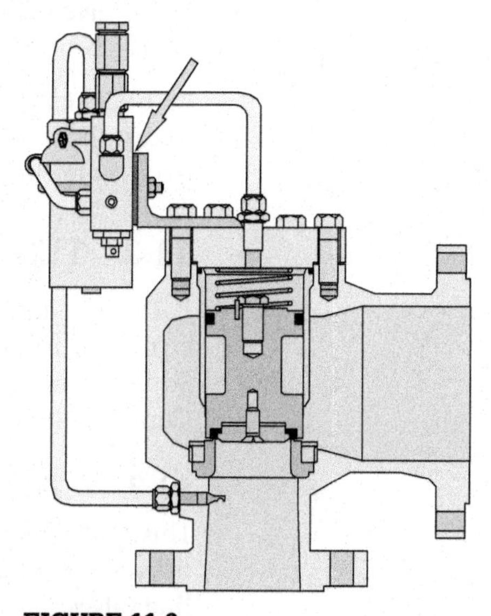

FIGURE 11.6
Thermal isolation between pilot and mounting bracket or main valve

If good care is taken to isolate the pilot from the main valve, the results are very satisfactory, as can be seen in Figure 11.7. While the main valve is flowing, the valve body and inlet piping become very cold and covered with ice, whereas the pilot does not show any icing.

A final consideration when installing a POSRV on an LNG application is the exhaust of the pilot. On some installations, the exhaust can be done to the atmosphere, either directly or via a mast. But on other installations, even the small volume relieved per opening cycle by the pilot (the volume of the main valve dome) cannot be accepted. Therefore, the pilot exhaust must be piped to a safe place.

FIGURE 11.7
Thermally isolated non-flowing type pilot does not freeze up

If the pilot is fully balanced against the backpressure and the valve is connected to a dry system, then the pilot exhaust can be safely piped to the main valve outlet so that all emissions go to the exhaust system. As the pilot is kept out of the cold, the problems which occur on balanced bellows spring-loaded SRV in the same situation will not occur.

Usually, the pilot valve types which are balanced against backpressure are the modulating pilots (to my knowledge, no snap action pilot currently on the market will accept backpressure with the pilot exhaust piped to the outlet of the valve). This modulating pilot will actually regulate the pressure in the dome to control the position of the main valve piston. It therefore regulates the flow relieved in order to match the needs of the system so that if the cause of overpressure is minimal, the main valve will be opened slightly, but if the pressure increase is created by a larger phenomenon, then the valve will be fully opened.

An undeniable advantage of POSRV is that it is relatively easy and inexpensive to test them *in situ* with accuracy. This field-testing capability enables more frequent testing to better monitor the critical valves of LNG installations.

11.3 TESTING FOR CRYOGENIC SERVICE

Because of the problems of relieving cryogenic fluids, many points of concern need to be verified before a valve is installed on an LNG application. So far the best way to do that is to test the valve under cryogenic conditions. Currently there are two main tests used.

11.3.1 The 'submerged test'

This test is mostly specified by end users and design engineers, and mimics the well-documented cryogenic tests usually done on control valves and other line valves.

In this test, the SRV is immersed into a cold or cryogenic medium (alcohol solution with dry ice or liquid nitrogen) usually up to its seat level, with temperature probes located in three or four places on the body. The inlet of the valve is blanked by a flange with a connection to supply a gas which will not condense at the testing temperature (helium, for example), while the outlet of the tested valve is blanked by another flange with a connection used to measure the leakage rate of the seat. When the valve is at the correct testing temperature, the pressure is increased at the inlet up to 90% of the set pressure and the leakage rate measured and compared to the maximum leakage rate acceptable.

Then the valve is actuated. The flow (volume) of helium being usually too small to lift the valve by itself, the valve to be tested must be previously fitted with a lift lever which will be used to actuate the valve. After an agreed number of actuations, the leak rate is measured once again at 90% of the set pressure. The valve is considered acceptable if the leak rate is below a certain level (e.g. according to API 527).

What information does this test provide on the reliability and operability of the valve under the actual operating conditions? Very little actually! This test is specified as a copy from control valves or other isolation valves but in fact does not apply to SRVs.

Since, first of all, the valve has to be manually operated to make sure it fully opens, no conclusion can be made on the actual opening pressure and behaviours under the test temperature. The gas used to measure the leak rate is normally not at the testing temperature, and therefore the critical seating parts will be warmed each time the gas passes through the valve, which is exactly the opposite of what is happening in the field.

Furthermore, the leakage rate is measured on helium (a far more searching gas than the methane or other gases the valve will see on site); it is somewhat difficult to transpose into relevant information the actual leakage rate the valve will have on, for instance, natural gas, which has a molecular weight much higher than helium.

On top of that, the valve body will be cooled down from outside in. So, the body will see the opposite thermal expansion behaviour than in process conditions where the cold will act inside out. Therefore, the thermal stresses that exist on the valve do not reflect the real conditions.

The only thing that this frequently specified test will prove is that the valve assembly can withstand the cold temperature, but not that the valve can actually operate on the field in cold or cryogenic circumstances.

The mistake which is made here, because of lack of specific SRV test specifications for cryogenic service, is that test specifications from line valves are simply transposed or used on SRVs.

However, the European normalization committee is in the process of looking at the issue and is working on the EN-13648-1: 'Cryogenic vessels – Safety devices for protection against excessive pressure – Part 1: Safety valves for cryogenic service'. The tendency is to go to a much more reliable test as described hereafter which will simulate the real process conditions to a greater degree.

11.3.2 The 'boil-off test'

This method of testing an SRV on cryogenic service has become more and more popular since the mid-1980s when it was first developed.

In this test, first the SRV is tested on a normal ambient temperature test bench, and its characteristics are recorded as 'normal'.

It is then installed on a tank filled with the cooling fluid. On many occasions. for lower pressures the natural 'boil-off' of the cryogenic liquid will increase the pressure in the test tank up to the set pressure of the valve. If not, a gas (identical to the fluid in the tank – for example, nitrogen gas and liquid nitrogen) is sent into the tank through the liquid. The gas will obviously condense into the liquid phase and will 'heat' it, causing boil-off to increase the pressure in the tank.

The pressure rises until the valve opens. Because the tank creates a sufficient test volume and the liquid creates an important boil-off gas volume when the valve starts to open, the valve actually lifts so as to give a correct indication of its set pressure. The valve is operated in this way a sufficient number of times until the temperature at the inlet flange is at the test value. The capacity is, however, usually not large enough to obtain a full lift of a large spring-loaded valve, for example. Therefore this test, like most of the tests on spring-loaded valves, cannot give an accurate measurement of the reseat or blowdown pressure. But then again, nothing except tests on site can provide large enough volumes.

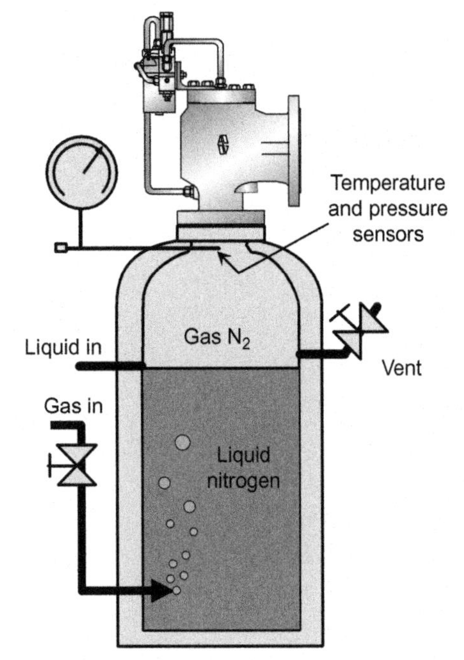

FIGURE 11.8
Boil-off test set-up

Then the valve is actuated again for an agreed upon number of times (5 or 10 times, depending on the type of valve and the application) while the temperature remains at its test value or below. The leak rate is finally measured after a few minutes (typically 2.5 minutes), at 90% of the set pressure, as for a classical API Standard 527 test with a set-up as shown in Figure 11.8.

After this test, the valve is tested again on a normal test bench to verify that the various pressure reference characteristics have not drifted from the original testing, which could indicate deformations due to thermal expansions or contractions.

The main advantage of this kind of test is that it tends to simulate as close as possible the actual conditions in which the valve will operate on site. The critical area, the seat, is cooled down by the cold/cryogenic fluid, as it would be in an actual upset. Because the valve operates on its own, the actual indication of its performances in the field is obtained.

11.4 CONCLUSION

While conventional spring-loaded SRVs can provide an acceptable level of reliability, particularly when fitted with a soft seat of proven design, the balanced bellows spring-loaded SRV can create a highly hazardous situation and should therefore be avoided on cold or cryogenic services.

With the proper attention to installation and the correct design, POSRVs can greatly improve the safety of the installation.

There must be a proper evaluation of their performances in the actual operating conditions, which can be quite well modelled for LNG applications by a 'boil-off test'.

The proven track record of a specific product is also a good factor to consider before any decision.

CHAPTER 12

Specifying Safety Relief Valves

In this chapter, we will provide just one safe example of how end users can specify their requirements for safety relief valves (SRVs) so that the risk for having wrong valves installed is minimized. This example is, of course, not exhaustive but provides guidance on what sort of information could be provided to the SRV manufacturer in order to ensure a safe situation on the site.

Note that this specification relates only to the compliance of the SRV component in itself and does not take into account installation or sizing/selection criteria which have already been discussed separately in this book.

Most parts referenced have been covered in this book except for piping, painting and packing codes, mentioned hereafter, which can be found in the relevant codes.

This practice covers the specification for the overall design, inspection, testing and preparation for shipment of SRVs. SRVs within the scope of ASME Code Section 1, Power Boilers, are not covered by this particular practice.*

Note: An asterisk (*) indicates that a specific decision by the purchaser is required or that additional information must be furnished by the purchaser.

12.1 SUMMARY OF OVERALL REQUIREMENTS

Table 12.1 lists the codes and standards, which can be used with this specific practice, depending where in the world the valves will be installed.

Table 12.2(*) lists practices and standards which are acceptable and/or are specific for the specific end user. They can be internationally accepted codes or recommendations, or can be internal codes and practices of the corporation. The given tables are non-exhaustive and are used as an example only, but are typical as encountered regularly.

Table 12.1 Codes and standards
Standards
API standards
526 – flanged steel safety relief valves
527 – seat tightness of pressure relief valves
ASME standard
B1.20.1 – pipe threads, general purpose (inch)
Section VIII – pressure vessels, Division I
Section VIII – pressure vessels, alternative rules, Division 2
PED standard
PED 97/23/EC – European CE mark
EN standard
EN/ISO 4126 – safety devices for protection against excessive pressure

Table 12.2 Practices and standards
Internal practices
General requirements for valves
Specifications for flanged joints, gaskets and bolting
Piping practices
Specific standards
NACE standard
MR-0175 – sulphide stress corrosion cracking-resistant metallic material for oil field equipment
ASME standard
ANSI B16.5 – pipe flanges and flanged fittings

Compliance with any local rules or regulations specified by the user* is mandatory.

12.2 MATERIALS

Materials shall be per API 526 and ASTM. Disc, nozzle and stem material shall be at least as follows:

Valve components	Material or material standard	Remarks
Disc and nozzle	High corrosion-resistant alloy of chromium-nickel or nickel-chromium	Ferritic steels not acceptable

(Continued)

Valve components	Material or material standard	Remarks
Stem	316SS Wrought 13 Cr 18 Cr 8 Ni Steel	 If body is ferritic If body is austenitic
Spring	316 SS Alloy steel with corrosion-resistant coating	Coatings subject to prior approval For temperatures between 215°C and 538°C

Impact testing: The need for impact testing shall be based on ASME code requirements and the minimum specified temperature on the specification sheets of the end user. Any impact testing criteria shall be in accordance with ASTM A370. Acceptable alternatives when ASME code compliance is not required include ISO 148, BS 131 Part II, or JIS B7722 and Z2242 5 (Japanese codes). Some specifications sheets also state: 'Impact testing may be done according to vendor's standards pending prior approval of end user'. The latter is definitely usually a more economic option, with major vendors being not very far from normal standards. The slight negative side is that it needs reviewing by the end user prior to acceptance.

12.3 DESIGN

Valves (*) which do not bear the ASME code stamp and/or the CE marking shall be calibrated and the capacity demonstrated in the manner prescribed by the ASME Code Section VIII.

If test certifications for spring-loaded valves on gas show the valve coefficient (API coefficient of discharge, K_d) to be less than 0.95, vendors' proposals for that particular valve design shall be submitted to the end user for verification and approval of valve sizes.

Valves not UV stamped, but which are identical with designs which have been previously certified and marked per ASME Code Section VIII, shall be considered acceptable without further qualifying tests. Vendors' certification to this effect shall be furnished.

Valve types shall be as follows:

- All *spring valves*, except thermal relief valves, shall be flanged, high lift, 'high-capacity type' with a top guided disc. Valves shall not be provided with a lifting device or test gags. All valves shall be provided with pressure tight bonnets except bellows type valves. The cap shall be of bolted design.
- The use of *pilot-operated valves* shall be considered for all backpressures above 30%. The pilot valves shall be of the non-flowing design.

Modulating pilot valves shall be exclusively used for two-phase flow conditions. All pilot-operated valves shall be equipped with on-site test connections and equipped with a supply filter with minimum 60 μm element. Pilot-operated valves can be of semi- or full nozzle design.

- *Dimensions* of both spring and pilot valves shall be in accordance with API 526.
- Flanged *spring valves* shall be of full nozzle design, arranged as such that the nozzle and the parts comprising the disc are the only parts exposed to the inlet pressure or to the corrosive action of the inlet fluid when the valve is closed.
- *Body flanges* accommodating a full nozzle may be modified to accommodate the nozzle except that the thickness shall not be less than the minimum thickness as specified by ASME B16.5.
- *Body flanges* should be by preference of integral design. If flanges are attached by welding, they shall have the complete circumference of each attachment weld 100% X-rayed and submitted to the end-user for approval. The radiography and acceptance criteria shall be per ASME Code Section VIII, Division 1.
- *Thermal relief valves* specified with threaded ends shall have internal, NPT-type, female-tapered threads.
- *Flange rating* and type of facing is specified in data sheets*. Flange facing finish shall be minimum per ASME B16.5.
- Individual *safety relief valve specifications* shall be communicated on API 526 specification sheets or equivalent substitute provided by the end user. When the relieving contingency is vapour and liquid, rates and properties for each at the allowable overpressure shall be provided.
- Materials for valves in *wet H_2S service* shall meet local plant standards (*). Where a plant standard is not available for wet H_2S, materials shall conform to NACE MR0175. Wet H_2S service is above 50 wppm H_2S in free water.
- *Test gags* (as defined by API 526) are not permitted.

12.4 IDENTIFICATION

- Valve marking shall be in accordance with ASME VIII requirements (see Section 4.2).
- Marking of springs with 0.125 in. (3 mm) wire diameter and larger shall be with symbols which identify, at a minimum, either the manufacturer and the spring characteristics (i.e. pressure–temperature range, material

and size) or the manufacturer and part number, which will allow characteristics to be determined. For springs less than 0.125 in. (3 mm) wire diameter, the designation shall be shown on an attached metal tag.

12.5 INSPECTION AND TESTING

Components of SRVs shall be hydrostatically tested before assembly, as given below. Parts made from forgings or bar stock are exempt.

Valve component	Hydrostatic test pressure
Both full and semi-nozzles, bodies for semi-nozzle valves, bodies with integral nozzles, primary pressure-containing components of pilot-operated main valve	1½ times the maximum allowable design pressure per the manufacturer's catalogue
Bodies for valves with full nozzles, closed bonnets with caps	1½ times lower of the maximum outlet flange rating pressure or maximum allowable backpressure per the manufacturer's catalogue
Bellows for balanced type valves	Minimum of 30 psig (2.1 barg)
Primary pressure-containing parts of pilot valves (minimum: main valve and cap)	1½ times the maximum inlet flange rating

The end user reserves the right to witness shop tests (*) and inspect valves at the manufacturer's plant, as specified in API 526.

Testing criteria for operation and set pressure testing must be in accordance with API 527.

Valves for cryogenic service below −60 °C shall be tested using the boil of test as per EN-13648-1: 'Cryogenic vessels – safety devices for protection against excessive pressure – Part1: Safety valves for cryogenic service'. Acceptance criteria to be in accordance with API 527.

For pilot-operated valves, set pressure testing on pilot only is allowed. Functional tests of the full unit are required.

12.6 PREPARATION FOR SHIPMENT

Unless otherwise specified (*), valves shall be painted using the manufacturer's standard paint specification using the following colors:

a. Conventional valves: white
b. Bellows type valves: body – yellow; bonnet – red
c. Pilot valves' main body: green.

*RAL numbers can be included according to end user standards.

Preparation for shipment shall conform to the requirements of the Protective Coatings and Valve Protection subsections of IP 3-12-9.

As already mentioned, this is just a typical general specification which could be sent to a manufacturer, and it will typically determine the general design and code requirements to which the SRVs must comply. This general specification is accompanied by a complete detailed technical data sheet per SRV, of which a typical example for the spring and pilot valve can be found in API 526 (see examples in Appendix K).

If, and only if, duly completed, the manufacturer will be able to make an appropriate sizing and recommend a selection of the required SRV for that specific application. However, in most cases it is recommended to ask more questions, in particular, related to the installation in order to cover all factors. The SRV manufacturer should then be able to provide a detailed sizing sheet per selected valve.

Most manufacturers use software to size and select their valves, but very little software is available which will calculate the valves independent of specific vendors' data and where data can objectively be used. In most, you unfortunately have to select a manufacturer's model number, and the selection gives specific valve model numbers with their specific configurations, K_d factors, etc. There are only a few software packages which are more independent.

An example of a complete (independent) sizing sheet can be found in Appendix L.

CHAPTER 13

Non-conformance of Existing Pressure Relief Systems

The safety relief valve (SRV) is considered the last silent sentinel to protect us from accidents which can cost us life, property and important economic losses. In fact, however, in practice in the process industry, they sometimes actually have two definitions:

Definition 1: *An automatic pressure-relieving device actuated by the static pressure upstream of the valve and characterized by a RAPID FULL OPENING or POP action. It is used for steam, gas or vapour service.*

Definition 2: *A useless appendage which should never have to work but which is capable of disrupting an operation and wasting a valuable product.*

Thirty-plus years of experience in the industry with leading manufacturers worldwide has taught me that people in the process industry are sometimes just not very familiar with SRVs. Many times, people install them because they have to by law, without giving a lot of thought or consideration on selection, installation or maintenance.

This has led to amazing experiences. We have been called many times about so-called 'non-functioning SRVs', and once on site, we found them buried in sand, mounted horizontally, upside down, mounted after closed isolation valves, installed with the test gag still blocking the spring or the remote sense line of the pilot valve not connected, and so forth. It is unimaginable how carelessly SRVs are sometimes treated within the industry, and this is mostly due to a lack of knowledge. At a certain point, as a manufacturer, we found that 75% of all the incoming complaints about so-called malfunctions on SRVs were due to careless handling during and after transportation, installation, testing or maintenance. Based on a period of 1 year, about 750 complaints were monitored and showed the following:

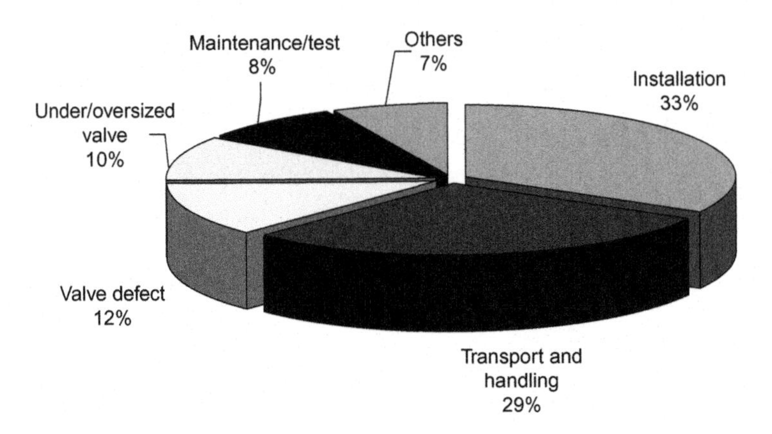

These are disturbing figures, showing once again that it is necessary to pay more attention to the safety in industrial plants when it comes to overpressure protection.

Looking beyond the incoming complaints about the valve itself, a lot of installations are just not protected at all. A statistical analysis shows that the pressure relief systems on nearly half of the equipment in the oil, gas and chemical industries lacks adequate overpressure protection as defined by recognized and generally accepted good engineering practices and by the codes, hence the law.

A detailed study was performed by Berwanger – 'Oil, gas & petrochemical consulting' – in order to evaluate how safe the installations were. It should be stated that the study was done in the United States, but with our experience I believe we can transpose the results to other parts of the world, perhaps taking into account about a 10% error margin to be on the safe side for Western Europe, where other (old) codes also still apply above the PED (TuV, ISPESL, GOST, etc.).

Nevertheless, the above figures on incoming complaints were from Europe, Middle East, Africa and Asia only, and are disturbing enough to warrant elaborating a little on the issue and the interesting study from Berwanger Consulting.

Why is this so important? First of all, it is the law, and management can be held accountable for any accidents. Oil and gas rigs, exploitation sites, refineries and petrochemical plants strain to increase capacity. While investments are made to increase this capacity, senior managers must ensure that the relief systems are also adequately upgraded as well, to avoid risk. Most disasters start with small, seemingly unimportant issues.

A strict focus on compliance only with OSHA's Process Safety Management Standard (29 CFR 1910.119) may not necessarily be enough to protect end users against pressure-relief failures. Although most of the units have been designed by reputable design firms worldwide, the vast majority of deviations from good practice are not identified during conventional process hazard analyses (PHAs), but either via external audits or when problems have already occurred. The problem is that design, instrument, process and piping engineers have such a variety of components to cover that there is little unique specialization in just SRVs or SRV systems by reputable design firms; these days, firms may also have some turnaround of personnel, depending on the workload and the economic situation.

Once installed and as the process changes, it might be wise to take a close look at the original selection of SRVs and evaluate whether this selection, made for different process criteria, is still valid. Then, organize compliance audits of pressure relief systems and particularly on individual valves and other devices and re-evaluate the potential overpressure scenarios described earlier in this book. The typical current standard internal audits might not catch the piece of equipment that, for instance, does not have a valve but should.*

Berwanger for instance, states it has performed 2000+ audits and that they have never encountered a plant that did not have a pressure relief issue. From my personal experience, I can confirm that only a few of the plants I visited had no problems in complying with good safety relief practices and/or codes, but more importantly many had regularly installed the wrong valve for a particular application or simply installed the valve incorrectly. This is of concern, to say the least. The extent of this risk merits the attention of those who are accountable to the company's stockholders and the safety of its personnel.

Many specialists in SRVs and systems agree that as a practical matter, conventional PHA methods are not always the most effective tools for evaluating pressure relief systems. They also conclude that the pressure relief system design process could be improved. Working closely with a lot of the design firms, I concluded that they merely try to comply with the codes at a minimum cost and care very little about LCC (life cycle cost) of the components. This puts pressure later on the end users' maintenance departments; usually, these employees have little or no input in the selection of design components whose problems ultimately end up in their laps. In my opinion, it is recommended that, in order to reduce this deficiency rate, the industry should adopt a more equipment-based approach to pressure relief system design and maintenance.

Besides the safety aspect another consideration is that about 30% of process industry losses can be at least partially attributed to deficient pressure relief systems. These losses alone could subsidize a closer look at the pressure relief systems in a process plant or a better upfront selection of the safety systems.

---

*Also the actual installation of the pressure relief devices deserve a detailed audit as this accounts for a large number of deficiencies.

An independent look at the pressure relief systems of 400 different operating units on 43,000 pieces of equipment and 27,000 pressure relief devices led to some very interesting results.

Types of equipment tested were a mix of the following:

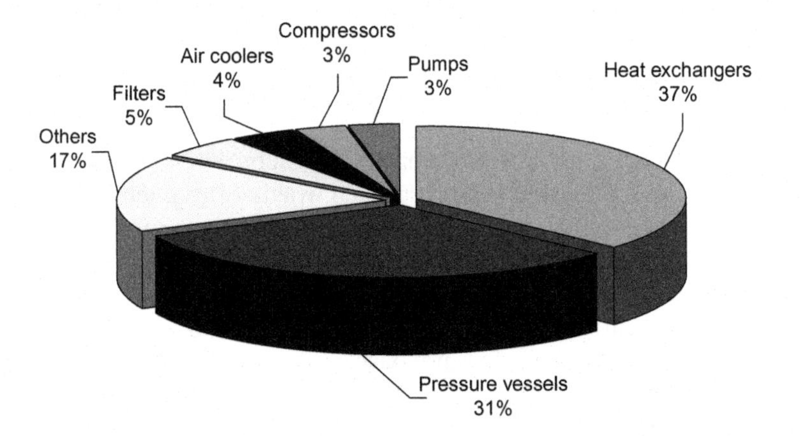

Almost all investigated companies had already undergone PHAs before investigation. Nevertheless, the conclusion was that one-third of the equipment in the oil, gas, petrochemical and chemical industries have some pressure relief system deficiency with reference to widely accepted engineering practices. This proves either a general lack of expertise of pressure relief systems, and in particular SRVs (and their application by both some end users and the reputable internationally renowned design companies and engineering companies), or too much focus on trying to cut costs, which should not be done on safety equipment. The types of deficiencies are roughly split between absent and/or undersized pressure relief devices and improperly installed pressure relief devices, which confirms the data seen from a manufacturer's point of view.

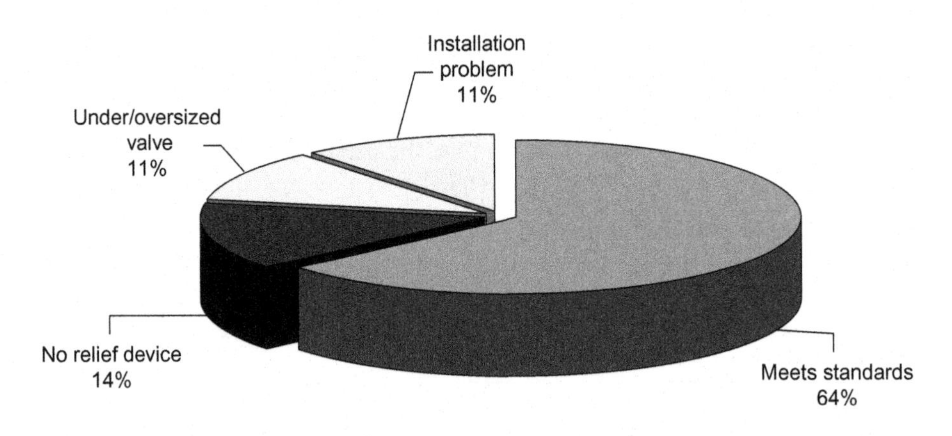

Not all possible deficiencies are included in the above statistical analysis. There are also concerns about items such as excessive flare radiation levels, inadequate knockout drums, poorly designed quench systems, discharge of toxic fluids to atmosphere, discharge of combustible or toxic liquids and gases to atmosphere and a general lack of process safety information upon which to base a safe pressure relief system design. Therefore, it could be stated that the actual total deficiency rates reported may even be understated.

Let's just look at the main categories as described in above pie chart.

13.1 NO RELIEF DEVICE PRESENT

In order to identify possible deficiencies, reference was made to API and then a full examination was performed on an equipment pool of 43,000. Of the investigated equipment, 12.5% did not have a relief valve where it should have one according to API recommendations.

Outside consultants used a very detailed method of investigation to check compliancy with API. Without going into the details on how this was done, the results are shown below:

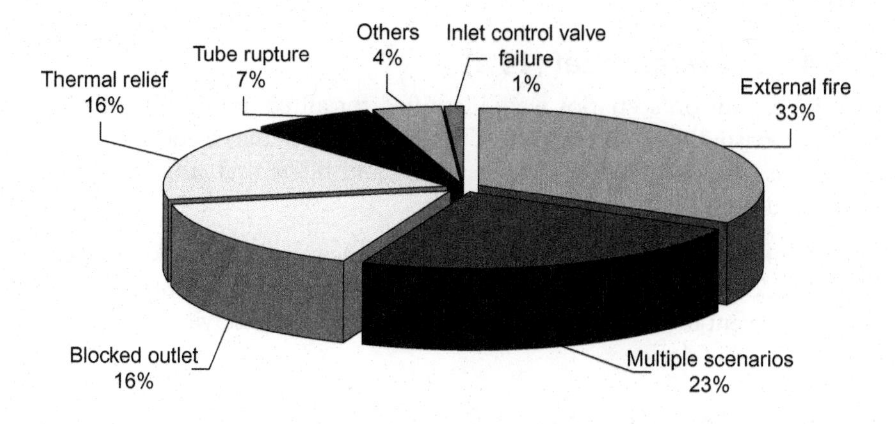

Discussing the scenarios in order of deficiency:

13.1.1 External fire (33%)

API RP 521 Section 3.15 contains extensive guidance relative to relief requirements for external fire (as discussed earlier in this book).

To identify external fire scenarios, all equipment with a liquid inventory located in an area of the facility potentially subject to an external fire was selected. External fire protection was not considered for vapour-filled vessels, as other methods of protection can alternatively be used. From an analysis standpoint, also pumps and compressors were not subjected to external fire scenarios. Thus, the total number of deficiencies was reported as a percentage of the vessels, filters, exchangers, air coolers and others.

The only solution here is installing an SRV that is suitable for this application after looking at the individual case to evaluate which SRV should be selected.

13.1.2 Multiple scenarios (23%)

This was identified as the number of deficiencies in which a piece of equipment had more than one potential overpressure scenario.

13.1.3 Blocked outlet (16%)

To identify blocked outlet scenarios, all maximum pressures of all inlet sources were compared to maximum allowable working pressure (MAWP) of the equipment under consideration. In addition, the potential for heat input from the process to result in vaporization at MAWP was considered.

Common solutions to this concern include installing a relief device, re-rating the equipment or removing the mechanism that results in the potential blockage (lock valves open, etc.).

13.1.4 Thermal relief (16%)

Thermal expansion scenarios were identified for all equipment in which the potential existed to isolate and heat a liquid-full system. In addition to heat exchangers, this included liquid-full vessels and filters that could potentially be isolated liquid-full.

Typical solutions to thermal expansion deficiencies include locking open of outlet block valves on the cold side of exchangers, reliance on shutdown procedures to prevent the scenario, draining and venting out of service equipment or installing a relief device.

13.1.5 Tube rupture (7%)

Here, instead of using API, it was referred to ASME VIII Division 13, Paragraph UG-133 (d) which states:

> Heat exchangers and similar vessels shall be protected with a relieving device of sufficient capacity to avoid overpressure in the case of an internal failure.

API RP 521 Section 3.18 states:

> Complete tube rupture, in which a large quantity of high-pressure fluid will flow to the lower pressure exchanger side, is a remote but possible contingency.

API has long used the 'two-thirds rule' to identify tube rupture scenarios. This rule states that tube rupture protection is not required when the ratio of the low pressure to high pressure side design pressure is greater than two-thirds. Tube rupture scenarios were identified only for shell and tube exchangers that did not meet the two-thirds rule.

Typical solutions include performing rigorous calculations of the relief capacity available via outflow from the low pressure side, re-rating the low pressure side to meet the two-thirds rule or installing a relief device.

13.1.6 Others (4%)

Any situation not covered by the rest described herein, but covered by the 16 overpressure scenarios in API RP 521 Section 3.1.

13.1.7 Control valve failure (1%)

To identify inlet control valve scenarios, the maximum expected pressure upstream of each inlet control valve was compared to the MAWP of the equipment under consideration.

Typical solutions include installing an SRV or re-rating the downstream equipment.

13.2 UNDER/OVERSIZED SAFETY VALVES

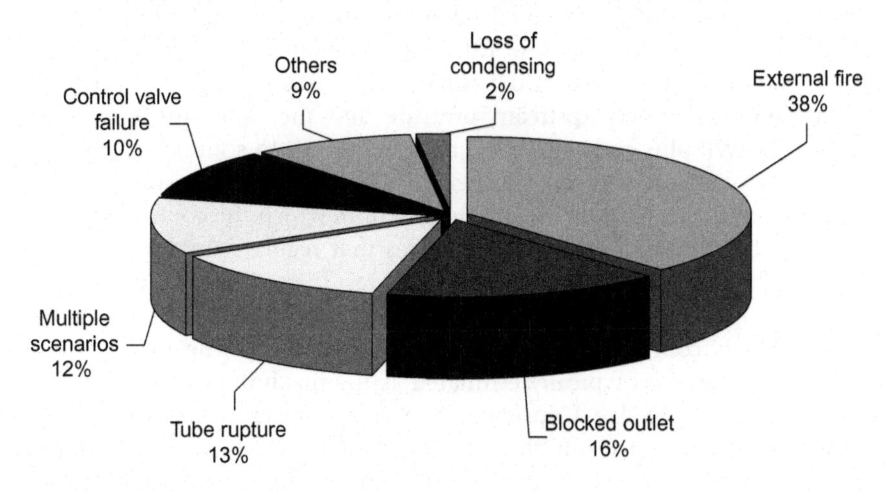

The scenarios are discussed in order of deficiency:

13.2.1 External fire (38%)

The external fire relief requirements were determined using equations presented in Appendix D of API RP 520, but which also depend on the adequacy of the firefighting and drainage system. Calculation of the wetted area was based on the high liquid level in the vessel up to 25 ft above grade (see Section 2.3). An environmental factor of 1.0 was generally used unless the insulation on the vessel (if present) was verified to be fireproof.

13.2.2 Blocked outlet (16%)

Due to the variety of blocked outlet situations, different methods were used to determine the required relief rates for different types of equipment. The quantity of the material to be relieved was generally determined at the relieving conditions (i.e. the MAWP plus the code-allowable overpressure) based on the capacity of upstream pressure sources or duty of process heaters.

13.2.3 Tube rupture (13%)

The two-thirds rule was used to determine if tube rupture protection is required for a heat exchanger.

13.2.4 Multiple scenarios (12%)

Same as above.

13.2.5 Control valve failure (10%)

Inlet control valve failure required relief rates that were based on the manufacturer's valve maximum flow calculations assuming a full open valve. The pressure differential across the control valve was the difference between the maximum expected upstream pressure and the downstream relieving pressure (MAWP plus code-allowable accumulation). In some cases, the flow through the control valve was limited by the capacity of upstream equipment, such as a pump, or by the piping system in which the control valve was installed. The failure of level control valves that regulate the flow of liquid from a higher to lower pressure system can result in a loss of liquid inventory in the upstream vessel and subsequent vapour flow to the low pressure system. This is commonly referred to as 'gas blow-by'. The relief requirement for this case was typically estimated as the maximum vapour flow rate through the control valve. Consideration was also given to the potential for the downstream vessel to fill above the normal level, which could result in the vapour flow from the control valve entering the liquid space in the downstream vessel and require two-phase relief.

13.2.6 Others (9%)

Same as above.

13.2.7 Loss of condensing (2%)

This is a very particular application where we have the potential for overpressure due to loss of overhead condensing or reflux failure. In the event the cooling medium in the condenser is lost, additional vapour may be present at the top of the column. This additional vapour may require pressure relief. In a typical distillation system, a cooling failure also results in a loss of reflux within a short period of time (typically about 15 minutes). API RP 521 states that the required relief rates before and after loss of reflux should be considered. The Berwanger audit method encompassed both of these calculations, as it was not intuitive, which case would require the larger required relief rate.

13.3 IMPROPER INSTALLATION

This is definitely the single most common problem encountered by manufacturers of so-called 'non-working SRVs'. Therefore in Chapter 6, we elaborated in length about the most common installation problems – inlet and outlet piping representing probably at least 60%. The outlet piping problems always result in uncontrolled backpressures which are not taken into account when selecting the valve. This is discussed in Section 3.4. Generally, these numbers are confirmed by the Berwanger audit on the mentioned investigated population.

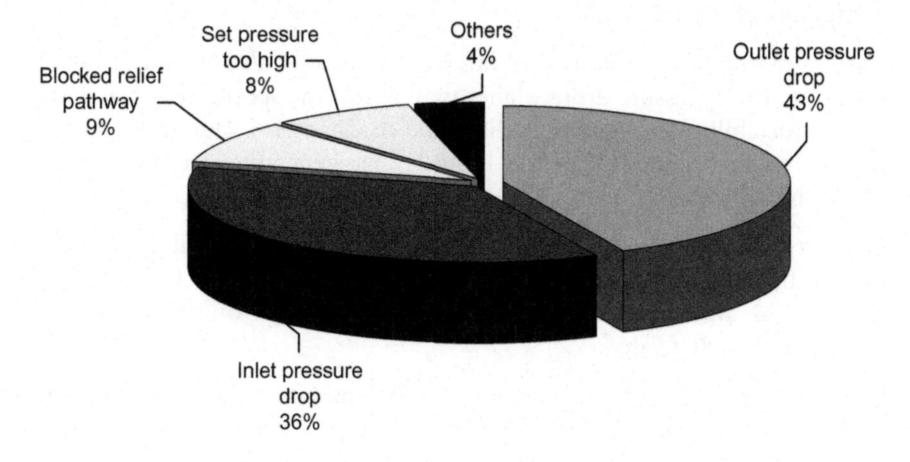

The scenarios are discussed in order of deficiency:

13.3.1 Outlet pressure drop too high (43%)

API RP 521 Section 5.4.1.3 states:

> Where conventional Safety Relief Valves are used, the relief manifold system should be sized to limit the built-up back pressure to approximately 10% of the set pressure...

While this is the general recommendation, compliance with this is not always possible. As seen earlier in this book, higher outlet pressure drops may result in reduced valve capacity and instability. In that case, other types of SRVs as discussed in Chapter 9 need to be taken into consideration. API RP 521 also states the outlet pressure drop for individual relief valves should be based on the actual rated valve capacity consistent with the inlet piping pressure drop as discussed in Section 6.1.

Outlet pressure drops were measured based on the frictional and kinetic losses from the outlet of the relief device to the discharge point, typically atmosphere or the entrance to a main relief header.

Typical solutions to meet the 10% rule include installation of a bellows kit or a pilot-operated safety relief valve. Alternatively, piping modifications can be considered, or a more rigorous analysis can be performed to further define the risk of chattering.

13.3.2 Inlet pressure drop too high (36%)

API RP 520 Part II Section 2.2.2 states:

> The total non-recoverable pressure loss between the protected equipment and the pressure relief valve should not exceed 3% of the set pressure of the valve...

The non-mandatory portion of ASME VIII Division I also includes the 3% limitation. Inlet pressure drops higher than 3% of the set pressure may result in valve instability and chatter as discussed in Chapter 6. Due to the lack of conclusive experimental evidence, industry has generally accepted the '3% rule' as the standard.

Inlet pressure drops were evaluated based on the frictional losses only between the adjacent equipment and the relief device. Per API RP 520 Part II Section 2.2.2, inlet pressure drops were calculated based on the actual rated capacity of the relief valve.

Typical solutions to meet the 3% rule include piping modifications, installing a pilot-operated relief device with remote sensing or performing more rigorous analysis to further define the risk of chattering.

Note: The API is commissioning a study to evaluate the dynamics of relief valve behaviour for various inlet and outlet pressure-drop scenarios. One should be cautioned that various experts in this area hold very strong yet contradictory opinions on this topic. However, it is the opinion of the author that with a good selection of the correct SRV, these differences of opinion are eliminated. It is wise to work on the safe side and allow the valve to overcome the possible pressure losses by selecting SRVs with good adjustable, known and tested blowdowns.

13.3.3 Blocked relief pathway (9%)

API RP 520 Part II Section IV states:

> All isolation valves in relief system piping shall meet the following guidelines:
>
> - Valves shall be full bore.
> - Valves shall have the capability of being locked open or carsealed open.

ASME VIII Section UG-135 specifically states:

> There shall be no intervening stop valves between the vessel and its protective device or devices, or between the protective device or devices and the point of discharge, except...

The ASME VIII requirements for block valves on relief system piping are similar in nature to API RP 520 Part II Section IV recommendations.

Deficiencies were identified by checking the entire relief pathway for each protected piece of equipment to ensure the existence of an open pathway. Any intervening block valves were required to be either carsealed or locked open.

The most common solution to this deficiency is to lock open all intervening block valves. It should be noted that a locked valve program should be in place to ensure the integrity of the locks.

13.3.4 Set pressure too high (8%)

Both API RP 520 and ASME VIII Division 1 state that the set or burst pressure of at least one relief device shall not exceed the MAWP of the associated equipment. In the case of multiple relief device installations, additional relief devices may be set at 105% or 110% of the MAWP, depending on the scenarios under consideration as discussed in Section 3.5.

Deficiencies were identified by comparing the current set or burst pressure to the MAWP of all protected equipment.

Common solutions to this concern include resetting the relief device or re-rating the equipment.

13.3.5 Others (4%)

API RP 520 Part II and ASME VIII Division 1 contain numerous other installation requirements. This category represents all other installation deficiencies that were identified for both relief valves and rupture discs.

13.4 WHAT CAN GO WRONG IN THE PROCESS SCENARIOS

As can be seen from the previous sections, undersized SRVs are a frequent occurrence for non-conformance of relief systems.

The best way to avoid non-conformance of existing pressure relief systems is to carefully study the 'what can go wrong' scenarios and hereby never assume that such a scenario will never happen. Also in the process industry Murphy does exist.

What can go wrong in a process? Plenty! A report in the August 2000 issue of CEP[1] (*Chemical Engineering Progress* magazine) shows that operator error or poor maintenance was the leading cause of accidents for unfired pressure vessels 8 years running.

Overpressure accidents can not only damage equipment but also cause injury or even death to plant personnel. In order to reduce the potential number of incidents or accidents, it is the job of the process engineer to analyse the process design and to determine the 'what can go wrong' scenarios and either find a way to 'design' out of them or provide protection against catastrophic failure in the event an accident does occur, that is, install an SRV and/or rupture disc.

A 'what can go wrong' scenario is defined as an action that could cause a vessel containing a gas or liquid to overpressure, leading to a catastrophic failure of that vessel if it were not for the presence of an SRV or rupture disc. To find these potentially deadly incidences, the process engineer should go through a detailed HAZOP (hazard and operability study), analysing the process to determine what these scenarios are. For each identified scenario, the process engineer can perform the calculations described in Section 2.3 to determine the amount of vapour (nominal relieving flow) or liquid that must be relieved from the vessel in a timely manner in order to prevent the overpressure from occurring, and then select the correct relieving device for the application.

Since there are many potential causes of failure, it would be nice to have a checklist to make the analysis organized and somewhat standard. As a guidance, a pretty good checklist is given by the Guide for Pressure-Relieving and Depressuring Systems, better known as API Recommended Practice 521 (API RP 521) table 1 in Section 3 (Table 13.1).

Since this is not a book on process engineering, here we will only establish a framework for analysing a given process. The ultimate goal is for the process engineer to identify credible 'what can go wrong' scenarios; perform relieving load calculations as described in Section 2.3 to prevent catastrophic failure; then size the relieving device and system as described in Chapter 8; and ultimately select the correct SRV for the application as described in Chapter 10.

Table 13.1 API RP 521 scenario checklist

API RP 521 item number	Overpressure cause
1	Closed outlets on vessels
2	Cooling water failure to condenser
3	Top-tower reflux failure
4	Side stream reflux failure
5	Lean oil failure to absorber
6	Accumulation of non-condensables
7	Entrance of highly volatile material
8	Overfilling storage or surge vessel
9	Failure of automatic control
10	Abnormal heat or vapour input
11	Split exchanger tube
12	Internal explosions
13	Chemical reaction
14	Hydraulic expansion
15	Exterior fire
16	Power failure (steam, electric or other)
	Others

In the selection of the right valve, it is always best to work in conjunction with one of the manufacturer's personnel or a consultant who is familiar with the different types of valve available on the market and who can advise the best solution for the application. Different types of valve are available for a reason. These reasons might sometimes be exclusively based on low cost, but also many times solve a particular application problem. Savings at the expense of safety is not a good idea and ultimately leads to increased LCC (life cycle cost of the valve), loss of valuable product, environmental pollution, damage to installations and, most importantly, potential loss of life.

It must be noted that API ignores failures that fall under the so-called 'double jeopardy' principle (see API 521, March 1997, 4th edition, paragraph 2.2). Double jeopardy basically means two unrelated failures occurring exactly at the same time, that is, simultaneously. This does not mean the failures occurred one minute, one second or even one millisecond apart. It means at exactly the same instant in time!

Let's consider an example:

A pump accidentally loses power, causing stoppage of cooling water to a condenser on a heating process. Because vapour from the heating process can no longer be condensed, vapour pressure builds up until it reaches the SRV's set pressure, that is, the system goes into relief. At the same time, an operator opens a steam flow control valve, adding more steam than normal to that same heating process and generating an additional excessive amount of vapour.

For sizing the SRV, should we take into account the excessive vapour produced by the wide-open steam valve, or should we consider only the normal amount of vapour exiting the heating process? Here, API 521, paragraph 2.3.2 says that the control valve should be considered to be in its normal operating position unless its function is affected by the primary initial cause of failure, the loss of power to a pump.

This is clearly a double jeopardy failure: two unrelated events occurring at exactly the same time. One has nothing to do with the other. Therefore, you need to calculate the relief capacity for one scenario at a time. For the loss of power to a pump scenario, the relief load would be based on the amount of vapour generated at the 'normal' rate of steam. For the steam control valve failure scenario, the relief capacity would be based on the amount of vapour generated by the heat provided by a wide-open steam valve; even accounting for the amount of vapour condensed in this failure, the condenser would still be in operation. So the SRV should be sized for the worst condition.

Let's look at the same situation again but with a different scenario. Suppose the pump stopped, so cooling water is lost to the condenser, causing the heating process to go into relief because of excess vapour. However, this time the operator realizes that the SRV has opened due to the pump being shut down and attempts to stop steam flow by closing the steam valve. The operator tries to put the steam control valve in manual and attempts to close it, but it won't respond because it is stuck. To free it, he strokes it wide open, shooting even more steam into the system and causing the generation of an excessive amount of vapour.

Now we have two related failures occurring at the exact same time. The power failure stops the pump and thus stops the cooling water to the column condenser. This causes the column to go into relief, which then causes the operator to react, initiating the second failure directly related to the first failure. This is a perfectly credible relieving scenario, and the calculation of relieving capacity should be based on the amount of vapour generated by the heat provided by a wide-open steam valve without taking into account the amount of vapour that can be condensed!

Note, however, that stuck-open control valves occurring simultaneously with a second failure does not constitute double jeopardy. That valve may have been stuck in its operating position for a significant amount of time before the second failure occurred. The first failure was the mechanical failure of the valve (sticking), and it did not happen at the same time as the second failure. These are unrelated failures and they do not occur simultaneously!

There are generally three approaches you can take when analysing your process. Taking a conservative approach is probably always the best. Following API 520 and 521 to the letter should be a minimum requirement, but be aware that, based on experience, some companies have consciously adopted internal rules that are even more conservative.

For example, API 521 for fire case calculations basically allows you to 'ignore' heights above a certain height when considering how much vessel surface to include in a fire zone calculation (see Section 2.3.2). Some companies go up to 8 m while others go up to 30 m for fire sizing and others simply have no height limit, considering that for a fire in a tank farm, for instance, it has been demonstrated that flames can reach over 100 m.

There is, however, an important economic factor when analysing for double jeopardy: Sometimes cost considerations by the end user dictate being less conservative. However, if there is a potential that double jeopardy failure can lead to loss of life or major equipment damage, it is wise to do the capacity calculations anyway.

When analysing a system for failures of control valves, it is best to assume all valves will fail as they are intended (fail to close will indeed fail to close, fail to open will indeed fail to open) except for the one control valve that will cause an overpressure hazard! This valve should be assumed to fail in the opposite direction (fails to close if it is intended to fail to open).

Once a credible scenario has been established, it is not recommended to take into account the use of instrumentation as a means of reducing the relieving load.

'What can go wrong' scenario analysis is a very important but complex process. It is impossible to cover every nuance associated with it and the scenarios can be open for interpretation, as is the whole API, ASME and PED. The only guidance here is to attempt at least a thorough scenario analysis and avoid major accidents and incidents that can cost money and life.

Appendix Section: Relevant Tables and References

A COMPARISONS BETWEEN THE SAFETY VALVES SIZING FORMULAS[1]

The following is a comparison between *Russian GOST Standards* and the *American API Recommended Practice 520* demonstrating that most calculation methods can look different according to different international codes, but that the results are very similar with API being usually on the conservative side.

Throughout we use the following terms and units, unless noted otherwise:

W, mass flow (kg/h)	V, volumetric flow (m^3/h)
P, pressures (bars) (0.1 Mpa)	T, temperatures (°C)
A, areas (cm^2)	ρ, density (kg/m^3)
Z, gas compressibility factor	R, constant of the gas (J/kgK)
K, nozzle flow coefficient of the valve	M, molar weight of the gas (kg/kmol)
Indices: 1 = inlet; 2 = outlet	

Gas flows

From *API RP 520,* we have:

$$W = \frac{ACKP_1K_b}{1.316}\sqrt{\frac{M}{TZ}} \quad \text{with} \quad C = 520\sqrt{k\left(\frac{2}{k+1}\right)^{(k+1)/(k-1)}}$$

and $K_b = 1$ for sonic flows.

[1] Written by Jean Paul Boyer.

For subsonic flows:

$$K_b = \frac{735F}{C} \quad \text{with} \quad F = \sqrt{\left(\frac{k}{k-1}\right)\left[\left(\frac{P_2}{P_1}\right)^{2/k} - \left(\frac{P_2}{P_1}\right)^{(k+1)/k}\right]}$$

From *GOST 12.2.085-82*, with P in kg/cm^2 and A in mm^2, we have:

$$W = BKA\sqrt{P_1\rho_1} \quad \text{with} \quad B = 1.59\sqrt{\frac{k}{k+1}}\left(\frac{2}{k+1}\right)^{1/(k-1)}$$

for sonic flows and

$$B = 1.59\sqrt{\left(\frac{k}{k-1}\right)\left[\left(\frac{P_2}{P_1}\right)^{2/k} - \left(\frac{P_2}{P_1}\right)^{(k+1)/k}\right]}$$

for subsonic flows.

Both formulas give the same results, as shown hereafter.

In the GOST formula, we first replace ρ_1 by its expression, and convert the units:

$$W = 100BKA\sqrt{1.02P_1\rho_1} \quad \text{and} \quad \rho_1 = \frac{p_1\,(\text{in Pa})}{rT_1Z} = \frac{P_1\,10^5}{rT_1Z} \Rightarrow W = 101BKA\sqrt{P_1\frac{P_1}{rT_1Z}10^5}$$

$$\text{and} \quad r = \frac{8314.3}{M} \Rightarrow W = 101BKAP_1\sqrt{\frac{M10^5}{8314.3ZT_1}} = 350.3BKAP_1\sqrt{\frac{M}{ZT_1}}$$

So now the only differences between the two formulas are:

$$GOST: 350.3B$$

$$API: \frac{CK_b}{1.316} \text{ or } 0.76CK_b$$

Critical (choked) flows

On critical flows, $K_b = 1$, so:

$$API : 0.76CK_b = 0.76 \times 520\sqrt{k\left(\frac{2}{k+1}\right)^{(k+1)/(k-1)}} = 395\sqrt{k\left(\frac{2}{k+1}\right)^{(k-1+2)/(k-1)}}$$

$$= 395\sqrt{k\left(\frac{2}{k+1}\right)^{(k-1)/(k-1)}\left(\frac{2}{k+1}\right)^{2/(k-1)}} = 395\sqrt{k\left(\frac{2}{k+1}\right)\left[\left(\frac{2}{k+1}\right)^{1/(k-1)}\right]^2}$$

$$= 395\left(\frac{2}{k+1}\right)^{1/(k-1)}\sqrt{2\left(\frac{k}{k+1}\right)} = 558.6\left(\frac{2}{k+1}\right)^{1/(k-1)}\sqrt{\frac{k}{k+1}}$$

$$= 351.3\left[1.59\left(\frac{2}{k+1}\right)^{1/(k-1)}\sqrt{\frac{k}{k+1}}\right] \approx 350.3B = \text{GOST formula}$$

Subcritical flows

$$API : 0.76CK_b = 0.76\ C\left(\frac{735F}{C}\right) = 558.6\sqrt{\left(\frac{k}{k-1}\right)\left[\left(\frac{P_2}{P_1}\right)^{2/k} - \left(\frac{P_2}{P_1}\right)^{(k+1)/k}\right]}$$

$$= 351.3\left(1.59\sqrt{\left(\frac{k}{k-1}\right)\left[\left(\frac{P_2}{P_1}\right)^{2/k} - \left(\frac{P_2}{P_1}\right)^{(k+1)/k}\right]}\right) \approx 350.3B$$

$$= \text{GOST formula}$$

In the same way we could demonstrate that the steam formulas are identical.

Note: For sonic conditions, some GOST standard copies show wrongly:

$$B = 1.59\sqrt{\frac{k}{k+1}\left(\frac{2}{k+1}\right)^{1/(k-1)}} \quad \text{instead of} \quad B = 1.59\sqrt{\frac{k}{k+1}}\left(\frac{2}{k+1}\right)^{1/(k-1)}$$

As critical flow occurs when

$$\frac{P_2}{P_1} = \beta_{cr} = \left(\frac{2}{k+1}\right)^{k/(k-1)}$$

the mistake can be shown from the coefficient for subcritical flow, by replacing P_2/P_1 by the critical ratio:

$$B = 1.59\sqrt{\frac{k}{k-1}}\sqrt{\beta_{cr}^{2/k} - \beta_{cr}^{(k+1)/k}}$$

$$= 1.59\sqrt{\frac{k}{k-1}}\sqrt{\left[\left(\frac{2}{k+1}\right)^{k/(k-1)}\right]^{2/k} - \left[\left(\frac{2}{k+1}\right)^{k/(k-1)}\right]^{(k+1)/k}}$$

Simplifying the powers by k:

$$B = 1.59\sqrt{\frac{k}{k-1}}\sqrt{\left(\frac{2}{k+1}\right)^{2/(k-1)} - \left(\frac{2}{k+1}\right)^{(k+1)/(k-1)}}$$

$$= 1.59\sqrt{\frac{k}{k-1}}\sqrt{\left(\frac{2}{k+1}\right)^{2/(k-1)} - \left(\frac{2}{k+1}\right)^{2/(k-1)}\left(\frac{2}{k+1}\right)^{(k-1)/(k-1)}}$$

$$= 1.59\sqrt{\frac{k}{k-1}}\sqrt{\left(\frac{2}{k+1}\right)^{2/(k-1)}\left(1 - \frac{2}{k+1}\right)}$$

$$= 1.59\left(\frac{2}{k+1}\right)^{1/(k-1)}\sqrt{\frac{k}{k-1}}\sqrt{\frac{k+1-2}{k+1}}$$

$$= 1.59\left(\frac{2}{k+1}\right)^{1/(k-1)}\sqrt{\frac{k}{k-1}\frac{k-1}{k+1}}$$

$$= 1.59\left(\frac{2}{k+1}\right)^{1/(k-1)}\sqrt{\frac{k}{k+1}}$$

Liquid flows

From *API RP 520* liquid formula, we have

$$V = \frac{AK}{0.19631}\sqrt{\frac{(P_1 - P_2)}{G}} \quad \text{with } G = \text{specific gravity} = \frac{\rho}{1000}$$

where V is expressed in m^3/h.

Or, in terms of kg/h:

$$W = \rho\frac{AK}{0.19631}\sqrt{\frac{(P_1 - P_2)1000}{\rho}} = \frac{AK\sqrt{1000}}{0.19631}\sqrt{(P_1 - P_2)\rho}$$

$$= 161.1AK\sqrt{(P_1 - P_2)\rho}$$

From *GOST 12.2.085-82*, gas formula, with P in kg/cm^2 and A in mm^2, we have:

$$W = 1.59KA\sqrt{(P_1 - P_2)\rho}$$

where A is in mm^2 and P is in kg/cm^2.

With A in cm^2 and P in bar:

$$W = 1.59K100A\sqrt{(P_1 - P_2)1.02\rho} = 159AK\sqrt{1.02}\sqrt{(P_1 - P_2)\rho}$$

$$W = 160.6AK\sqrt{(P_1 - P_2)\rho} \approx 161.1AK\sqrt{(P_1 - P_2)\rho} = \text{API formula}$$

B BACKPRESSURE CORRECTION FACTORS

The approximate backpressure correction factor for vapours and gases, K_b for most conventional valves with constant backpressure and backpressure-compensated pilot-operated valves with backpressures exceeding critical pressure (generally taken as 53%–55% of accumulated inlet pressure absolute), is:

$$\begin{aligned} P_b/P_1 &= \text{Backpressure percentage} \\ &= \frac{\text{Backpressure (absolute)}}{\text{Relieving pressure (absolute)}} \times 100 \end{aligned}$$

Backpressure correction factors depend on the design of the valve and it is impossible to provide general figures. Each manufacturer should provide their own data, and per EN 4126 they should be able to demonstrate them by tests.

Contrary to the European norms, API 520 has always published 'typical' backpressure correction factors in its code. These curves only serve as guidance and represent a sort of average for a number of manufacturers. API states that they can be used eventually when the make of the valve is unknown (which is rather unlikely) or, for gases and vapours, when the critical flow pressure point is unknown. What has happened is that a lot of manufacturers just adopted the API table in their sizing programs but are not able to demonstrate the numbers by test. It is interesting to see that when you start comparing backpressure correction factors in many manufacturers' catalogues or sizing programs, you will notice many will just use the same numbers used in API 520. Of course this may be correct but therefore it is wise to double-check the test data as also suggested by EN 4126. This is why we are issuing them hereafter as a reference only but not necessarily as definitive numbers to use.

Since this number is design related, the European EN 4126 has taken a more conservative approach in this particular case by asking confirmation of the numbers by tests. However, to be fair, same as in EN 4126, API 520 also recommends to contact the manufacturer for this data and, contrary to what happens, not just use the published numbers in the API document.

As a side note, it needs to be noticed that the backpressure correction factors given by API 520 are for pressures above 3.45 Barg (50 psig) only and that they are limited to backpressures below critical flow pressure for a given pressure. For everything below 3.45 Barg, the manufacturer should in any case be contacted.

For fire case applications where 21% overpressure is allowed $K_b = 1$ may be used up to a backpressure of 50%.

Figure B.1 shows the backpressure correction factor for balanced bellows safety relief valves (SRVs) on gases and vapours as published in API.

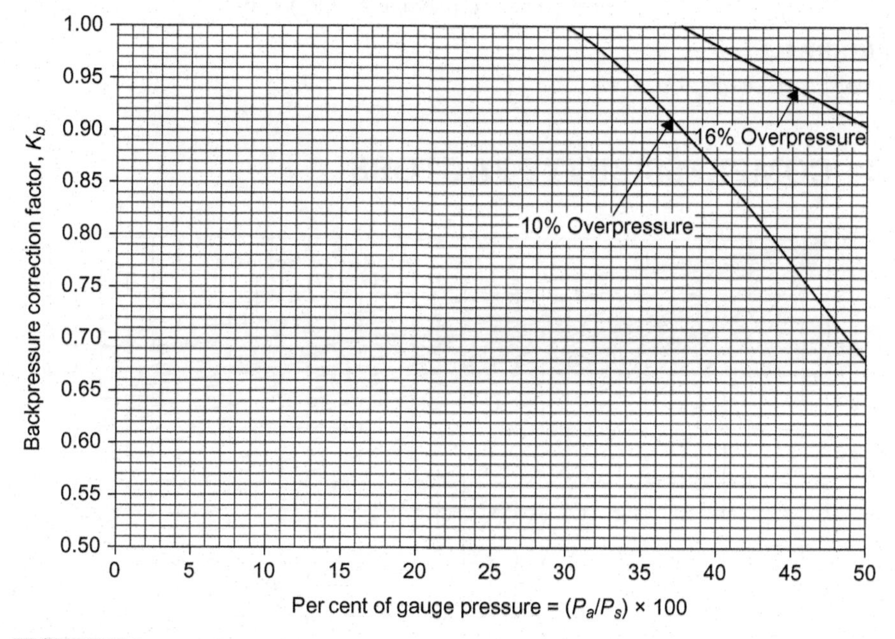

FIGURE B.1
API 520 backpressure correction factors for gases and vapours

Figure B.2 shows the capacity correction factor K_w due to backpressure on balanced bellows SRVs in liquid service.

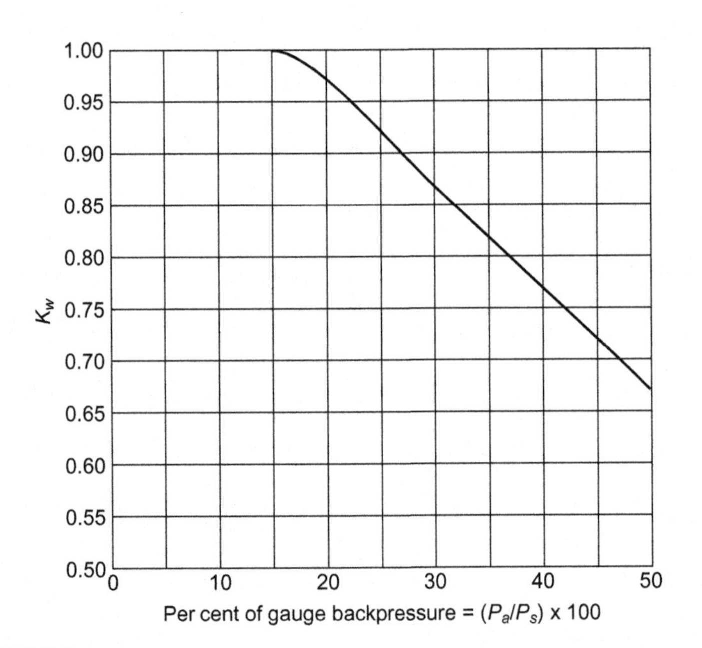

FIGURE B.2

Capacity correction graph for liquids

C COMPRESSIBILITY FACTORS

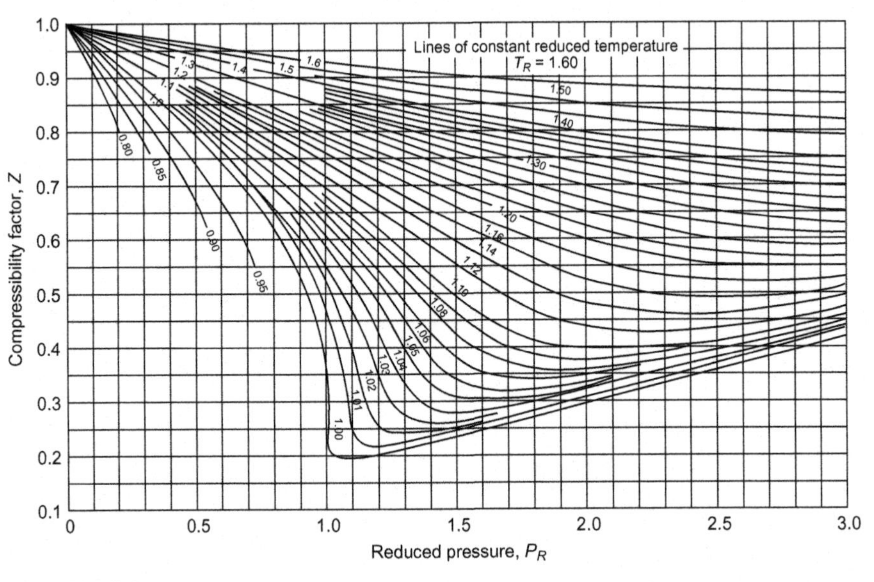

FIGURE C.1

Compressibility factors

D RATIO OF SPECIFIC HEATS k AND COEFFICIENT C

The following formula equates the ratio of specific heats to the coefficient C used in sizing methods for gases and vapours. Figure D.1(b) provides the calculated solution to this formula, where k is the ratio of specific heats.

$$C = 520\sqrt{k\left(\frac{2}{k+1}\right)^{(k+1)/(k-1)}}$$

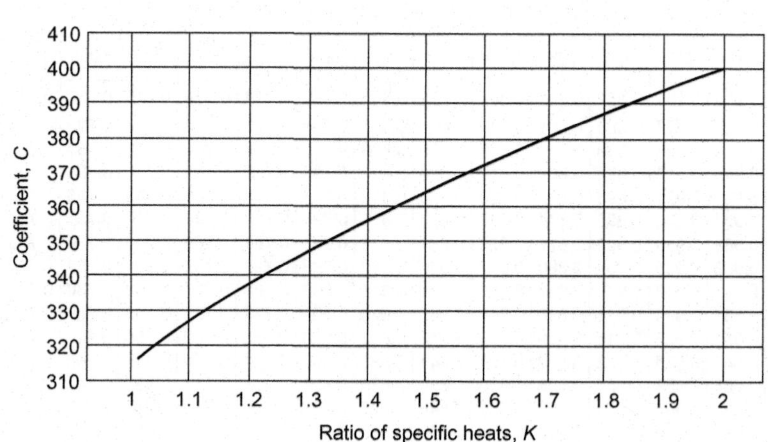

(a)

k	c	k	c	k	c	k	c	k	c
1.01	317	1.21	338	1.41	357	1.61	373	1.81	388
1.02	318	1.22	339	1.42	358	1.62	374	1.82	389
1.03	319	1.23	340	1.43	359	1.63	375	1.83	389
1.04	320	1.24	341	1.44	360	1.64	376	1.84	390
1.05	321	1.25	342	1.45	360	1.65	376	1.85	391
1.06	322	1.26	343	1.46	361	1.66	377	1.86	391
1.07	323	1.27	344	1.47	362	1.67	378	1.87	392
1.08	325	1.28	345	1.48	363	1.68	379	1.88	393
1.09	326	1.29	346	1.49	364	1.69	379	1.89	393
1.10	327	1.30	347	1.50	365	1.70	380	1.90	394
1.11	328	1.31	348	1.51	365	1.71	381	1.91	395
1.12	329	1.32	349	1.52	366	1.72	382	1.92	395
1.13	330	1.33	350	1.53	367	1.73	382	1.93	396
1.14	331	1.34	351	1.54	368	1.74	383	1.94	397
1.15	332	1.35	352	1.55	369	1.75	384	1.95	397
1.16	333	1.36	353	1.56	369	1.76	384	1.96	398
1.17	334	1.37	353	1.57	370	1.77	385	1.97	398
1.18	335	1.38	354	1.58	371	1.78	386	1.98	399
1.19	336	1.39	355	1.59	372	1.79	386	1.99	400
1.20	337	1.40	356	1.60	373	1.80	387	2.00	400

(b)

FIGURE D.1

C coefficients table

E CAPACITY CORRECTION FACTOR FOR SUPERHEAT, K_{sh}

The steam sizing formulas are based on the flow of dry saturated steam. To size for superheated steam, the superheat correction factor is used to correct the calculated saturated steam flow to superheated steam flow.

For saturated steam, $K_{sh} = 1$. When the steam is superheated, use Figure E.1 and read the superheat correction factor under the total steam temperature column.

Flowing pressure (psia)	Total temperature of superheated steam, (°F)																
	400	450	500	550	600	650	700	750	800	850	900	950	1000	1050	1100	1150	1200
50	0.987	0.957	0.930	0.905	0.882	0.861	0.841	0.823	0.805	0.789	0.774	0.759	0.745	0.732	0.719	0.708	0.696
100	0.998	0.963	0.935	0.909	0.885	0.864	0.843	0.825	0.807	0.790	0.775	0.760	0.746	0.733	0.720	0.708	0.697
150	0.984	0.970	0.940	0.913	0.888	0.866	0.846	0.826	0.808	0.792	0.776	0.761	0.747	0.733	0.721	0.709	0.697
200	0.979	0.977	0.945	0.917	0.892	0.869	0.848	0.828	0.810	0.793	0.777	0.762	0.748	0.734	0.721	0.709	0.698
250	—	0.972	0.951	0.921	0.895	0.871	0.850	0.830	0.812	0.794	0.778	0.763	0.749	0.735	0.722	0.710	0.698
300	—	0.968	0.957	0.926	0.898	0.874	0.852	0.832	0.813	0.796	0.780	0.764	0.750	0.736	0.723	0.710	0.699
350	—	0.968	0.963	0.930	0.902	0.877	0.854	0.834	0.815	0.797	0.781	0.765	0.750	0.736	0.723	0.711	0.699
400	—	—	0.963	0.935	0.906	0.880	0.857	0.836	0.816	0.798	0.782	0.766	0.751	0.737	0.724	0.712	0.700
450	—	—	0.961	0.940	0.909	0.883	0.859	0.838	0.818	0.800	0.783	0.767	0.752	0.738	0.725	0.712	0.700
500	—	—	0.961	0.946	0.914	0.886	0.862	0.840	0.820	0.801	0.784	0.768	0.753	0.739	0.725	0.713	0.701
550	—	—	0.962	0.952	0.918	0.889	0.864	0.842	0.822	0.803	0.785	0.769	0.754	0.740	0.726	0.713	0.701
600	—	—	0.964	0.958	0.922	0.892	0.867	0.844	0.823	0.804	0.787	0.770	0.755	0.740	0.727	0.714	0.702
650	—	—	0.968	0.958	0.927	0.896	0.869	0.846	0.825	0.806	0.788	0.771	0.756	0.741	0.728	0.715	0.702
700	—	—	—	0.958	0.931	0.899	0.872	0.848	0.827	0.807	0.789	0.772	0.757	0.742	0.728	0.715	0.703
750	—	—	—	0.958	0.936	0.903	0.875	0.850	0.828	0.809	0.790	0.774	0.758	0.743	0.729	0.716	0.703
800	—	—	—	0.960	0.942	0.906	0.878	0.852	0.830	0.810	0.792	0.774	0.759	0.744	0.730	0.716	0.704
850	—	—	—	0.962	0.947	0.910	0.880	0.855	0.832	0.812	0.793	0.776	0.760	0.744	0.730	0.717	0.704
900	—	—	—	0.965	0.953	0.914	0.883	0.857	0.834	0.813	0.794	0.777	0.760	0.745	0.731	0.718	0.705
950	—	—	—	0.969	0.958	0.918	0.886	0.860	0.836	0.815	0.796	0.778	0.761	0.746	0.732	0.718	0.705
1000	—	—	—	0.974	0.959	0.923	0.890	0.862	0.838	0.816	0.797	0.779	0.762	0.747	0.732	0.719	0.706
1050	—	—	—	—	0.960	0.927	0.893	0.864	0.840	0.818	0.798	0.780	0.763	0.748	0.733	0.719	0.707
1100	—	—	—	—	0.962	0.931	0.896	0.867	0.842	0.820	0.800	0.781	0.764	0.749	0.734	0.720	0.707
1150	—	—	—	—	0.964	0.936	0.899	0.870	0.844	0.821	0.801	0.782	0.765	0.749	0.735	0.721	0.708
1200	—	—	—	—	0.966	0.941	0.903	0.872	0.846	0.823	0.802	0.784	0.766	0.750	0.735	0.721	0.708
1250	—	—	—	—	0.969	0.946	0.906	0.875	0.848	0.825	0.804	0.785	0.767	0.751	0.736	0.722	0.709
1300	—	—	—	—	0.973	0.952	0.910	0.878	0.850	0.826	0.805	0.786	0.768	0.752	0.737	0.723	0.709
1350	—	—	—	—	0.977	0.958	0.914	0.880	0.852	0.828	0.807	0.787	0.769	0.753	0.737	0.723	0.710
1400	—	—	—	—	0.982	0.963	0.918	0.883	0.854	0.830	0.808	0.788	0.770	0.754	0.738	0.724	0.710
1450	—	—	—	—	0.987	0.968	0.922	0.886	0.857	0.832	0.809	0.790	0.771	0.754	0.739	0.724	0.711
1500	—	—	—	—	0.993	0.970	0.926	0.889	0.859	0.833	0.811	0.791	0.772	0.755	0.740	0.725	0.711
1550	—	—	—	—	—	0.972	0.930	0.892	0.861	0.835	0.812	0.792	0.773	0.756	0.740	0.726	0.712
1600	—	—	—	—	—	0.973	0.934	0.894	0.863	0.836	0.813	0.792	0.774	0.756	0.740	0.726	0.712
1650	—	—	—	—	—	0.973	0.936	0.895	0.863	0.836	0.812	0.791	0.772	0.755	0.739	0.724	0.710
1700	—	—	—	—	—	0.973	0.938	0.895	0.863	0.835	0.811	0.790	0.771	0.754	0.738	0.723	0.709
1750	—	—	—	—	—	0.974	0.940	0.896	0.862	0.835	0.810	0.789	0.770	0.752	0.736	0.721	0.707
1800	—	—	—	—	—	0.975	0.942	0.897	0.862	0.834	0.810	0.788	0.768	0.751	0.735	0.720	0.705
1850	—	—	—	—	—	0.976	0.944	0.897	0.862	0.833	0.809	0.787	0.767	0.749	0.733	0.718	0.704
1900	—	—	—	—	—	0.977	0.946	0.898	0.862	0.832	0.807	0.785	0.766	0.748	0.731	0.716	0.702
1950	—	—	—	—	—	0.979	0.949	0.898	0.861	0.832	0.806	0.784	0.764	0.746	0.729	0.714	0.700
2000	—	—	—	—	—	0.982	0.952	0.899	0.861	0.831	0.805	0.782	0.762	0.744	0.728	0.712	0.698

FIGURE E.1

Superheat correction factors

F CAPACITY CORRECTION FACTOR FOR HIGH PRESSURE STEAM, K_n

The high pressure steam correction factor K_n is used when the steam pressure P_1 is greater than 1500 psia [10,340 kPaa] and up to 3200 psia [22,060 kPaa]. This factor has been adopted by ASME to account for the deviation between steam flow as determined by Napier's equation and actual saturated steam flow at high pressures. K_n can also be calculated by the following equation or may be taken from Figure F.1.

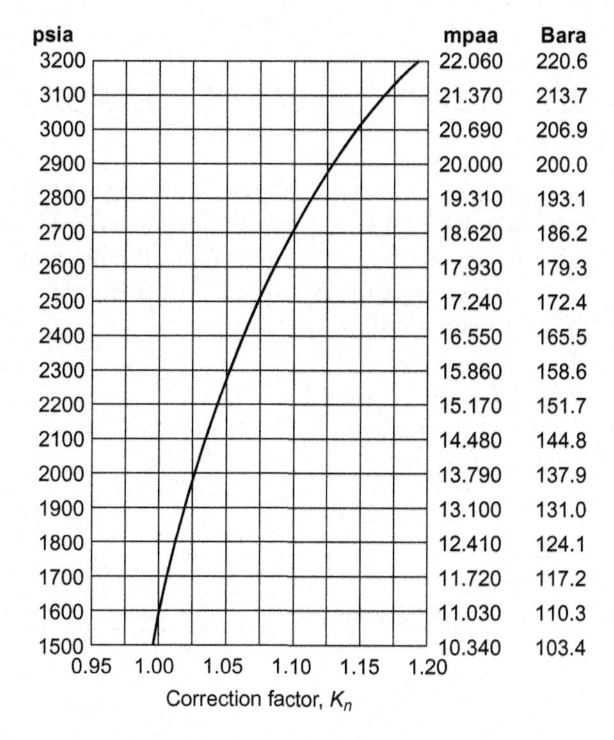

FIGURE F.1

Correction factor for high pressure steam, k_n

U.S.C.S. units:

$$K_n = \frac{0.1906P_1 - 1000}{0.2292P_1 - 1061}$$

Metric units:

$$K_n = \frac{0.02763P'_1 - 1000}{0.03324P'_1 - 1061}$$

where:

K_n = High pressure steam correction factor
P_1 = Relieving pressure (psia). This is the set pressure + overpressure + atmospheric pressure
P'_1 = Relieving pressure (kPaa).

G CAPACITY CORRECTION FACTOR FOR VISCOSITY, K_v

When an SRV is sized for viscous liquid service, it is suggested that it would be sized first as for a non-viscous type application in order to obtain a preliminary required effective discharge area (A). From the manufacturer's standard effective orifice sizes, select the next larger orifice size and calculate the Reynolds' number, *Re*, per the following formula:

English units:

$$Re = \frac{W(2800G)}{\mu\sqrt{A}} \qquad Re = \frac{12,700W}{U\sqrt{A}}$$

Metric units:

$$Re = \frac{Q(18,800G)}{\mu\sqrt{A'}} \qquad Re = \frac{85,225Q}{U\sqrt{A'}}$$

where:

W = Flow rate at the flowing temperature (USGPM)
G = Specific gravity of the liquid at the flowing temperature referred to water = 1.00 at 70 °F or 21 °C
A = Effective discharge area (in^2, from manufacturers' standard orifice areas)
U = Viscosity at the flowing temperature, Saybolt Universal Seconds (SSU).

μ = Absolute viscosity at the flowing temperature (cp = centipoise)
Q = Flow rate at the flowing temperature (l/min)
A' = Effective discharge area (mm^2).

After the value of *Re* is determined, the factor K_V is obtained from Figure G.1. Factor K_V is applied to correct the 'preliminary required discharge area'. If the corrected area exceeds the 'chosen effective orifice area', the above calculations should be repeated using the next larger effective orifice size as the required effective orifice area of the valve selected cannot be less than the calculated required effective area.

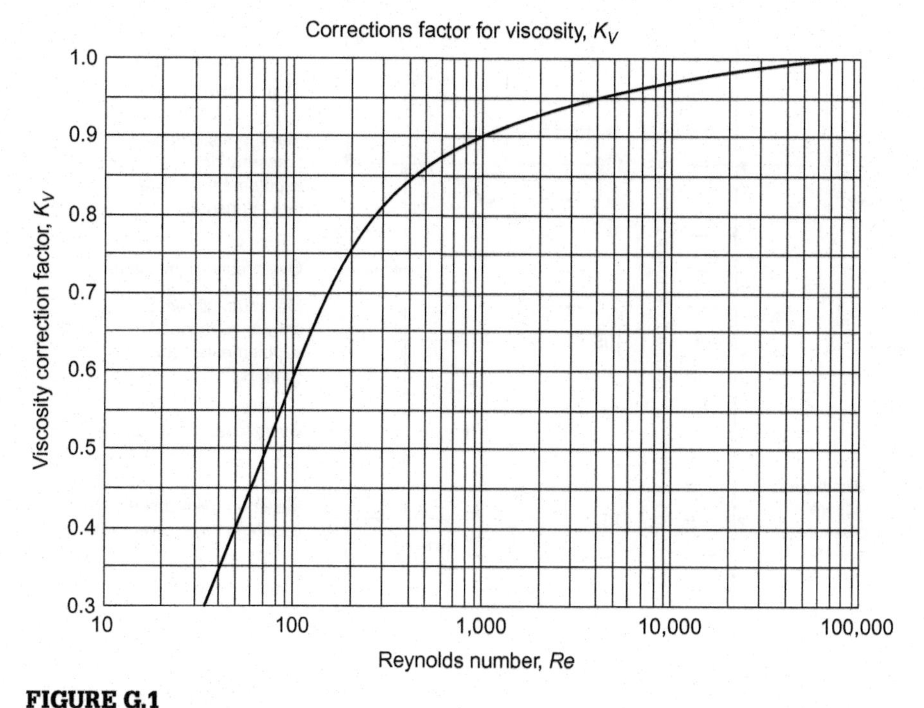

FIGURE G.1
Viscosity correction factor

H ALLOWABLE OPERATING, WORKING, RELIEF, SET AND BLOWDOWN PRESSURES

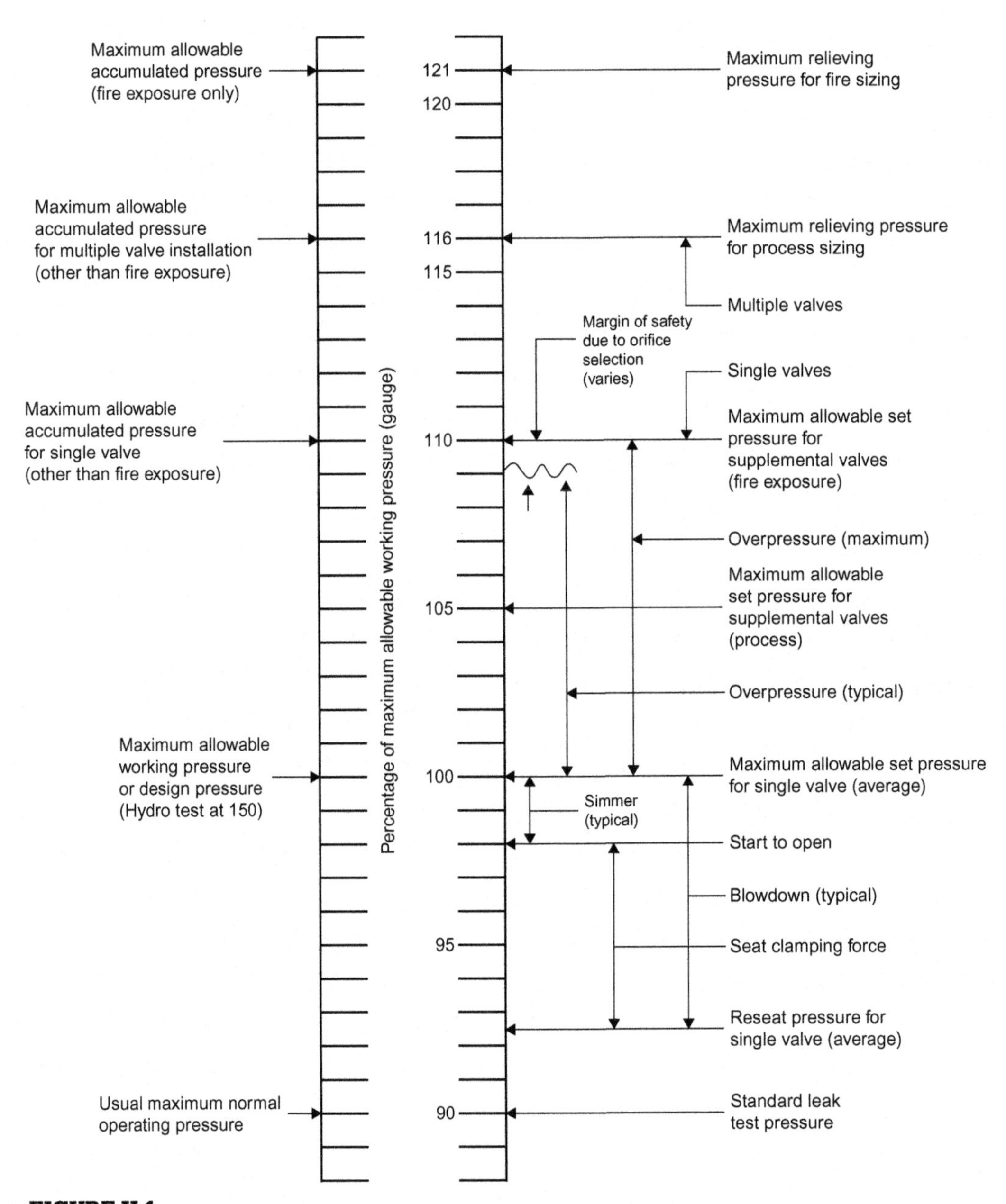

FIGURE H.1

Pressure table

I CODES AND STANDARDS ORGANIZATIONS

Organization	Publications
American Petroleum Institute Washington www.api.org	*API Recommended Practice 520 Part 1* – Sizing and Selection *API Recommended Practice 520 Part 2* – Installation *API Recommended Practice 521* – Guide for Pressure Relief and Depressurizing Systems *API Standard 526* – Flanged Steel Pressure Relief Valves *API Standard 527* – Seat Tightness of Pressure Relief Valves *API Recommended Practice 576* – Inspection of Pressure Relieving Devices
ASME International New York www.asme.org	*ASME PTC 25* – Pressure Relief Devices *ASME B16.34* – Valves – Flanged, Threaded and Welding End *ASME B31.1* – Power Piping *ASME B31.3* – Process Piping *ASME B31.8* – Gas Transmission and Distribution Piping Systems *ASME Boiler and Pressure Vessel Code* Section I – Power Boilers Section II – Materials Section III – Nuclear Power Stations Section IV – Heating boilers Section V – Non-destructive Examination Section VIII – Pressure Vessels Section IX – Welding and Brazing Qualifications
ISO – International Organization for Standardization Geneva www.iso.org	*EN/ISO 4126 – Safety Devices for Protection Against Excessive Pressure* Part 1 – Safety Valves Part 2 – Bursting Disc Safety Devices Part 3 – Safety Valves and Bursting Disc Safety Devices in Combination Part 4 – Pilot Operated Safety Valves Part 5 – Controlled Safety Pressure Relief Systems Part 6 – Application, Selection and Installation of Bursting Disc Safety Devices Part 7 – Common Data Part 9 – Application and Installation of Safety Devices Excluding Standalone Bursting Disc Safety Devices ISO 9001 – 2000 – Quality Management System

(Continued)

Organization	Publications
CEN – European Committee for Standardization Brussels www.cenorm.be	*EN/ISO 4126 – Safety Devices for Protection Against Excessive Pressure* Part 1 – Safety Valves Part 2 – Bursting Disc Safety Devices Part 3 – Safety Valves and Bursting Disc Safety Devices in Combination Part 4 – Pilot Operated Safety Valves Part 5 – Controlled Safety Pressure Relief Systems Part 6 – Application, Selection and Installation of Bursting Disc Safety Devices Part 7 – Common Data
NACE – National Association of Corrosion Engineers Houston www.nace.org	*NACE MR0175-2002* – Sulphide stress cracking resistant metallic materials for oilfield equipment *NACE MR0175-2003* – Metals for sulphide stress cracking and stress corrosion cracking resistance in sour oilfield environments *NACE MR0175/ISO15156-2003 Parts 1, 2 and 3* – Petroleum and natural gas industries materials for use in H_2S containing environments in oil and gas production *NACE MR0103-2003* – Materials resistant to sulphide stress cracking in corrosive petroleum environments
National Board of Boiler and Pressure Vessel Inspections Columbus www.nationalboard.org	*NB-18* – National Board Pressure Relief Device Certifications *NB-23* – National Board Inspection Code (NBIC)

J API 526 DATA SHEET RECOMMENDATION

SPRING-LOADED PRESSURE RELIEF VALVE SPECIFICATION SHEET	
	Page of
	Requisition No.
	Job No.
	Date
	Revised
	By

GENERAL	BASIS OF SELECTION
1. Item Number:	5. Code: ASME VIII [ ] Stamp Req'd: Yes [ ] No [ ]
2. Tag Number:	Other [ ] Specify:
3. Service, Line, or Equipment Number:	6. Comply With API Std 526: Yes [ ] No [ ]
4. Number Required:	7. Fire [ ] Other [ ] Specify:
	8. Rupture Disk: Yes [ ] No [ ]
VALVE DESIGN	**MATERIALS**
9. Design Type:	17. Body
Conventional [ ] Bellows [ ] Balanced Piston [ ]	18. Bonnet:
10. Nozzle Type: Full [ ] Semi [ ]	19. Scat (Nozzle): Disk:
Other [ ] Specify:	20. Resilient Seat:
11. Bonner Type: Open [ ] Closed [ ]	21. Guide
12. Seat Type: Metal to Metal [ ] Resilient [ ]	22. Adjusting Ring(s):
13. Scat Tightness: API Std 527 [ ]	23. Spring: Washer:
Other [ ] Specify:	24. Bellows:
CONNECTIONS	25. Balanced Piston:
	26. Comply With NACE MRO175: Yes [ ] No [ ]
14. Inlet Size Rating Facing	27. Other (Specify):
15. Outlet Size Rating Facing	
16. Other (Specify):	
SERVICE CONDITIONS	**ACCESSORIES**
33. Fluid and State:	28. Cap: Screwed [ ] Bolted [ ]
34. Required Capacity Per Valve & Units:	29. Lifting Lever: Plain [ ] Packed [ ] None [ ]
35. Molecular Weight or Specific Gravity:	30. Test Gag: Yes [ ] No [ ]
36. Viscosity at Flowing Temperature & Units:	31. Bug Screen: Yes [ ] No [ ]
37. Operating Pressure & Units:	32. Other (Specify):
38. Set Pressure & Units:	
39. Blowdown: Standard [ ] Other [ ]	
40. Latent Heat of Vaporization & Units:	
41. Operating Temperature & Units:	**SIZING AND SELECTION**
42. Relieving Temperature & Units:	49. Calculated Orifice Area (in square in.):
43. Built-up Back Pressure & Units:	50. Selected Orifice Area (in square in.):
44. Superimposed Back Pressure & Units:	51. Orifice Designation (letter):
45. Cold Differential Test Pressure & Units:	52. Manufacturer:
45. Allowable Overpressure in Percent or Units:	53. Model Number:
47. Compressibility Factor, Z:	54. Manufacturer's Orifice Area (in square in):
48. Ratio of Specific Heats:	55. Manufacturer's Coefficient of Discharge:
	56. Vendor Calculations Required: Yes [ ] No [ ]

Note: Indicate items to be filled in by the manufacturer with an asterisk (*).

PILOT- OPERATED
PRESSURE RELIEF VALVE
SPECIFICATION SHEET

Page of
Requisition No.
Job No.
Date
Revised
By

GENERAL	BASIS OF SELECTION
1. Item Number:	5. Code: ASME VIII [ ] Stamp Req'd: Yes [ ] No [ ]
2. Tag Number:	Other [ ] Specify:
3. Service, Line, or Equipment Number:	6. Comply With API Std 526: Yes [ ] No [ ]
4. Number Required:	7. Fire [ ] Other [ ] Specify:
	8. Rupture Disk: Yes [ ] No [ ]
VALVE DESIGN	**MATERIALS**
9. Design Type: Piston [ ] Diaphragm [ ] Bellows [ ]	20. Body:
10. Number of Pilots:	21. Scat (Nozzle): Piston:
11. Pilot Type: Flowing [ ] Non-flowing [ ]	22. Resilient Seat: Seals:
12. Pilot Action: Pop [ ] Modulating [ ]	23. Piston Seal:
13. Pilot Senses Internal [ ] Remote [ ]	24. Piston Liner/Guide:
14. Seat Type: Metal to Metal [ ] Resilient [ ]	25. Diaphragm/Bellows:
15. Seat Tightness: API Std 527 [ ]	**MATERIALS, PILOT**
Other [ ] Specify:	
16. Pilot Vent: Atmosphere [ ] Outlet [ ]	26. Body/Bonnet:
Other [ ] Specify:	27. Internals:
	28. Seat: Seals:
CONNECTIONS	29. Diaphragm:
	30. Tubing/Fillings:
17. Inlet Size Rating Facing	31. Filter Body: Cartridge:
18. Outlet Size Rating Facing	32. Spring:
19. Other (Specify):	33. Comply With NACE MRO175: Yes [ ] No [ ]
	34. Other (Specify):
SERVICE CONDITIONS	**ACCESSORIES**
43. Fluid and State:	35. External Filter Yes [ ] No [ ]
44. Required Capacity Per Valve & Units:	36. Lifting Lever: Plain [ ] Packed [ ] None [ ]
45. Molecular Weight or Specific Gravity:	37. Field Test Connection: Yes [ ] No [ ]
46. Viscosity at Flowing Temperature & Units:	38. Field Test Indicator: Yes [ ] No [ ]
47. Operating Pressure & Units:	39. Backflow Preventer: Yes [ ] No [ ]
48. Set Pressure & Units:	40. Manual Blowdown Valve: Yes [ ] No [ ]
49. Blowdown: Standard [ ] Other [ ]	41. Test Gag: Yes [ ] No [ ]
50. Latent Heat of Vaporization & Units:	42. Other (Specify):
51. Operating Temperature & Units:	
52. Relieving Temperature & Units:	**SIZING AND SELECTION**
53. Built-up Back Pressure & Units:	
54. Superimposed Back Pressure & Units:	59. Calculated Orifice Area (in square in.):
55. Cold Differential Test Pressure & Units:	60. Selected Orifice Area (in square in.):
56. Allowable Overpressure in Percent or Units:	61. Orifice Designation (letter):
57. Compressibility Factor, Z:	62. Manufacturer:
58. Ratio of Specific Heats:	63. Model Number:
	64. Manufacture's Orifice Area (in square in.):
	65. Manufacture's Coefficient of Discharge:
	66. Vendor Calculations Required: Yes [ ] No [ ]

Note: Indicate items to be filled in by the manufacturer with an asterisk (*).

K GENERIC SIZING PROGRAM

Page : of

JOB Nr: 3006750

Spring Loaded Safety Rellet valves

	CLIENT :						REVISIONS	
	CLIENT REF :				DATE:		1	3
	PROJECT :						2	4
1	ITEM Nr	QUANTITY						
2	TAG Nr							
3	NOZZLE TYPE		SEMI		FULL		FULL	
4	DESIGN	BONNET TYPE	CONVENT.	CLOSED	CONVENT.	CLOSED	CONVENT.	CLOSED
5	**CONNECTIONS**	TYPE						
6	SIZE INLET	OUTLET						
7	RATING INLET	OUTLET						
8	FACING INLET	OUTLET						
9	**MATERIAL**	BODY | BONNET						
10	DISC	SOFT SEAT						
11	NOZZLE	SEALS / GASKETS						
12	GUIDE	RINGS						
13	SPRING							
14	BELLOWS		N/A		N/A		N/A	
15	**OPTIONS**							
16	CAP. SCREWED or BOLTED		YES					
17								
18								
19								
20	PAINTING		STANDARD		STANDARD		STANDARD	
21	**BASIS**	SIZING	API 520		API 520		API 520	
22		CASE						
23	**UNITS** (Press: GAUGE)	AREA (1=cm² 2=in² 3=mm²)					2	sq in
24	P. (1=Bar 2=PSI 3=kg/cm² 4=kPa 5=oz/in² 6=Pa 7=mBar 8=in wc 9=g/cm² 10=mm wc 11=mmHg)						1	BarG
25	GAS / STEAM FLOW (1=kg/hr 2=lb/hr 3=t/hr 4=kg/s 5=SCFM 6=Nm³/hr 7=Sm³/hr 8=SCFH 9=SCFD)						6	Nm³/hr
26	LIQUID FLOW (1=m³/hr 2=GPM-US 3=l/min 4=kg/hr 5=lb/hr 6=kg/s 7=Brl/Day 8=GPM-Imp)						1	m³/hr
27	TEMPERATURE (1=ºC 2=ºF 3=ºK 4=ºR)						1	ºC
28	**FLUID DATAS**	FLUID DESCRIPTION						
	INLET LOSS							
29	CAPACITY	GAS | STEAM		0.				
30		LIQUID	0.		0.			
31	MOLEC. WEIGHT	SPEC. GRAVITY		1,000	1,00	1,000	1,00	1,000
32	Cp/Cv	Const. C	1,000	315	1,000	315	1,000	315
33	Ksh or Z	VISCOSITY	1,000		1,000		1,000	
34	SERVICE TEMP	RELIEF TEMP				0.		0.
35	CONST. B.-PRESS.	VARIABLE B.-P.						
36	TOTAL BACK-PRESSURE		0.		0.		0.	
37	SERVICE PRESS.	SET PRESSURE						
38	COLD or DIFFER. SET PRESSURE		0.		0.		0.	
39	OVER-PRESSURE	BLOWDOWN	10%	Std	10%	Std	10%	Std
40	**ORIFICE**							
41	K Gas / Steam	K Liquid	0,000	0,000	0,000	0,000	0,000	0,000
42	Kb (Back-P, gas)	Kw (Back-P, liquid)	1,000	1,000	1,000	1,000	1,000	1,000
43	Kp (Over-P, liquid)	Kv (Visc factor)	1,000	1,000	1,000	1,000	1,000	1,000
44	CALC. AREA GAS	LIQUID	0.	0.	0.	0.	0.	0.
45	TOTAL CALCULATED AREA		0.		0.		0.	
46	SELECTED AREA	DESIGNATION						
47	FULL CAPACTITY	GAS | LIQUID	0	0	0	0	0	0
48	REACTION FORCE	daN | lbs	N/A	N/A	N/A	N/A	N/A	N/A
49	MAX. NOISE AT 30. M, OPEN OUTLET		N/A	dBa	N/A	dBa		dBa
50	**VALVE MODEL Nr**							
51	**SERIAL NUMBER**							
52	**Dimensions, Weight**	A / B						
53	(mm / kg)	C / WEIGHT						
54	Formulas used : $A(gas)=1.316W\sqrt{(TZ)}/(CKP_1Kb\sqrt{M})$ $A(steam)=W/(52.5KP_1Ksh)$ $A(liquid)=0.19631W\sqrt{G}/[KKpKvKw\sqrt{(p_1-BP)}]$							
55	with $P_1=p_1+1.013$ (absolute) and $p_1=SET(1+OverPress)$ (gauge) in: kg/hr - m³/hr - BarG - ºK							
56	**NOTES** • Dimensions & Weight for information. To be confirmed on certified drawings.							
57	• Sizing per formulas from API RP 520 & 521 (noise level). Noise & Force given at Full Opening of valve.							
58	• Given reaction force does not take in account the effects of outlet static pressure if any							
59								
60								
61								
62								

C

A

B

L WORLDWIDE CODES AND STANDARDS (MOST COMMON)

Country	Standard no.	Description
Germany	A.D. Merkblatt A2 TRD 421 TRD 721	Pressure vessel equipment safety devices against excess pressure – safety valves Technical equipment for steam boilers safeguards against excessive pressure – safety valves for steam boilers of groups I, III and IV Technical equipment for steam boilers safeguards against excessive pressure – safety valves for steam boilers of group II
United Kingdom	BS 6759	Part 1 specification for safety valves for steam and hot water Part 2 specification for safety valves for compressed air or inert gas Part 3 specification for safety valves for process fluids
France	AFNOR NFE-E 29-411 to 416 NFE-E 29-421	Safety and relief valves Safety and relief valves
Korea	KS B 6216	Spring-loaded safety valves for steam boilers and pressure vessels
Japan	JIS B 8210	Steam boilers and pressure vessels – spring-loaded safety valves
Australia	SAA AS1271	Safety valves, other valves, liquid level gauges and other fittings for boilers and unfired pressure vessels
United States	ASME I ASME III ASME VIII ANSI/ASME PTC 25.3	Boiler applications Nuclear applications Unfired pressure vessel applications Safety and relief valves – performance test codes
	API RP 520	Sizing selection and installation of pressure-relieving devices in refineries Part 1 Design Part 2 Installation
	API RP 521 API STD 526 API STD 527	Guide for pressure-relieving and depressurizing systems Flanged steel pressure relief valves Seat tightness of pressure relief valves
Europe	prEN ISO 4126*	Safety devices for protection against excessive pressure
International	ISO 4126	Safety valves – general requirements

**pr = pre-ratification. This harmonized European standard is not officially issued.*

M PROPERTIES OF COMMON GASES

Gas or vapour	Molecular weight, *M*	Ratio of specific heats, *k* (14.7 psia)	Coefficient, C^b	Specific gravity	Critical pressure (psia)	Critical temperature (°R) (°F + 460)
Acetylene	26.04	1.25	342	0.899	890	555
Air	28.97	1.40	356	1.000	547	240
Ammonia	17.03	1.30	347	0.588	1638	730
Argon	39.94	1.66	377	1.379	706	272
Benzene	78.11	1.12	329	2.696	700	1011
N-butane	58.12	1.18	335	2.006	551	766
Iso-butane	58.12	1.19	336	2.006	529	735
Carbon dioxide	44.01	1.29	346	1.519	1072	548
Carbon disulphide	76.13	1.21	338	2.628	1147	994
Carbon monoxide	28.01	1.40	356	0.967	507	240
Chlorine	70.90	1.35	352	2.447	1118	751
Cyclohexane	84.16	1.08	325	2.905	591	997
Ethane	30.07	1.19	336	1.038	708	550
Ethyl alcohol	46.07	1.13	330	1.590	926	925
Ethyl chloride	64.52	1.19	336	2.227	766	829
Ethylene	28.03	1.24	341	0.968	731	509
Freon 11	137.37	1.14	331	4.742	654	848
Freon 12	120.92	1.14	331	4.174	612	694
Freon 22	86.48	1.18	335	2.965	737	665
Freon 114	170.93	1.09	326	5.900	495	754
Helium	4.02	1.08	377	0.139	33	10
N-heptane	100.20	1.05	321	3.459	397	973
Hexane	86.17	1.06	322	2.974	437	914
Hydrochloric acid	36.47	1.41	357	1.259	1198	584
Hydrogen	2.02	1.41	357	0.070	188	80
Hydrogen chloride	36.47	1.41	357	1.259	1205	585
Hydrogen sulphide	34.08	1.32	349	1.176	1306	672

(Continued)

Gas or vapour	Molecular weight, *M*	Ratio of specific heats, *k* (14.7 psia)	Coefficient, C[b]	Specific gravity	Critical pressure (psia)	Critical temperature (°R) (°F + 460)
Methane	16.04	1.31	348	0.554	673	344
Methyl alcohol	32.04	1.20	337	1.106	1154	924
Methyl butane	72.15	1.08	325	2.491	490	829
Methyl chloride	50.49	1.20	337	1.743	968	749
Natural gas (typical)	19.00	1.27	344	0.656	671	375
Nitric oxide	30.00	1.40	356	1.038	956	323
Nitrogen	28.02	1.40	356	0.967	493	227
Nitrous oxide	44.02	1.31	348	1.520	1054	557
N-octane	114.22	1.05	321	3.943	362	1025
Oxygen	32.00	1.40	356	1.105	737	279
N-pentane	72.15	1.06	325	2.491	490	846
Iso-pentane	72.15	1.06	325	2.491	490	829
Propane	44.09	1.13	330	1.522	617	666
Sulphur dioxide	64.04	1.27	344	2.211	1141	775
Toluene	92.13	1.09	326	3.180	611	1069

**If 'C' is not known, then use C = 315. If the ratio of specific heats 'k' is known, refer to Annex D to calculate 'C'.*

N RELEVANT CONVERSION FACTORS

A Multiply	B By	C Obtain
Atmospheres	14.70	Pounds per square inch
Atmospheres	1.033	Kilograms per square centimetre
Atmospheres	29.92	Inches of mercury
Atmospheres	760.0	Millimetres of mercury
Atmospheres	407.5	Inches of water
Atmospheres	33.96	Feet of water
Atmospheres	1.013	Bara
Atmospheres	101.3	Kilo Pascals
Barrels	42.00	Gallons (US)
Bara	14.50	Pounds per square inch
Bars	1.020	Kilograms per square centimetre
Bara	100.0	Kilo Pascals
Centimetres	0.3937	Inches
Centimetres	0.03281	Feet
Centimetres	0.010	Metres
Centimetres	0.01094	Yards
Cubic centimetres	0.06102	Cubic inches
Cubic feet	7.481	Gallons
Cubic feet	0.1781	Barrels
Cubic feet per minute	0.02832	Cubic metres per minute
Cubic feet per second	448.8	Gallons per minute
Cubic inches	16.39	Cubic centimetres
Cubic inches	0.004329	Gallons
Cubic metres	264.2	Gallons
Cubic metres per hour	4.403	Gallons per minute
Cubic metres per minute	35.31	Cubic feet per minute
Standard cubic feet per minute	60.00	Standard cubic feet per hour
Standard cubic feet per minute	1440.	Standard cubic feet per day

(Continued)

A	B	C
Standard cubic feet per minute	0.02716	Nm^3/min (0°C, 1 Bara)
Standard cubic feet per minute	1.630	Nm^3/h (0°C, 1 Bara)
Standard cubic feet per minute	39.11	Nm^3/day (0°C, 1 Bara)
Standard cubic feet per minute	0.02832	Nm^3/min
Standard cubic feet per minute	1.699	Nm^3/h
Standard cubic feet per minute	40.78	Nm^3/day
Feet	0.3048	Metres
Feet	0.3333	Yards
Feet	30.48	Centimetres
Feet of water (68°F)	0.8812	Inches of mercury (0°C)
Feet of water (68°F)	0.4328	Pounds per square inch
Gallons (US)	3785.	Cubic centimetres
Gallons (US)	0.1337	Cubic feet
Gallons (US)	231.0	Cubic inches
Gallons (Imperial)	277.4	Cubic inches
Gallons (US)	0.8327	Gallons (Imperial)
Gallons (US)	3.785	Litres
Gallons of water (60°F)	8.337	Pounds
Gallons of liquid	500 × Sp. Gr.	Pounds per hour liquid per minute
Gallons per minute	0.002228	Cubic feet per second
Gallons per minute (60°F)	227.0 × SG	Kilograms per hour
Gallons per minute	0.06309	Litres per second
Gallons per minute	3.785	Litres per minute
Gallons per minute	0.2271	M^3/h
Grams	0.03527	Ounces
Inches	2.540	Centimetres
Inches	0.08333	Feet
Inches	0.0254	Metres
Inches	0.02778	Yards
Inches of mercury (0°C)]	1.135	Feet of water (68°F)
Inches of mercury (0°C)	0.4912	Pounds per square inch

(Continued)

A	B	C
Inches of mercury (0°C)	0.03342	Atmospheres
Inches of mercury (0°C)	0.03453	Kilograms per square centimetre
Inches of water (68°F)	0.03607	Pounds per square inch
Inches of water (68°F)	0.07343	Inches of mercury (0°C)
Kilograms	2.205	Pounds
Kilograms	0.001102	Short tons (2000 lbs)
Kilograms	35.27	Ounces
Kilograms per minute	132.3	Pounds per hour
Kilograms per square centimetre	14.22	Pounds per square inch
Kilograms per square centimetre	0.9678	Atmospheres
Kilograms per square centimetre	28.96	Inches of mercury
Kilograms per cubic metre	0.0624	Pounds per cubic foot
Kilo Pascals	0.1450	Pounds per square inch
Kilo Pascals	0.0100	Bars
Kilo Pascals	0.01020	Kilograms per square centimetre
Litres	0.03531	Cubic feet
Litres	1000.0	Cubic centimetres
Litres	0.2842	Gallons
Litres per hour	0.004403	Gallons per minute
Metres	3.281	Feet
Metres	1.094	Yards
Metres	100.0	Centimetres
Metres	39.37	Inches
Pounds	0.1199	Gallons H_2O @ 60F (US)
Pounds	453.6	Grams
Pounds	0.0005	Short tons (2000 lbs)
Pounds	0.4536	Kilograms
Pounds	0.0004536	Metric tons
Pounds	16.00	Ounces

(Continued)

A	B	C
Pounds per hour	6.324/M.W.	SCFM
Pounds per hour	0.4536	Kilograms per hour
Pounds per hour liquid	0.002/Sp.Gr.	Gallons per minute liquid (at 60°F)
Pounds per square inch	27.73	Inches of water (68°F)
Pounds per square inch	2.311	Feet of water (68°F)
Pounds par square inch	2.036	Inches of mercury (0°C)
Pounds per square inch	0.07031	Kilograms per square centimetre
Pounds per square inch	0.0680	Atmospheres
Pounds per square inch	51.71	Millimetres of mercury (0°C)
Pounds per square inch	0.7043	Metres of water (68°F)
Pounds per square inch	0.06895	Bar
Pounds par square inch	6.895	Kilo Pascals
Specific gravity (of gas or vapours)	28.97	Molecular weight (of gas or vapours)
Square centimetre	0.1550	Square inch
Square inch	6.4516	Square centimetre
Square inch	645.16	Square millimetre
SSU	0.2205 × SG	Centipoise
SSU	0.2162	Centistoke
Water (cubic feet @ 60°F)	62.37	Pounds
Temperature: Centigrade Kelvin Fahrenheit Fahrenheit Fahrenheit		 = 5/9 (Fahrenheit −32) = Centigrade + 273 = 9/5 [Centigrade] +32 = Rankine – 460 (9/5 Kelvin) – 460

Further Reading

API (www.api.org) Recommended Practice 520, "Sizing, Selection, and installation of Pressure-Relieving Device in Refineries, Part 1 – Sizing and Selection", 7th Edition (January 2000).

ASME (www.asme.org) "Boiler and Pressure Vessel Code, Section VIII, Division 1" (1998).

PED 97/23/EC (Pressure Equipment Directive).

European Normalisations : EN/ISO 4126.

Guidelines on noise. Medical Research Report EA 7301. API 1973.

Design principles: Working environment. NORSOK Standard S-DP-002 Rev 1, Dec. 1994 (PO Box 547, N-4001 Stavanger, Norway. Fax (47) 51562105).

86/188/EEC Council Directive of 12 May 1986 on the protection of workers from the risks related to the exposure to noise at work.

The noise at work regulations. SI No. 1790, 1989.

IEC 534-8-3: 1995 Industrial process control valves Part 8 Noise Considerations

Section 3 Control valve aerodynamic noise prediction method

API RP 521. Guide for pressure relieving and depressuring systems (March 1997)

Noise Reduction. L. L. Beranek (Ed.) McGraw-Hill (Pub. 1960) Institute of Sound and Vibration Research, University of Southampton, Hampshire, UK

Presentation at the Institute of Acoustics Conference, Windermere, Cumbria, UK in November 1997.

Presentation of Eur Ing. MDG Randall, from Foster Wheeler, Energy on PSV noise – criteria, limits and prediction.

Berwanger – Oil, Gas & Petrochemical consulting – Houston, Tx : Patrick C. Berwanger, Robert A. Kreder, Wai-Shan Lee.

BP Amoco Exploration and E. Smith and J. McAleese from the city University of London.

University of Wisconsin – Madison : D. Reindl, Ph.D. and T. Jekel Ph.D.

BVAA – British Valve and Actuator Association

Guidelines on the Application of Directive 94/9/ec of 23 March 1994.

Index